URBAN WARFARE

Emergence, Evolution, Strategies and Mastery of the Modern Conflict Landscape

Richard Skiba

AFTER MIDNIGHT
PUBLISHING

Skiba, Richard (author)

Urban Warfare: Emergence, Evolution, Strategies and Mastery of the Modern Conflict Landscape

ISBN 978-1-7638046-1-6 (Paperback) 978-1-7638046-2-3 (eBook) 978-1-7638046-3-0 (Hardcover)

Non-fiction

CONTENTS

INTRODUCTION

This book does not advocate for urban warfare or war in any form. Instead, it seeks to take an analytical approach to understanding the complexities, challenges, and realities of urban warfare. The intent is not to glorify or promote conflict but to explore the what, why, how, and where of urban warfare in a way that informs understanding, strategy, and planning for those who may need to address its consequences or engage with it in professional, academic, or policy-making capacities.

The "what" focuses on defining urban warfare, examining its characteristics, and understanding its unique challenges compared to other forms of combat. The "why" explores the reasons conflicts are drawn into urban settings, looking at historical, sociopolitical, and strategic factors that make cities pivotal in warfare. The "how" delves into the tactics, technologies, and methodologies used by militaries and combatants to navigate and operate within urban environments. Lastly, the "where" examines specific locations and contexts where urban warfare has occurred or could occur, drawing lessons from past conflicts and considering future implications in an increasingly urbanized world.

By adopting this analytical perspective, the book aims to provide a deeper understanding of urban warfare's dynamics without endorsing violence or conflict. The emphasis is on learning from history and current events to develop strategies that prioritize minimizing harm, protecting civilian populations, and fostering peace and stability in the aftermath of war. It is a resource for military professionals, policymakers, academics, and anyone seeking to comprehend the complexities of urban warfare with the hope of mitigating its devastating impacts.

Urban warfare represents a significant evolution in military operations, characterized by engagements in urbanized terrains such as cities and towns. This form of warfare diverges from conventional combat, which typically occurs on open battlefields, as it unfolds within complex environments defined by intricate structures, including buildings, streets, and underground systems. The unique challenges presented by urban warfare include limited manoeuvrability, reduced visibility, and an elevated risk of collateral damage to civilian populations and infrastructure [1-3]. The confined nature of urban settings necessitates a shift in military strategies, emphasizing close-quarters combat, ambush tactics, and the critical need for precision in distinguishing between combatants and civilians [3, 4].

The complexity of urban environments forces military forces to adapt their strategies continuously. Each building, alley, and rooftop can serve as potential cover or vantage points for adversaries, complicating operational planning [1, 2]. This dynamic landscape requires soldiers to maintain heightened situational awareness, as the presence of civilians introduces ethical considerations and operational constraints. Military operations must prioritize minimizing harm to non-combatants and adhere to the laws of armed conflict, which adds layers of complexity to urban engagements [1, 2]. Furthermore, logistical challenges are amplified in urban warfare, as the supply of essential resources such as food, water, ammunition, and medical aid becomes increasingly difficult in densely populated areas [1, 2, 5].

Figure 1: The Asymmetric Warfare Group (AWG) assist U.S. Soldiers, assigned to 101st Airborne Division (Air Assault) in breach training for Dense Urban Terrain, focusing on Subterranean aspects at Fort Campbell, Ky., July 20, 2018. This Home Station Training allows Soldiers to prepare for potential environments that they may see while deployed over seas. (U.S. Army Photo by Spc. Jordan Buck). Defense Visual Information Distribution Service, Public Domain, via NARA & DVIDS Public Domain Archive.

Moreover, the psychological and physiological responses of soldiers in urban combat scenarios are markedly different from those in traditional warfare. Studies indicate that urban combat environments can lead to increased anxiety and cardiovascular responses due to the multitude of stimuli that soldiers must process in a short timeframe [6, 7]. This heightened stress underscores the necessity for specialized training and preparation for military personnel engaged in urban operations [2]. The implications of urban warfare extend beyond immediate tactical concerns; they also influence broader military doctrines and strategies, as the urbanization of conflict necessitates a re-evaluation of how military forces operate within civilian contexts [1, 8].

Learning about urban warfare serves various purposes depending on the individual's or organization's interests, roles, and responsibilities. Understanding urban warfare is crucial not only for military professionals but also for policymakers, humanitarians, academics, and even civilians in conflict-prone areas.

Urban warfare is one of the most challenging types of combat, requiring specialized knowledge and training. Military personnel, including soldiers, commanders, and strategists, study urban warfare to prepare for scenarios where they must operate in densely populated and complex environments. Training focuses on tactics, minimizing collateral damage, and effectively navigating the urban landscape. Understanding the psychological and logistical challenges of urban warfare ensures military readiness.

Government officials, policymakers, and defence analysts study urban warfare to develop strategies, allocate resources, and make informed decisions about military operations in urban settings. Understanding the implications of urban combat can help in planning humanitarian aid, reconstruction efforts, and post-conflict governance.

Humanitarian workers and organizations like the Red Cross or UN agencies often operate in conflict zones affected by urban warfare. Learning about its dynamics helps them anticipate challenges, such as evacuating civilians, delivering aid, and setting up medical or food supply chains in war-torn cities. Understanding urban combat scenarios also aids in advocating for the protection of civilians and adherence to international humanitarian laws.

Historians, political scientists, and defence researchers study urban warfare to analyse its causes, strategies, and consequences. They aim to identify patterns, improve military doctrines, and develop innovative solutions to mitigate civilian harm. Research can also inform public policy and contribute to a better understanding of the evolving nature of modern conflicts.

Urban planners and emergency service providers, such as firefighters and medics, benefit from understanding urban warfare to prepare for disasters resulting from conflicts. Insights into the destruction of infrastructure, evacuation challenges, and the needs of displaced populations can help cities develop resilience strategies for potential crises.

Civilians living in regions at risk of conflict or urban warfare may seek knowledge about it for personal safety and preparedness. Understanding evacuation routes, survival strategies, and how to avoid danger zones during urban combat can be life-saving.

War correspondents and media professionals covering conflicts need a working knowledge of urban warfare to report accurately and safely. Understanding military tactics, civilian impacts, and the geopolitical implications of urban battles ensures responsible reporting that informs the public and policymakers.

Military Terminology

Historically, the terminology used by the United States Armed Forces to describe urban warfare has evolved significantly. Initially, the term "UO" (urban operations) was prevalent, but it has largely been supplanted by "MOUT" (military operations in urban terrain) as the preferred nomenclature in military doctrine and training [9]. This shift reflects a broader understanding of the complexities and unique challenges associated with conducting military operations in urban environments, which require specialized tactics and strategies tailored to the dense and multifaceted nature of urban settings [10].

In contrast, the British armed forces utilize different terminologies to describe similar operations. The terms "OBUA" (operations in built-up areas) and "FIBUA" (fighting in built-up areas) are commonly employed, with colloquial expressions such as "FISH" (fighting in someone's house) and "FISH and CHIPS" (fighting in someone's house and causing havoc in people's streets/public spaces) also in use [10]. These terms highlight the focus on close-quarters combat and the implications of urban warfare on civilian populations and infrastructure, reflecting the operational realities faced by military personnel in urban combat scenarios [11].

The term "FOFO" (fighting in fortified objectives) is another critical concept in urban warfare, referring specifically to operations aimed at clearing enemy personnel from fortified positions such as bunkers and trenches. This includes the dismantling of mines and securing footholds in enemy territory, which are essential tasks in urban combat operations [12]. The emphasis on such operations underscores the need for specialized training and tactics to effectively engage in environments where the enemy is entrenched and fortified.

The Israel Defense Forces (IDF) have developed their own terminology and strategies for urban warfare, referred to in Hebrew as "לש"ב" (LASHAB), which stands for warfare on urban terrain. This approach incorporates large-scale tactics, including the use of heavy armoured personnel carriers, armoured bulldozers, and UAVs for intelligence gathering. The IDF has emphasized close-quarters battle (CQB) training for infantry units, which is crucial for operations in densely populated urban areas [13]. The evolution of LASHAB tactics has been significantly influenced by experiences in conflicts such as the 1982 Lebanon War and the Second Intifada, where urban warfare played a pivotal role [14]. The IDF has also established advanced training facilities specifically designed for urban combat, reflecting the importance of preparing soldiers for the unique challenges posed by urban environments [15].

The terminology and strategies associated with urban warfare have evolved across different military organizations, reflecting the complexities of conducting operations in urban settings. The United States Armed Forces have transitioned from UO to MOUT, while the British forces have adopted terms like OBUA and FIBUA. The IDF's LASHAB approach highlights the need for specialized tactics and training in urban warfare, shaped by historical experiences and the operational environment.

Overview of Urban Warfare's Historical Context

Urban warfare has been a pivotal aspect of military history, evolving alongside the growth of cities and their strategic significance in conflict. The historical context of urban warfare can be traced back to ancient civilizations, where fortified cities served as critical centres of power and conflict. Empires such as the Romans and Greeks developed sophisticated siege tactics and technologies, including siege engines, to breach city defences, underscoring the importance of urban centres in military strategy [16]. The ancient city, as a focal point of culture and politics, was often the site of intense military engagements, where the control of urban spaces was synonymous with power [16].

During the Middle Ages, the prominence of urban warfare continued, particularly through the use of castles and walled cities. The Siege of Constantinople in 1453 exemplified the evolution of siege warfare, combining traditional tactics with the innovative use of gunpowder artillery, which marked a significant shift in urban combat methods [17]. This period highlighted the strategic importance of urban centres, as they became battlegrounds for both military and political control, reflecting the intertwined nature of urbanization and warfare [18].

Figure 2: Infantrymen of the 44th Division, Seventh U.S. Army, fire at snipers concealed in a building in Mannheim, Germany, during mopping-up operations on the day of its capture March 29, 1945. On the far side of the street, a soldier with a bazooka gun is about to fire a rocket at the position. The city was subjected to such heavy shelling and bombing that hardly a building remained intact. Mannheim is on the Rhine River 35 miles south of Mainz. OWI Staff Photo by Richard Boyer. Bundesarchiv, Bild 146-1971-053-59 / Boyer, Richard / CC-BY-SA, Public domain, via Wikimedia Commons.

The 19th and early 20th centuries saw urban warfare take on new dimensions, particularly with the advent of industrialization and urbanization. The Napoleonic Wars illustrated the strategic significance of cities, as seen in the Battle of Moscow, where urban environments shaped military tactics and outcomes [19]. The two World Wars further transformed urban warfare, with battles like Stalingrad becoming emblematic of the brutal and destructive nature of combat in urban settings. The intense, block-by-block fighting in Stalingrad resulted in staggering civilian and military casualties, illustrating the complexities and challenges of urban warfare [20].

Figure 3: Soviet troops attacking German positions, armed with rifles and machine guns, note the fallen soldier in the foreground. Soviet archives, CC BY-SA 4.0, via Wikimedia Commons.

In the latter half of the 20th century, urban warfare increasingly became a feature of asymmetric conflicts, where insurgents utilized guerrilla tactics within urban environments. The Battle of Hue during the Vietnam War and the Siege of Sarajevo during the Bosnian War highlighted the difficulties of urban combat, particularly in protecting civilian populations while engaging in military operations [21]. More recently, conflicts in the Middle East, such as the battles for Fallujah and Mosul, have demonstrated the ongoing challenges militaries face in urban warfare, including the need to clear urban areas of entrenched insurgents while minimizing civilian casualties [17].

Figure 4: Hashed Shaabi in Fallujah after Retaking The City from ISIS. Tasnim News Agency, CC BY 4.0, via Wikimedia Commons.

Urban warfare has played a pivotal role in military history, as cities have long been strategic and symbolic centres of power. Wars have frequently revolved around sieges and urban battles, with large open-field engagements being relatively less common. From ancient conflicts to modern asymmetrical wars, urban settings have consistently posed unique challenges, demonstrating that technological military superiority does not guarantee success. Historical examples such as the battles of Hue and Fallujah underscore that even tactical victories in urban warfare may not translate to broader strategic success in the larger conflict [22].

Cities like Kadesh, Constantinople, and Carthage have historically been focal points of military campaigns due to their economic, political, and cultural significance. For example, the fall of Constantinople in 1453 marked the end of the Byzantine Empire and a turning point in world history [22]. The idea of cities as central to warfare is reflected in the writings of Carl von Clausewitz, who viewed war as a continuation of politics, a term derived from the Greek "polis," meaning city. This centrality persisted into modern conflicts, where urban environments continued to serve as battlegrounds for ideological, political, and territorial disputes [22].

During the Cold War, urban guerrilla warfare escalated, especially in Latin America, forcing conventional forces to adapt strategies for counterinsurgency in dense, populated areas. This period also gave rise to notable urban confrontations, such as the Battle of Hue in 1968 during the Vietnam War, where the extensive destruction and civilian displacement eroded political support despite a tactical military victory. Similarly, in Mogadishu in 1993, advanced U.S. military technology was outmanoeuvred by local fighters in the urban chaos, highlighting the limitations of firepower and the importance of cultural and tactical adaptability [22].

The First Chechen War (1994–1996) offers another stark lesson in urban warfare. Russian forces, equipped for conventional combat, suffered significant losses when they underestimated the complexities of urban combat in Grozny. Chechen fighters utilized urban terrain to their advantage, employing tunnels and fortified positions to outlast and counteract Russian offensives. By the Second Chechen War, Russian forces adapted, using encirclement and heavy bombardments to achieve victory, but at a devastating cost to civilians and infrastructure [22].

Modern urban warfare has also highlighted the evolving nature of technology and its applications in conflict. For instance, in Gaza and similar densely populated areas, combat vehicles and tanks have faced challenges in manoeuvrability, visibility, and vulnerability to concealed anti-tank units [22]. Enhancements like stabilized panoramic sights and advanced infrared systems have been developed to address these issues, yet the balance between operational effectiveness and minimizing civilian harm remains a critical concern [22].

Urban warfare also underscores broader strategic challenges, particularly in post-conflict governance and reconstruction. In Fallujah and Grozny, even after military victories, the lack of trust and prolonged instability hindered sustainable peace. The concept of localized governance, as seen in the "Chechenization" strategy during the Second Chechen War, highlights the importance of integrating cultural and socio-political considerations into military and post-conflict planning [22].

The lessons of urban warfare demonstrate that while cities are critical to military campaigns, their complexity demands a holistic approach that combines tactical innovation, cultural understanding, and long-term planning. Each city's unique geography, population dynamics, and socio-political context require tailored strategies to achieve not only battlefield success but also enduring stability and peace [22].

As urban areas continue to grow and evolve, urban warfare is expected to remain a critical aspect of military doctrine. Advances in technology, such as drones and precision-guided munitions, are reshaping the landscape of urban combat, providing new tools for military forces [4]. However, the inherent challenges of urban warfare—such as the protection of civilians, the complexity of operations, and the destruction of infrastructure—will continue to test the adaptability and ingenuity of military forces worldwide [23]. The historical context of urban warfare thus not only informs current military strategies but also underscores the enduring significance of urban centres in the dynamics of conflict.

Why Urban Warfare Has Overtaken Traditional Warfare

Urban warfare has increasingly supplanted traditional warfare due to a confluence of factors, including rapid urbanization, the strategic significance of cities, shifts in warfare dynamics, global conflict trends, and the inherent challenges of urban defence. This transformation reflects the evolving nature of conflicts in the contemporary world.

The global trend towards urbanization is a significant driver of the prevalence of urban warfare. The United Nations projects that nearly 70% of the world's population will reside in urban areas by 2050, up from over 50% today [24]. This demographic shift makes cities critical centres for governance, commerce, and infrastructure, rendering them attractive targets in conflicts. As urban areas grow, they become focal points for military operations, as controlling these regions can disrupt an adversary's logistical and economic capabilities [25]. Furthermore, the urbanization of military operations has been noted, with increasing populations in cities correlating with heightened vulnerability to political unrest and armed conflict [24].

Cities are not only population centres but also hubs of transportation and communication networks essential for military operations. The strategic value of urban areas is underscored by their role in providing psychological advantages in warfare; capturing a city can symbolize dominance, while losing it can demoralize both the military and civilian populations [17, 25]. The complexity of urban environments often necessitates a shift in military strategy, as traditional tactics may be less effective in densely populated areas where insurgents can blend in with civilians [17, 25].

The nature of warfare has evolved towards asymmetrical conflicts, where non-state actors and insurgents often engage in urban settings to exploit the advantages offered by these environments. Urban areas provide cover and concealment, allowing these groups to utilize guerrilla tactics effectively against more conventional military forces [25, 26]. This shift is exacerbated by the increasing difficulty of conducting traditional military operations in urban landscapes, where the presence of civilians complicates engagement rules and operational strategies [25, 27].

Contemporary conflicts are characterized by a decline in direct confrontations between state armies in open fields, influenced by factors such as international scrutiny and advanced surveillance technologies [25, 27]. Instead, urban warfare has become more common, as combatants can operate within civilian populations, complicating military responses and prolonging engagements [27]. This trend is evident in recent conflicts, such as those in Ukraine and Syria, where urban centres have become battlegrounds for both state and non-state actors [17, 27].

Defending urban areas presents unique challenges that often favour defenders over attackers. The dense infrastructure, narrow streets, and underground networks in cities limit the manoeuvrability of attacking forces and allow defenders to establish fortified positions [17, 25]. This dynamic has led to a preference among insurgents and non-state actors to engage in urban warfare, where they can leverage their knowledge of the terrain and the presence of civilians to their advantage [25, 27].

Urban warfare typically arises from a combination of political, ideological, and territorial conflicts. Political uprisings and civil wars often target cities as centers of power, as seen in conflicts in Kyiv and Baghdad [17, 25]. Insurgent groups frequently retreat to urban areas to evade conventional military forces, using the urban landscape to conduct hit-and-run attacks and challenge state authority [17, 25]. Additionally, ethnic and sectarian tensions can ignite urban warfare, particularly in diverse cities where multiple groups vie for control [27]. Resource scarcity and economic drivers also play a role, as competition for essential resources can lead to violent confrontations in urban settings [27].

Devastating Consequences of Urban Warfare for Civilians

Urban warfare poses severe and multifaceted consequences for civilian populations, leading to significant loss of life, psychological trauma, destruction of infrastructure, and lingering dangers from unexploded ordnance (UXO). The unique characteristics of urban environments exacerbate these effects, making the need for effective conflict mitigation and post-conflict recovery strategies paramount.

The close proximity of combatants to civilian populations in urban warfare significantly increases civilian casualties. Studies have shown that urban combat leads to high rates of collateral damage, even with the use of precision weaponry. For instance, Bilukha et al. [28] highlight that unexploded ordnance (UXO) in conflict zones, particularly in Afghanistan, has resulted in more injuries than landmines, with a notable proportion of these injuries occurring among children. Similarly, Khan [29] discusses the severe injuries and fatalities resulting from explosive devices, emphasizing the vulnerability of civilians, especially children, during military operations. The chaotic nature of urban combat often results in civilians being caught in crossfire or affected by indirect fire, leading to a tragic loss of life and severe injuries [30]. Furthermore, the lack of accessible medical care in besieged urban areas exacerbates the situation, as treatable injuries can become fatal due to insufficient resources [25].

The psychological impact of urban warfare on civilians is profound and long-lasting. Constant exposure to violence, loss of loved ones, and the destruction of familiar environments contribute to a pervasive atmosphere of fear and uncertainty. Research indicates that children are particularly susceptible to developing post-traumatic stress disorder (PTSD), anxiety, and depression following exposure to violent conflict [31]. Adults also experience significant psychological distress, often compounded by economic instability and the loss of community support structures [25]. The enduring nature of these psychological scars can affect multiple generations, leading to a cycle of trauma that persists long after the cessation of hostilities [31].

Urban warfare frequently results in the destruction of critical infrastructure, including homes, schools, hospitals, and transportation networks. This destruction disrupts essential services, leaving civilians without access to clean water, electricity, and healthcare [25]. Bilukha et al. [28] note that the collapse of healthcare systems during prolonged conflicts leads to a humanitarian crisis, as the remaining medical facilities struggle to cope with the influx of casualties from both combatants and civilians. The rebuilding of infrastructure post-conflict is often slow and resource-intensive, delaying recovery and exacerbating civilian suffering [25].

The remnants of urban warfare, particularly UXO, pose significant dangers to civilian populations long after the fighting has ceased. UXO can cause injuries or fatalities, particularly among children who may unknowingly handle these dangerous remnants [28]. The presence of UXO complicates reconstruction efforts, as areas cannot be safely cleared for rebuilding until demining operations are completed [32]. This delay can leave entire neighbourhoods uninhabitable for years, prolonging displacement and hindering economic recovery [33]. The ongoing threat of UXO underscores the importance of comprehensive post-conflict recovery strategies that include effective clearance operations and community education on the dangers of unexploded munitions [32].

The cumulative effects of urban warfare create a cycle of suffering and instability for civilian populations. The loss of family members, destruction of homes, and disruption of social and economic systems often force civilians to flee, resulting in large-scale displacement and refugee crises [31]. Internally displaced persons (IDPs) and refugees face additional challenges, including lack of access to resources, education, and employment opportunities in host countries [25]. The long-term implications of urban warfare highlight the urgent need for adherence to international humanitarian law and the implementation of effective conflict mitigation strategies to protect civilian lives and facilitate recovery [25].

Importance of Urban Warfare in Modern Conflict Zones

Urban warfare has emerged as a critical aspect of modern conflict zones, influenced by various factors including the strategic significance of cities, increasing urbanization, and the complexities inherent in urban environments. This synthesis will explore these dimensions, supported by relevant literature.

Cities serve as vital centres of power, often functioning as administrative, economic, and cultural hubs. Their control can significantly impact the outcome of conflicts. Urban centres are not only strategic military targets but also symbols of political dominance, as controlling a major city can demoralize adversaries and consolidate power for the victor [17, 34]. Furthermore, cities provide access to essential infrastructure such as communication networks and transportation systems, which are crucial for military operations and logistics [5]. The ability to control these resources can dictate the flow of supplies and information, thereby influencing broader conflict dynamics [23].

The trend of urbanization is undeniable, with over half of the global population now residing in urban areas. This demographic shift has transformed cities into inevitable battlegrounds in contemporary conflicts [5]. The dense population and complex architecture of urban environments necessitate specialized military tactics, as traditional warfare strategies may not be effective in such settings [35]. The urban landscape complicates military operations, requiring forces to adapt to three-dimensional combat scenarios that involve fighting in buildings, streets, and underground systems [36].

Urban warfare presents unique challenges, primarily due to the presence of civilians and the intricate nature of urban infrastructure. The need to distinguish between combatants and non-combatants complicates military operations, necessitating adherence to international humanitarian law to minimize civilian casualties (Andrade et al., 2023). The dense urban environment also poses logistical challenges, as narrow streets and alleys can hinder movement and visibility, demanding high levels of adaptability and specialized training for military personnel [35].

Modern conflicts often feature asymmetric warfare, where non-state actors utilize urban areas as strongholds. This shift has led to the adoption of guerrilla tactics, which leverage the urban terrain for ambushes and surprise attacks [34]. The urban environment favours smaller, decentralized forces, making it essential for conventional military units to develop countermeasures, including advanced technologies such as drones and precision-guided munitions [5]. The integration of smart technologies into military operations further enhances situational awareness and operational effectiveness in urban settings [37, 38].

Figure 5: Ukrainian 25th Sicheslavska bde showing their improvised FPV strike drones. АрміяInform, CC BY 4.0, via Wikimedia Commons.

Urban warfare raises significant humanitarian concerns, particularly regarding civilian protection and the psychological impact of conflict. High civilian casualties and displacement are common outcomes of urban combat, necessitating careful planning to mitigate harm [23, 39]. Additionally, the extensive destruction of urban areas requires long-term reconstruction efforts, which can strain resources and complicate post-conflict recovery [23]. The psychological scars left on both civilians and combatants highlight the need for comprehensive strategies that address the aftermath of urban warfare [39].

Recent conflicts, such as those in Fallujah, Grozny, and the Gaza Strip, have underscored the complexities and challenges of urban warfare. The Battle of Fallujah demonstrated the difficulty of achieving strategic objectives despite tactical successes, while the conflict in Grozny illustrated the high costs associated with indiscriminate tactics in urban settings [17, 34]. The ongoing situation in the Gaza Strip highlights the challenges of conducting military operations in densely populated areas, where the presence of civilians complicates military objectives and raises ethical concerns [23].

Looking ahead, urban warfare is likely to remain central to future conflicts, particularly as hybrid warfare becomes more prevalent. Urban areas will continue to serve as operational bases for insurgent and terrorist groups, necessitating effective counterinsurgency and counterterrorism strategies [35]. The evolution of military technology, including unmanned systems and AI-driven decision-making, will further shape the landscape of urban combat, enabling forces to adapt to the complexities of modern warfare [37, 38].

The importance of urban warfare in modern conflict zones cannot be overstated. It shapes military strategies, influences political dynamics, and has profound implications for civilian populations. The unique challenges posed by urban environments demand a comprehensive approach that balances military effectiveness with humanitarian considerations and long-term recovery efforts.

Differences Between Conventional and Urban Warfare

The distinctions between conventional and urban warfare are significant, particularly in terms of environment, forces involved, tactics, civilian impact, logistics, technology, psychological implications, and the duration of engagements. Each of these aspects reflects the evolving nature of conflict in modern warfare.

Environment and Terrain: Conventional warfare typically occurs in open fields, deserts, and forests, characterized by fewer obstacles and clear lines of sight, which facilitate the movement of large forces and mechanized units [40]. In contrast, urban warfare unfolds in cities and densely populated areas, presenting a complex terrain filled with buildings, streets, and underground structures. This environment limits visibility and manoeuvrability, complicating operations due to the presence of civilians [41, 42]. The urban landscape not only obstructs traditional military strategies but also requires adaptations in tactics to navigate confined spaces effectively [41].

Forces Involved: Conventional warfare generally involves organized state armies with established hierarchies, often engaging in battles between evenly matched forces [43]. Urban warfare, however, frequently features a mix of state militaries, insurgents, and non-state actors, where irregular forces employ guerrilla tactics and blend into civilian populations, complicating the identification of combatants [44, 45]. This shift in the nature of combatants necessitates a re-evaluation of military strategies to address the asymmetrical dynamics present in urban settings [46].

Tactics and Combat Style: The tactics employed in conventional warfare focus on large-scale troop movements and massed formations, heavily utilizing mechanized forces such as tanks and artillery [40, 43]. Conversely, urban warfare emphasizes close-quarters combat, requiring soldiers to secure individual buildings and infrastructure while employing ambushes, snipers, and improvised explosive devices (IEDs) [41, 42]. The decentralized nature of urban combat necessitates small-unit tactics that prioritize flexibility and rapid response to changing conditions [44].

Civilian Presence and Impact: In conventional warfare, civilians are typically located away from the battlefield, resulting in less frequent collateral damage [43]. Urban warfare, however, sees civilians directly in the combat zone, leading to higher risks of casualties and displacement. The ethical and legal challenges of distinguishing between combatants and non-combatants are pronounced in urban settings, where the destruction of civilian infrastructure significantly hampers post-conflict recovery efforts [45, 47]. The psychological trauma experienced by both soldiers and civilians is exacerbated in urban warfare due to the constant threat of violence and the ethical dilemmas faced by combatants [48].

Logistics and Supply Chains: Establishing and maintaining supply chains in conventional warfare is relatively straightforward due to predictable routes over open terrain [40]. In contrast, urban warfare presents logistical challenges, as narrow streets and destroyed infrastructure can disrupt supply lines, necessitating innovative solutions for resource delivery [41]. The complexity of urban environments requires military planners to develop adaptable strategies to ensure the continuous flow of supplies to combatants [49].

Technology and Equipment: Conventional warfare relies heavily on tanks, artillery, and large-scale weapon systems designed for open terrain [40]. Urban warfare, however, demands a shift toward infantry-based weapons and equipment tailored for close combat, including modifications to armoured vehicles to navigate urban landscapes [4, 41]. The use of drones and precision-guided munitions becomes critical in urban settings to minimize collateral damage and enhance situational awareness [4].

Psychological and Morale Implications: The psychological challenges faced by combatants in conventional warfare are often less intense due to the more predictable nature of engagements [43]. In urban warfare, however, soldiers experience heightened stress from constant threats of ambushes and the ethical dilemmas associated with civilian casualties. Civilians in urban conflict zones endure significant psychological trauma, exacerbated by their proximity to violence and destruction [48, 50].

Duration and Intensity: Conventional warfare battles are often large-scale but may be resolved quickly through decisive engagements [43]. Urban warfare, on the other hand, tends to involve prolonged engagements with no clear frontlines, where victory in a specific battle does not necessarily translate to success in the broader conflict [49]. This complexity can lead to a drawn-out struggle for control over urban areas, as seen in historical examples like the battles of Hue and Fallujah [49].

Key Challenges: High-Density Population, Infrastructure, and Terrain Complexities

Urban warfare presents unique challenges due to the high-density population, complex infrastructure, and intricate terrain. These factors create an environment that is far more unpredictable and demanding than conventional battlefields, requiring tailored strategies and tactics.

One of the most significant challenges in urban warfare is the presence of large civilian populations. Cities are often densely populated, which complicates military operations and heightens the risk of civilian casualties. Combatants must navigate ethical and legal responsibilities to distinguish between civilians and combatants, making precision and restraint critical. The displacement of civilians during urban conflict also creates humanitarian crises, as people flee their homes, overwhelming relief efforts and infrastructure. Additionally, insurgents or adversaries may exploit civilian populations by using them as human shields or blending in to avoid detection, further complicating operations.

Urban areas are characterized by critical infrastructure such as power grids, water systems, transportation networks, and communication systems, which are often damaged or destroyed during conflict. The destruction of these systems disrupts civilian life and complicates military logistics. Accessing and securing infrastructure, such as bridges, tunnels, or highways, becomes a strategic priority to ensure operational success. Moreover, damaged infrastructure can hinder the mobility of military forces and restrict supply chains, requiring innovative approaches to resource management and troop movement.

Figure 6: A Georgian soldier from the 1st Infantry Brigade's NATO Response Force company surveys the terrain during an urban warfare training exercise with the 173rd Airborne Brigade and the Georgian 1st Infantry Brigade as part of Exercise Noble Partner here May 17. Noble Partner is a combined U.S. Army Europe-Georgian army exercise designed to increase interoperability between Georgia's contribution to the NATO Response Force and allied militaries. (U.S. Army photo by Sgt. A.M. LaVey). U.S. Army Europe Images from Wiesbaden, Germany, CC BY 2.0, via Wikimedia Commons.

The urban terrain itself poses significant challenges. Cities are three-dimensional battlefields with streets, buildings, underground tunnels, and rooftops creating multiple levels of engagement. Narrow alleys, barricaded streets, and concealed positions allow adversaries to stage ambushes, set traps, and evade pursuit. The dense concentration of buildings provides ample cover and concealment, making it difficult for conventional forces to maintain situational awareness. Additionally, modern mapping and navigation tools may be inadequate in the face of rapidly changing urban landscapes, forcing forces to rely on real-time intelligence and reconnaissance.

Urban warfare's challenges stem from the interplay of high-density populations, critical infrastructure, and complex terrain. These factors demand a multidisciplinary approach that combines tactical innovation, precision technology, and careful planning to minimize civilian harm, maintain operational effectiveness, and adapt to the complexities of urban environments.

Chapter 1

CHARACTERISTICS OF THE URBAN BATTLEFIELD

Structure and Layout of Urban Areas: Buildings, Streets, and Networks

The structure and layout of urban areas significantly influence the dynamics of urban warfare, creating both opportunities and challenges for military forces. Urban environments are characterized by densely packed buildings, intricate street layouts, and interconnected networks of infrastructure, all of which affect how combatants operate, manoeuvre, and engage. Understanding these elements is essential for devising effective strategies in urban combat scenarios.

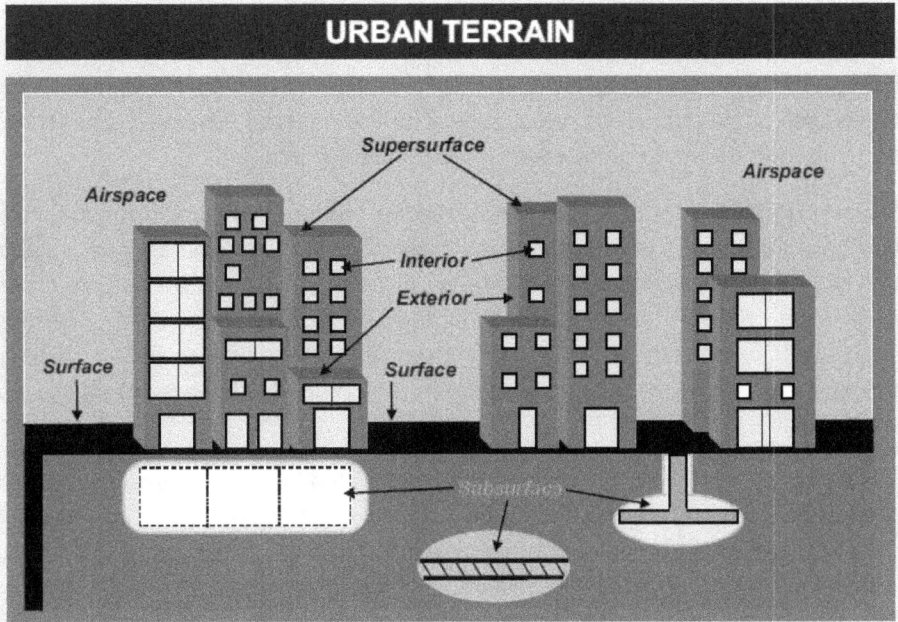

Figure 7: Urban Terrain Concept. Department of Defense, Public domain, via Wikimedia Commons.

Figure 7 provides a illustration of the various layers and dimensions of urban terrain, emphasizing their critical role in understanding the complexities of urban warfare. Urban terrain serves as a multi-dimensional battlefield, where each zone presents unique challenges and opportunities. These zones must be strategically analysed and addressed during military operations to ensure operational success.

Airspace: The airspace above urban areas is a vital component of urban warfare, facilitating aerial reconnaissance, drone operations, and airstrikes. It provides opportunities for situational awareness and surveillance, allowing forces to monitor enemy movements and coordinate attacks effectively. However, airspace can also be heavily contested, with adversaries employing anti-aircraft systems or small arms fire to target helicopters and drones, making aerial operations inherently risky.

Surface: The surface level includes streets, roads, sidewalks, parks, and other open spaces visible above ground. This zone serves as the primary area for troop movements, vehicle navigation, and direct combat. It also connects the urban environment to its subsurface and supersurface layers, making it a critical battlefield for maintaining control. Surface-level operations face numerous challenges, including debris, blockades, and the constant threat of ambushes, particularly in narrow streets or areas with concealed positions.

Supersurface: The supersurface encompasses the vertical dimension of buildings and structures, such as rooftops and upper floors. This layer introduces the concept of vertical combat, where forces may engage at different heights, from snipers on rooftops to defenders barricaded on upper levels. Controlling the supersurface is essential for maintaining visibility over the urban terrain, gaining strategic vantage points, and denying enemy forces the ability to use buildings as strongholds.

Interior and Exterior: Urban terrain also includes the interior and exterior spaces of buildings. Interior spaces, such as rooms and hallways, are often used for concealment, close-quarters combat, and logistical operations. Exterior spaces, including building facades, courtyards, and adjacent open areas, are critical during assaults, as they serve as entry points for clearing operations. Interior operations require meticulous planning to counter risks like traps and ambushes, while exterior zones often expose combatants to sniper fire and aerial observation.

Subsurface: The subsurface layer consists of underground tunnels, subways, basements, sewers, and bunkers. These spaces are invaluable in urban warfare for enabling covert movement, staging surprise attacks, and storing supplies. However, subsurface areas are often difficult to map, complicating intelligence gathering and operational planning. Specialized equipment and tactics are necessary to navigate and clear these hidden zones effectively.

Figure 8: A 10 Kilometre-Long Terror Tunnel Passing Underneath a Hospital and a University: IDF Troops Located an Underground Tunnel Network Connecting the North and South of the Gaza Strip. IDF Spokesperson's Unit, Public Domain, via Wikimedia Commons.

Strategic Implications

Figure 7 underscores the three-dimensional nature of urban terrain, highlighting the need for military forces to address threats from above (airspace and supersurface), below (subsurface), and at eye level (surface). Securing each layer is essential to prevent adversaries from exploiting them as vantage points, escape routes, or defensive positions. The interconnectedness of these layers requires synchronized operations to ensure comprehensive control of the battlefield.

Urban warfare demands a multi-layered approach to planning and combat. Success relies on understanding the interplay between airspace, surface, supersurface, and subsurface zones, as well as coordinating effectively between infantry, mechanized units, and aerial assets. This image serves as a critical tool for grasping the complexities of urban environments and their profound impact on military strategies.

Buildings

Urban warfare presents unique challenges and opportunities, largely defined by the built environment. Buildings in urban areas serve as both obstacles and assets for combatants, providing cover, concealment, and strategic vantage points. The diversity of building types, from high-rises to residential homes, complicates military operations. High-rise buildings, in particular, facilitate vertical combat, allowing forces to engage from rooftops and basements, which can be advantageous for snipers and observation posts [17, 25]. The urban landscape's complexity necessitates a nuanced understanding of how these structures influence combat dynamics, as they can enhance the effectiveness of certain combat strategies while also presenting significant challenges.

However, the presence of buildings also introduces considerable difficulties in urban warfare. Collapsing structures can obstruct access routes, posing risks to both civilians and combatants [23, 51]. The labour-intensive nature of clearing buildings, often requiring door-to-door searches and specialized tactics, can slow military operations significantly [17, 25]. Moreover, the density of buildings can obscure sightlines and limit the effectiveness of heavy weaponry, necessitating the use of precision-guided munitions or close-quarters combat techniques [52]. The interconnected nature of urban structures, such as underground passages and shared basements, enables adversaries to evade detection and launch surprise attacks, further complicating military engagements [17, 25].

The implications of urban warfare are profound, as modern technologies and rapid urbanization have fundamentally altered the nature of conflict [17, 25]. The strategic use of precision-guided munitions has become increasingly important in urban settings, where the risk of collateral damage is heightened due to the proximity of civilian populations and infrastructure [52]. As military strategies evolve, understanding the interplay between urban environments and combat operations remains critical for effective planning and execution in contemporary warfare.

Streets

The street layout of urban areas plays a crucial role in determining how military forces can manoeuvre and engage in combat scenarios. The design of streets, whether wide and organized or narrow and maze-like, significantly influences the operational capabilities of both mechanized units and infantry. Wide streets facilitate easier access for vehicles and supply chains, allowing for rapid movement and logistics support. However, this accessibility also increases vulnerability to ambushes from concealed positions, as attackers can exploit the open nature of such environments to launch surprise assaults [53-55].

Conversely, narrow streets and alleys present unique challenges for military operations. These confined spaces restrict vehicle movement, complicating logistics and troop deployment. The limited manoeuvrability increases the risk of ambushes, particularly for infantry units, as they may find themselves cornered or exposed to enemy fire in tight quarters [56]. The irregularity of urban street networks, characterized by intersections, dead ends, and overpasses, creates a complex three-dimensional battlefield. Adversaries can utilize these features to establish choke points, set traps, or conduct hit-and-run attacks, thereby leveraging the urban landscape to their advantage [57-59].

Moreover, the urban street layout complicates logistical operations. Debris from combat or blockades can render certain routes impassable, forcing military commanders to identify alternative pathways or rely on aerial supply drops to sustain their forces. This necessitates detailed mapping of the street network to coordinate troop movements and maintain situational awareness [60-62]. The need for precise navigation is underscored by the fact that urban environments can significantly alter the effectiveness of communication and reconnaissance technologies, as dense structures may obstruct signals and hinder operational efficiency [62, 63].

Networks and Infrastructure

Urban networks, encompassing transportation, utilities, and communication systems, are integral to the functionality of urban areas, significantly influencing both civilian life and military operations. Transportation networks, including subways, railways, and highways, provide essential mobility for civilians and military forces alike. However, these networks can also become sites of ambush or logistical bottlenecks during combat operations. The complexity of urban environments often leads to increased vulnerability for military units, as they must navigate through densely populated areas where transportation routes can be easily monitored or obstructed by adversaries [25]. The strategic importance of these networks is underscored by their dual role in facilitating movement and serving as potential targets for disruption [64].

Moreover, underground transportation systems, such as subways and tunnels, add another layer of complexity to urban warfare. These infrastructures allow for covert movement, enabling combatants to stage attacks or evade capture effectively. The ability to utilize these hidden networks can provide a tactical advantage, as they can be used to bypass surface-level threats and engage in surprise manoeuvres [20]. This dynamic is particularly relevant in urban combat scenarios, where the control of such networks can dictate the outcome of engagements [4].

Utility networks, including power grids, water supplies, and sewage systems, are often weaponized or disrupted in urban warfare. Combatants may target these critical infrastructures to undermine the opposing force's logistics or to demoralize the civilian population. For example, attacks on water supply systems can severely impact the operational capabilities of military forces, as access to clean water is essential for sustaining troops [20]. Conversely, military forces may rely on intact utility networks for their operations, using electrical grids for communication and water systems for logistical support. The interdependence of military operations and urban infrastructure highlights the strategic significance of these utility networks in both combat and civilian contexts.

Urban communication networks also play a crucial role in modern warfare. Civilian infrastructures, such as cell towers and internet lines, can be exploited for intelligence gathering or disrupted to isolate adversaries. The dual-use nature of these systems means that while they can facilitate coordination and propaganda for combatants, they can also be targeted to disrupt enemy communications [65]. This necessitates robust countermeasures from military forces to protect their own communication lines while attempting to disrupt those of their adversaries [66]. The complexity of urban communication networks, particularly in the context of network-centric warfare, emphasizes the need for effective strategies to manage and secure these vital systems [65].

Cities in Conflict

Cities hold immense cultural, psychological, and strategic value in warfare, influencing the outcomes of conflicts through their geographical and sociological significance. Urban centres often sit astride critical infrastructure, such as major ground routes, rail hubs, water crossings, and power projection sites, making their control pivotal in military campaigns. Historically and in modern conflicts, including the ongoing war in Ukraine, cities have repeatedly demonstrated their importance [67]. Despite this, urban warfare receives limited attention in military theory and doctrine, particularly in the U.S. Army, where it is often characterized as inherently difficult and best avoided. However, as demonstrated in both historical and contemporary examples, conflicts are frequently decided in the cities due to their symbolic, logistical, and operational significance [67].

Cities are more than just geographical locations; they are symbols of sovereignty, resistance, and national pride. The battle for Kyiv during the Ukraine conflict exemplifies this, as Russia's attempt to capture the capital was intended to delegitimize Ukrainian sovereignty and impose regime change [67]. Similarly, Mariupol emerged as a symbol of defiance and resilience, holding psychological and morale significance far beyond its physical terrain. Historical parallels include Stalingrad during World War II, where the city became a turning point in the conflict due to its symbolic importance, and Manila during the Philippines liberation campaign, which drew fierce urban combat. These examples illustrate how cities can become "hero cities," whose resistance embodies national pride and morale, drawing combatants into prolonged and costly urban fights [67].

While symbolic value drives battles for cities, the inability of militaries to successfully wage urban warfare can have profound psychological and strategic consequences. Cities like Moscow, Washington, D.C., and Richmond have historically fallen without ending their respective wars, emphasizing that the symbolic loss of a city does not necessarily equate to defeat. Nevertheless, the symbolic power of urban centres remains a potent factor in shaping both morale and the broader trajectory of conflicts [67].

Cities often develop around critical infrastructure, including railheads, road networks, ports, and airfields, making them essential logistical hubs in military campaigns. The failure to secure these infrastructures can cripple operational success, as seen in Russia's inability to capture key rail hubs in cities like Sumy, Chernihiv, and Kharkiv during its northeastern offensive in Ukraine. These logistical failures contributed to Russia's inability to advance on Kyiv and forced a strategic pivot toward other regions [67].

Historical examples underscore the importance of urban infrastructure in warfare. During the American Civil War, the capture of the railroad hub city of Corinth was critical to isolating Vicksburg, while in World War II, strategic bombing campaigns targeted urban logistical hubs to disrupt Axis supply lines. Similarly, during Operation Iraqi Freedom, the seizure of major bridges and road networks was vital for advancing coalition forces. These examples highlight that controlling urban centres is often less about the cities themselves and more about the critical infrastructure they contain [67].

Ports and airfields, often surrounded by urban environments, are equally vital for sustaining military operations and projecting power. The battles for Mariupol and Odesa in Ukraine emphasize this point, as these cities provide access to the Sea of Azov and the Black Sea, respectively, enabling logistical and operational dominance. The fierce fighting over airfields like Hostomel near Kyiv further illustrates the strategic importance of urban-adjacent infrastructure in modern warfare [67].

Urban terrain provides a natural advantage for resistance movements, enabling guerrilla fighters to exploit the anonymity and complexity of densely populated environments. In the current Ukrainian conflict, cities like Kherson have become hotbeds of resistance against Russian occupation [67]. Urban environments neutralize many of the technological advantages of counter-guerrilla forces, such as surveillance and reconnaissance, by offering concealment and opportunities for information operations. Resistance fighters in cities can manipulate civilian congestion to amplify their messaging, create disinformation, and provoke overreactions from occupying forces, further complicating urban operations.

Figure 9: Destructions in Kherson after Russian shelling and strike with guided air-dropped bomb on 2 February 2024. National Police of Ukraine, CC BY 4.0, via Wikimedia Commons.

Navigating Urban Terrain and Its Population

Cities have always held strategic significance as centres of power, governance, and human civilization, making them central to both historical and contemporary conflicts. In the context of modern warfare, urban terrain presents unique and complex challenges. As the U.S. military shifts its focus from counterinsurgency to preparing for large-scale combat operations against near-peer adversaries, the intricacies of urban combat remain both highly relevant and increasingly difficult to address. Urban warfare demands enormous manpower, resources, and planning due to the unique features of cities, including their density, scale, and the presence of large civilian populations. These challenges were evident in battles like Mosul and are now unfolding in real time in Ukraine, demonstrating that urban combat is a critical area of study and preparation for modern militaries [68].

One of the most significant challenges of urban warfare is the presence of civilians. Adhering to the principle of distinction—separating combatants from non-combatants—is fundamental under international law, yet it becomes extremely difficult in cities where civilians and enemy forces occupy the same space. This difficulty complicates the use of airpower and artillery, as the risk of collateral damage and civilian casualties is magnified. For example, NATO's *Protection of Civilians Handbook* emphasizes the need to minimize harm to civilians, as protecting the population is essential to achieving long-term peace. However, this goal is difficult to achieve in densely populated urban environments, where distinguishing between civilians and combatants becomes an almost insurmountable challenge [68].

To mitigate these issues, militaries have developed tactics to isolate enemy forces geographically within a city, thereby limiting their ability to blend in with civilian populations. This tactic was effectively used during the Battle of Sadr City in 2008, where U.S. and coalition forces built the "Gold Wall" to geographically contain Jaish-al-Mahdi (JAM) forces [68]. By forcing the insurgents to emerge from their urban strongholds to fight in more open environments, coalition forces were able to diminish the enemy's defensive urban advantage. This approach, while effective in certain scenarios, highlights the inherent tension in urban operations: balancing military objectives with the need to minimize civilian harm and preserve urban functionality [68].

While isolating enemy forces within an urban area can yield tactical advantages, the long-term effectiveness of such strategies remains questionable, especially in larger, more interconnected urban centres. Sadr City, for example, was a specific district within Baghdad with a relatively homogenous population and limited external connections. In contrast, applying similar isolation tactics to a megacity— such as those in the Asia-Pacific region—would be far more complex due to the diversity, scale, and interconnectedness of modern cities [68].

Moreover, the rise of digital communications complicates the idea of isolation in urban warfare. The ongoing conflict in Ukraine demonstrates that even with disrupted communication networks, urban populations and defenders can still exchange information, undermining the effectiveness of isolation tactics. Cities are dynamic and adaptive systems, and any attempt to disrupt their normal operations—whether by cutting off movement, communication, or resources— risks significant second- and third-order effects. For instance, during the Afghan campaign, disruptions to civilian life and livelihoods undermined the Afghan government's legitimacy and fuelled support for insurgent forces like the Taliban. Similarly, in Vietnam, programs like the Strategic Hamlets and "free-fire zones" alienated civilians by severely interrupting their daily lives, ultimately proving counterproductive [68].

Urban operations must carefully consider the socio-economic and cultural framework of a city. Military actions that severely disrupt civilian life—such as the prolonged degradation of infrastructure or the imposition of restrictive measures—can erode trust in governing or occupying forces. In Sadr City, for example, while the Gold Wall was tactically successful in isolating JAM forces, it also became a controversial symbol of U.S. military intervention, eliciting mixed reactions from the local population. These disruptions can create long-term grievances that insurgents or resistance movements can exploit, as seen in Iraq, Afghanistan, and other theatres of urban conflict [68].

To effectively navigate urban terrain and its population, military planners must prioritize minimizing collateral damage and preserving the city's identity and socioeconomic functions. This is particularly critical in the post-conflict stabilization phase, where irregular threats and local grievances can undermine control even after initial military objectives are achieved. Stability operations must address these challenges by engaging with local communities, maintaining essential services, and ensuring that military actions align with the long-term goal of securing and rebuilding the city [68].

Environmental Impact on Visibility, Manoeuvrability, and Engagement

Urban environments significantly influence visibility, manoeuvrability, and engagement in warfare, creating unique challenges that differ greatly from traditional open-field combat. The dense and layered structure of cities introduces complexities that require specialized tactics, equipment, and planning to overcome.

The impact of urban terrain on visibility is a critical factor in modern combat operations, particularly in densely populated areas. The unique challenges posed by urban environments significantly complicate situational awareness for combatants, affecting both offensive and defensive operations.

Restricted Observation: Urban environments inherently restrict line-of-sight communication, which is vital for situational awareness. Combatants often face difficulties in detecting enemy movements due to the presence of buildings, narrow streets, and other structural obstructions. This is particularly evident in conflict zones like Syria, where a significant proportion of explosive weapon use has occurred in densely populated civilian areas, leading to high risks for civilians and complicating the operational landscape for combatants [69]. The ability to conduct effective reconnaissance is severely hampered as adversaries can easily conceal themselves behind walls or within buildings, making it challenging to assess threats in real-time [69].

Three-Dimensional Challenges: The verticality of urban landscapes introduces additional complexities. Combatants must monitor not only the horizontal plane but also the vertical dimensions, including rooftops and basements. This three-dimensional visibility challenge allows snipers and ambush teams to exploit vantage points that are often overlooked, thereby increasing the risk to ground forces [70]. The need for comprehensive situational awareness in three dimensions is underscored by studies that emphasize the importance of visibility analysis in urban planning and military operations [70].

Reduced Effectiveness of Traditional Reconnaissance: Traditional reconnaissance methods, such as aerial surveillance and ground-based observation, are often less effective in urban settings due to structural obstructions. While modern technologies like drones and thermal imaging have been developed to mitigate these challenges, they cannot fully compensate for the inherent limitations of urban environments [70]. The effectiveness of these technologies is often reduced in areas with high building density, where the potential for obstructions is significant [71]. This limitation necessitates the development of new strategies for reconnaissance and intelligence gathering in urban warfare.

Civilian Presence: The presence of civilians in urban combat zones further complicates visibility and situational awareness. The intermingling of combatants and non-combatants makes it difficult to distinguish between friend and foe, particularly in scenarios where insurgents blend into the civilian population [72]. This challenge is exacerbated by the increasing involvement of civilians in hostilities, which complicates the application of international humanitarian law principles, such as the distinction between combatants and civilians [72]. The difficulty in maintaining situational awareness in such contexts can lead to tragic outcomes, including civilian casualties, which have been documented in various conflict scenarios [69].

Urban environments present significant challenges to the manoeuvrability of both infantry and mechanized units, necessitating adaptations to navigate confined spaces and unpredictable layouts. The complexities of urban warfare are multifaceted, with several key challenges impacting operational effectiveness.

Narrow streets and dense layouts are primary obstacles for mechanized units, such as tanks and armoured personnel carriers (APCs). The limited width of roadways and the presence of alleyways restrict the movement of these larger vehicles, complicating their ability to manoeuvre, turn, or deploy effectively. This vulnerability is exacerbated in urban settings where ambushes and anti-tank weapons are prevalent, as mechanized units become easy targets when they are unable to reposition quickly or effectively [73, 74]. The constraints imposed by urban geography necessitate a re-evaluation of tactics and strategies to mitigate these risks.

Obstructions and debris further complicate manoeuvrability in urban environments. Collapsed buildings, wrecked vehicles, and makeshift barricades can block routes, forcing military units to seek alternative pathways or engage in obstacle-clearing operations under hostile fire. Such scenarios not only slow operational tempo but also disrupt coordination between units, as they may become isolated or delayed in their movements [75, 76]. The need for rapid adaptability in response to these challenges is critical for maintaining operational effectiveness in urban warfare.

Moreover, complex underground systems, such as subways, sewers, and tunnels, provide adversaries with covert routes for movement and resupply. These systems allow enemy forces to evade detection and bypass frontline engagements, complicating the ability of advancing forces to secure areas comprehensively. The presence of these underground networks necessitates a more nuanced approach to urban combat, where situational awareness and intelligence gathering become paramount [77, 78].

Finally, constrained supply lines present a significant challenge for logistics in urban warfare. The manoeuvring of logistics vehicles through urban terrain is fraught with risks, as supply routes can be easily targeted by snipers or improvised explosive devices (IEDs). This vulnerability complicates the sustainment of forces operating in urban environments, as timely resupply becomes a critical factor for operational success [79, 80]. The interplay between urban geography and military logistics underscores the importance of strategic planning and resource allocation in urban combat scenarios.

The dynamics of combat engagement in urban settings present unique challenges and operational requirements that differ significantly from traditional battlefields. The close-quarters combat (CQC) inherent in urban environments necessitates specialized training and equipment, as soldiers frequently engage in confined spaces such as hallways and narrow streets. This type of combat leads to heightened physiological responses, including increased heart rates and muscle strength, as soldiers navigate the intense stressors of CQC scenarios. Research indicates that sympathetic nervous system activation is significantly more pronounced in CQC compared to traditional combat settings, which can adversely affect operational performance and increase the risk of psychological conditions such as PTSD among soldiers [7, 81, 82].

Urban environments also favour ambush and guerrilla tactics, allowing defenders to exploit their intimate knowledge of the terrain. This strategic advantage enables them to stage ambushes, set traps, and establish sniper positions, complicating the attackers' efforts and often leading to hit-and-run engagements. The asymmetric nature of these confrontations can create a dynamic where attackers must adapt quickly to unexpected threats, further emphasizing the need for specialized training that prepares soldiers for the unpredictability of urban warfare [51]. The complexity of urban terrain can lead to operational challenges, as soldiers must remain vigilant against sudden attacks while managing the risks associated with civilian presence in these environments [51].

Collateral damage is another critical concern in urban combat, as the proximity of civilians and critical infrastructure necessitates more precise engagement methods. Heavy artillery and airstrikes, while effective in traditional warfare, are often restricted in urban settings due to the potential for civilian casualties, which can undermine strategic objectives and lead to broader humanitarian crises [23, 51]. The need for precision in targeting is underscored by the psychological impact of collateral damage on both military personnel and civilian populations, which can exacerbate tensions and complicate post-conflict recovery efforts [23, 51].

Moreover, the intensity of urban warfare typically results in increased ammunition usage. Combatants may fire suppressive rounds to clear rooms or engage enemies hiding behind cover, leading to a significant escalation in the consumption of ammunition compared to rural operations. This heightened demand for ammunition reflects the sustained engagements characteristic of urban combat, where soldiers must be prepared for prolonged firefights [83]. The challenges of providing effective fire support in urban environments are compounded by the difficulties in accurately targeting concealed enemies, raising the risk of friendly fire incidents and further complicating operational effectiveness [23, 51].

Psychological and Morale Challenges for Troops in Urban Warfare

Urban warfare is one of the most gruelling forms of combat, not just because of its physical demands but also due to the intense psychological and morale challenges it imposes on troops. The unique environment of cities—dense populations, confined spaces, and constant uncertainty—creates stressors that test soldiers' mental resilience, decision-making, and teamwork.

The unpredictable nature of urban warfare significantly contributes to the psychological strain experienced by soldiers. Urban environments are characterized by confined spaces such as narrow streets, hallways, and underground tunnels, where threats can emerge unexpectedly. This unpredictability is exacerbated by the prevalence of ambushes, sniper fire, booby traps, and improvised explosive devices (IEDs), which necessitate a state of heightened alertness among troops for prolonged periods. The constant vigilance required in these scenarios can lead to both physical exhaustion and mental fatigue, impairing decision-making and reaction times [7].

Research indicates that the asymmetrical combat scenarios typical of urban warfare can provoke heightened physiological responses, such as increased heart rates, due to the multitude of stimuli that soldiers must process in a short timeframe [7]. This overstimulation can result in anxiety and stress, further compounding the psychological burden faced by military personnel. Additionally, the difficulty in distinguishing between combatants and civilians adds another layer of stress, as soldiers must remain vigilant for potential threats hidden within crowds or buildings [7]. The psychological toll of this uncertainty is significant, as soldiers are forced to navigate complex social dynamics while maintaining operational effectiveness.

Moreover, the implications of prolonged mental fatigue in combat situations are profound. Studies have shown that mental fatigue can impair performance in critical tasks, such as marksmanship, due to the shared cognitive resources required for both mentally taxing and physically demanding tasks [84]. This suggests that the psychological strain induced by urban warfare not only affects soldiers' mental health but also their operational capabilities. The inability to effectively manage stress and fatigue can lead to detrimental outcomes in high-stakes environments, where quick and accurate decision-making is essential [84].

The emotional toll of civilian presence in urban warfare is a complex issue that intertwines moral dilemmas, psychological impacts on military personnel, and the challenges posed by international humanitarian law. The principle of distinction, which mandates that combatants differentiate between military targets and civilians, becomes increasingly challenging in urban environments where combatants often blend into civilian populations or utilize them as shields. This complexity can lead to significant emotional distress among soldiers, who may grapple with feelings of guilt and helplessness when faced with the suffering of civilians, including injuries, deaths, and displacement [51, 85].

The increasing prevalence of urban warfare has been linked to the rise of non-state actors, which complicates adherence to international humanitarian principles. These actors often operate outside the traditional frameworks of warfare, making it difficult for soldiers to apply the principle of distinction effectively [85]. As noted by John-Hopkins, the nature of asymmetric conflicts can lead to a situation where militarily stronger sides inadvertently escalate violence against civilian populations, further complicating the moral landscape for soldiers engaged in such conflicts [51]. The emotional burden of witnessing civilian suffering can lead to long-lasting psychological effects, including moral injury, which is characterized by feelings of guilt and shame stemming from actions or inactions during combat.

Moreover, the psychological impact of urban warfare extends beyond immediate combat experiences. Soldiers often face the dual challenge of fulfilling operational objectives while being acutely aware of the humanitarian implications of their actions. This constant balancing act can exacerbate stress and anxiety, as soldiers may feel pressured to protect civilians while also achieving military goals. The emotional toll is further compounded by the realities of modern warfare, where the use of explosive weapons in densely populated areas can lead to high civilian casualties, creating a cycle of trauma for both civilians and military personnel [23].

Urban warfare is characterized by close-quarters combat (CQC) in confined spaces, such as stairwells, alleys, and buildings, which significantly heightens the intensity of violence experienced by soldiers. This environment exposes combatants to traumatic sights and sounds, including the screams of wounded comrades and the chaos of explosions, which can lead to sensory overload. The cacophony of gunfire, collapsing structures, and civilian distress creates a tumultuous atmosphere that can overwhelm troops, impairing their ability to focus and maintain composure during critical moments of engagement [3, 7].

The psychological ramifications of urban warfare are profound, often resulting in long-term mental health issues such as post-traumatic stress disorder (PTSD). Studies indicate that the high-stress levels associated with urban combat scenarios, particularly those involving asymmetrical warfare, can lead to increased cardiovascular responses and heightened anxiety among soldiers [3, 7]. The unique challenges posed by urban environments, where combatants must navigate complex terrains and manage multiple stimuli, exacerbate these psychological effects [3, 7]. Furthermore, the nature of urban warfare necessitates a highly professionalized infantry capable of adapting to the intense demands of CQC, which further underscores the psychological toll on military personnel [3].

In addition to the immediate psychological impacts, the long-term consequences of exposure to such traumatic environments can lead to chronic mental health issues. The prevalence of PTSD among veterans of urban warfare has been documented, highlighting the need for effective mental health support systems for soldiers returning from combat [3, 7]. The interplay between the physical and psychological stressors of urban combat necessitates a comprehensive understanding of the urban battlespace, as it shapes not only the tactics employed but also the well-being of those engaged in such conflicts [3].

Urban warfare presents a unique set of challenges for military units, particularly due to the fragmented nature of urban environments. The maze-like structure of cities often forces military personnel to operate in small groups or even as individuals, leading to a significant risk of isolation from their larger units or commanders. This fragmentation can severely impair communication, resulting in troops feeling vulnerable and unsupported during critical moments of combat. Research indicates that such isolation can exacerbate psychological stress, contributing to a decline in morale among soldiers who are already facing the intense pressures of urban combat scenarios [86, 87].

The psychological ramifications of operating in isolated conditions are profound. Studies have shown that morale plays a crucial role in how military personnel cope with the stresses of combat. High morale can serve as a buffer against the negative effects of combat exposure, while low morale can lead to increased feelings of isolation and vulnerability [86]. The lack of coordination and support in urban warfare not only heightens these feelings but can also lead to interpersonal conflicts within units. The close confines of urban environments, combined with the high-stress nature of combat, can strain relationships among troops, further undermining teamwork and cohesion at a time when such unity is essential for operational success [86, 88].

Moreover, the psychological impact of urban warfare is compounded by the stressors inherent in combat situations. Troops often face unpredictable threats, such as ambushes and improvised explosive devices (IEDs), which can lead to acute stress reactions and long-term psychological issues, including post-traumatic stress disorder (PTSD) [89, 90]. The interplay of these factors creates a challenging environment where the mental health of soldiers can deteriorate, impacting their ability to function effectively as a cohesive unit. The importance of maintaining unit cohesion and morale cannot be overstated, as these elements are critical for operational effectiveness and the well-being of military personnel [86, 91].

Urban warfare is characterized by prolonged engagements, necessitating the systematic clearing of urban spaces, which can lead to significant physical and mental strain on military personnel. The complexity of urban environments often results in battles that extend over weeks or even months, as troops must navigate and secure each block and building methodically. This drawn-out nature of combat can lead to chronic fatigue among soldiers, exacerbated by the constant threat of enemy attacks and the challenges of conducting operations during nighttime, where the risk of ambush is heightened. Such conditions contribute to a pervasive state of sleep deprivation, which has been documented to severely impair soldiers' cognitive functions and emotional stability, increasing the likelihood of errors and lapses in judgment during critical operations [92, 93].

The psychological toll of urban warfare is profound, as the repetitive and gruelling nature of these conflicts can foster feelings of futility and despair among troops. When progress is slow or the objectives of missions become unclear, soldiers may experience a decline in morale and motivation, further complicating their operational effectiveness. This phenomenon is particularly evident in contemporary conflicts, such as the ongoing urban warfare in Ukraine, where the dynamics of urban combat have been observed to differ significantly from traditional warfare paradigms, leading to unexpected challenges for military strategists [17]. The emotional and psychological impacts of such prolonged engagements are critical considerations for military leadership, as they directly affect operational effectiveness.

Urban warfare presents unique challenges for military personnel, often leading to profound ethical dilemmas that significantly impact their psychological well-being. Soldiers are frequently faced with decisions that can result in civilian casualties, such as whether to call in airstrikes on buildings suspected of housing both combatants and non-combatants. This uncertainty can create intense moral conflicts, as troops grapple with the potential consequences of their actions. The psychological burden associated with these decisions is substantial, as evidenced by research indicating that soldiers in urban combat situations experience heightened stress and fatigue, which can impair their cognitive processing and decision-making abilities [94].

The moral injuries resulting from these ethical dilemmas can have long-lasting effects on soldiers' mental health. Studies have shown that experiences of civilian death or the inability to protect civilians can lead to feelings of guilt and betrayal, contributing to the development of moral injury [95, 96]. This condition is characterized by psychological distress stemming from actions that violate an individual's moral or ethical standards, which is particularly prevalent in combat scenarios where the lines between right and wrong are often blurred [97]. Furthermore, the perception of contributing to civilian suffering can exacerbate feelings of guilt, leading to a cycle of moral injury that may manifest as post-traumatic stress disorder (PTSD) or suicidal ideation among veterans [98, 99].

The impact of these moral dilemmas on troop morale cannot be overstated. When soldiers feel unsupported or uncertain about the ethical guidelines governing their actions, their morale can erode, further complicating their ability to cope with the psychological aftermath of combat (Britt et al., 2013). Research indicates that morale plays a crucial role in moderating the relationship between combat exposure and PTSD symptoms, suggesting that a supportive environment can help mitigate the adverse effects of combat experiences [86]. Additionally, leadership that fosters moral awareness and open discussions about ethical challenges can enhance soldiers' psychological resilience and overall mental health [100, 101].

In urban warfare contexts, adversaries frequently utilize psychological tactics to undermine the morale of attacking forces. Propaganda and misinformation are pivotal tools in this strategy, as they can effectively sway public opinion and demoralize troops. For instance, the dissemination of false narratives can create an environment of distrust and insecurity among soldiers, making them question their mission and the legitimacy of their actions. This is particularly evident in modern conflicts where social media platforms are exploited to amplify fear and disinformation, thereby exacerbating the psychological strain on military personnel [102].

Figure 10: S Marine Corps (USMC) Marines watch the local Iraqi civilian population and guide the 1ST Marine Division (MAR DIV), Division Main, convoy through the city of Ash Shumali, Iraq, on Highway 27 in support of Operation IRAQI FREEDOM. The U.S. National Archives, Public Domain, via NARA & DVIDS Public Domain Archive.

Moreover, the psychological impact of being perceived as an occupying force can significantly affect soldiers' morale. When troops feel that their efforts are misunderstood or unappreciated by the local populace, it can lead to feelings of isolation and frustration. This phenomenon is supported by research indicating that soldiers often experience heightened stress and emotional turmoil when faced with the complexities of civilian interactions in conflict zones [103, 104]. The perception of being a destructive force, rather than a liberating one, can further sap the morale of soldiers, leading to increased rates of mental health issues such as depression and anxiety [105].

Targeting supply lines is another tactic that adversaries may employ to weaken military resolve. Disrupting logistics not only hampers operational effectiveness but also instils a sense of vulnerability among troops, who may feel that their support systems are under constant threat. This can lead to a decline in combat readiness and an increase in psychological stress, as soldiers grapple with the implications of being cut off from essential resources [106].

The cumulative effect of these psychological tactics can be profound, as they create an atmosphere where soldiers are not only physically challenged but also mentally strained. The interplay between operational stressors and psychological resilience is critical; soldiers who possess higher levels of resilience are better equipped to cope with these adversities, while those lacking such resilience may succumb to the negative psychological impacts of warfare [103, 107]. Thus, understanding and addressing the psychological dimensions of urban warfare is essential for maintaining troop morale and operational effectiveness.

The cumulative psychological strain of urban warfare significantly impacts the mental health of troops, often leading to long-term challenges such as post-traumatic stress disorder (PTSD), anxiety, depression, and substance abuse. Urban combat environments, characterized by their complexity and intensity, expose soldiers to traumatic experiences that can leave lasting psychological scars. Research indicates that veterans returning from urban warfare often grapple with vivid memories of violent encounters, civilian suffering, and personal losses, which can persist for years and severely hinder their ability to reintegrate into civilian life [108, 109].

The transition from military to civilian life is particularly challenging for veterans, as they face an identity struggle that exacerbates mental health issues. The contrasting demands of military life—where obedience and deindividuation are emphasized—against the expectations of civilian autonomy can create significant psychological distress [108]. This struggle is compounded by the lack of adequate mental health support and debriefing following combat, which is critical for helping soldiers process their experiences and cope with the emotional aftermath of urban warfare [109]. Without such support, many veterans find themselves isolated and unable to manage their mental health effectively, leading to increased rates of PTSD and other psychological disorders [110, 111].

Moreover, the environmental and social contexts of urban warfare further complicate the mental health landscape for veterans. Urban areas, often densely populated and chaotic during conflicts, can amplify the psychological toll on soldiers who witness civilian casualties and destruction [25]. The psychological impact of these experiences is not only individual but also collective, as communities affected by urban warfare may struggle with their own mental health challenges, further complicating reintegration efforts for returning veterans [23]. Thus, addressing the mental health needs of veterans requires a comprehensive approach that includes both individual therapy and community support systems to facilitate healing and reintegration [109].

Civilian Presence and Impact on Tactical Decisions

The presence of civilians in urban warfare is one of the most challenging and consequential factors for military operations. Unlike traditional battlefields, urban areas are densely populated, requiring military forces to navigate not only the physical complexities of the terrain but also the ethical, legal, and operational challenges posed by civilian populations. This dynamic significantly influences tactical decisions, as military forces must balance achieving their objectives with minimizing harm to non-combatants and maintaining compliance with international law.

Principle of Distinction and Ethical Considerations

The principle of distinction is a fundamental tenet of international humanitarian law (IHL) that mandates the differentiation between combatants and civilians during armed conflict. This principle becomes particularly challenging in urban warfare, where the close proximity of civilian populations to combatants can blur these lines. Insurgent forces often exploit this ambiguity by integrating themselves within civilian communities, utilizing tactics such as human shielding and launching attacks from populated areas. This manipulation of the urban environment places military forces in ethically and legally precarious positions, as they must navigate the complexities of engagement without causing civilian harm [51, 112].

In urban warfare, the tactical decisions made by military commanders are heavily influenced by the need to minimize civilian casualties. The use of heavy artillery, airstrikes, or other area-effect weapons poses significant risks to non-combatants, leading to a dilemma where commanders must weigh the safety of their troops against the potential for civilian harm. This ethical consideration is further complicated by the broader implications of civilian casualties, which can affect the legitimacy of military operations and influence public perception and political outcomes [51, 113]. The consequences of failing to adhere to the principle of distinction can extend beyond immediate tactical failures, potentially undermining strategic objectives and leading to long-term repercussions in the conflict [51, 114].

Moreover, the evolving nature of warfare, particularly with the rise of asymmetric conflicts, necessitates a re-evaluation of how the principle of distinction is applied. As noted by Gordon and Perugini [112], the phenomenon of human shields exemplifies a contemporary challenge where civilians, while still protected under IHL, become intertwined with hostilities, complicating their status and the obligations of combatants. This situation underscores the need for a robust legal framework that can adequately address the unique challenges posed by urban warfare and ensure the protection of civilians amidst the complexities of modern conflict [113].

Constraints on Firepower and Engagement Rules

The dynamics of urban warfare are profoundly influenced by the presence of civilians, which imposes strict limitations on the use of firepower. The necessity to protect civilian lives and infrastructure leads to the implementation of restrictive rules of engagement (ROEs) that mandate precision and restraint in targeting. This is particularly evident in urban environments where the risk of collateral damage is significantly heightened. Collateral damage not only results in civilian casualties but also threatens critical infrastructure such as hospitals, schools, and water supplies, exacerbating humanitarian crises [51, 115]. The implications of these realities are profound, as military forces are compelled to rely heavily on intelligence, surveillance, and reconnaissance (ISR) assets to accurately identify enemy positions, a task that is complicated by the limited visibility and situational awareness inherent in urban settings [4].

The use of precision-guided munitions (PGMs) is often seen as a solution to mitigate collateral damage. However, these munitions are resource-intensive and depend on accurate intelligence for effective targeting. The operational complexity increases as forces may be required to conduct surgical operations, such as systematically clearing buildings to minimize civilian harm [52]. While these methods are essential for protecting non-combatants, they also demand significant time, manpower, and increase the risk to military personnel, thereby influencing broader operational timelines and strategies [115]. The balance between operational effectiveness and civilian protection remains a critical challenge in urban warfare, as military planners must navigate the ethical and tactical implications of their actions [116].

Precision-guided munitions (PGMs) are an indispensable tool in urban warfare, where the dense environment and civilian presence demand a high level of accuracy. These advanced weapons, including guided bombs, missiles, and artillery shells, are specifically designed to strike targets with precision, minimizing collateral damage. Their use is not only an operational necessity in complex urban combat scenarios but also an ethical imperative, ensuring adherence to international humanitarian law by reducing harm to non-combatants and critical infrastructure.

One of the most significant advantages of PGMs in urban warfare is their ability to limit collateral damage. In densely populated areas, where civilians often live and work near combatants, traditional weapons pose significant risks. PGMs, however, enable military forces to strike precise targets, such as enemy strongholds within buildings, while minimizing destruction to surrounding areas and non-combatants.

PGMs also excel in neutralizing hardened or concealed targets, which are common in urban warfare. Enemies often fortify themselves within reinforced structures or hide in concealed positions. Weapons like bunker-buster bombs and guided artillery shells are specifically designed to penetrate these fortifications and eliminate threats with high accuracy. This capability is particularly crucial when adversaries create defensive strongholds within urban settings.

Another advantage is the reduced risk to friendly forces. PGMs allow for long-range, accurate strikes, minimizing the need for close-quarters combat, which is inherently dangerous in urban environments. By neutralizing heavily fortified or high-value targets from a distance, troops are less exposed to enemy fire and ambushes. Additionally, PGMs provide tactical flexibility, enabling commanders to respond effectively to dynamic battlefield conditions. Whether targeting a sniper nest on a rooftop or neutralizing a vehicle-borne explosive in a crowded street, PGMs deliver precision across a range of tactical scenarios.

Different types of PGMs serve distinct roles in urban warfare. Air-launched weapons, such as laser-guided bombs (e.g., Paveway) and GPS-guided bombs (e.g., JDAM), are commonly deployed to strike specific structures or vehicles with precision. Missiles like the Hellfire, often launched from drones or helicopters, are ideal for targeting small vehicles or individuals in urban settings.

Figure 11: Aviation Ordnancemen aboard USS Abraham Lincoln (CVN 72) load a Joint Direct Attack Munition (JDAM) onto a F/A-18C Hornet. U.S. NAVY, Public Domain, via Picryl.

Artillery and mortar shells, such as GPS-guided Excalibur rounds, are highly effective in urban operations. These munitions can strike within meters of their intended target, minimizing risks to nearby civilians and infrastructure. Precision mortar systems are also increasingly utilized to accurately target threats in areas where traditional mortars would pose excessive risks. Ground-launched missiles, including anti-tank guided missiles (ATGMs) like the Javelin or Spike, are invaluable in urban warfare for engaging vehicles and fortified positions with high precision.

Despite their advantages, PGMs face several challenges in urban environments. One of the primary limitations is their high cost. Precision-guided weapons are significantly more expensive than conventional munitions, with some costing tens of thousands of dollars per unit. This financial burden makes their large-scale or prolonged use difficult to sustain.

Figure 12: Aviation Ordnancemen prepare to load 500-pound laser guided bombs (GBU-12) onto weapon pylons under an F-14B Tomcat. U.S. NAVY, Public Domain, via Picryl.

Another challenge is the reliance on accurate intelligence to identify targets. In urban warfare, where adversaries blend into civilian populations or use human shields, obtaining reliable target data is a persistent difficulty. Misidentification or faulty intelligence can lead to unintended casualties or collateral damage, undermining the objectives of precision strikes.

Sophisticated adversaries may also employ electronic countermeasures, such as jamming or spoofing, to disrupt the guidance systems of PGMs, reducing their effectiveness. Additionally, urban environments themselves pose limitations. Dense infrastructure and narrow streets can interfere with GPS signals or obscure targets, complicating the deployment of guided weapons.

The use of PGMs aligns with the principles of international humanitarian law, particularly the obligation to minimize civilian harm. By enabling precision targeting, these weapons help uphold the principle of distinction between combatants and civilians. However, their use also carries strategic implications. Successful deployment of PGMs can enhance the legitimacy of military operations by demonstrating a commitment to reducing collateral damage. Conversely, failures or misuse of these weapons can damage public perception and undermine broader strategic goals.

PGMs also come with the risk of overreliance on their precision. The perception that these weapons enable "surgical" warfare can lower the threshold for military interventions, particularly in urban areas, where the consequences of conflict are profound.

PGMs have played a critical role in several urban conflicts. During the battle to retake Mosul from ISIS (2016-2017), precision-guided bombs and artillery were used extensively to target insurgents in fortified urban positions. Despite their effectiveness, the dense population and the use of human shields by ISIS highlighted the limitations of even the most advanced PGMs. Similarly, in the Gaza Strip, PGMs have been used to target militant positions while attempting to minimize harm to civilians. These operations demonstrate the potential and constraints of PGMs in complex urban environments.

Precision-guided munitions have transformed urban warfare by offering unparalleled accuracy and reducing collateral damage. Their ability to target threats precisely while protecting civilians and infrastructure makes them a crucial tool in addressing the ethical and operational challenges of urban combat. However, their effectiveness is tempered by high costs, the need for accurate intelligence, and the inherent complexities of urban environments. While not a perfect solution, PGMs remain an essential asset for modern militaries, enabling them to navigate the demands of urban warfare with greater precision and responsibility.

The urban landscape itself also presents unique challenges that further complicate military operations. The three-dimensional nature of urban environments provides enhanced concealment for combatants, making it difficult to distinguish between combatants and civilians [47]. This complexity necessitates a shift in military doctrine, where traditional firepower is often replaced by tactics that prioritize troop protection and civilian safety [115]. The evolving nature of urban warfare underscores the need for military forces to adapt their strategies to address the intricate realities of conducting operations in densely populated areas, ensuring that the principles of distinction and proportionality are upheld in accordance with international humanitarian law [51, 52].

Impact on Movement and Manoeuvrability

The presence of civilians in urban environments significantly impacts military operations, particularly troop movement and manoeuvrability. Soldiers are often required to navigate through crowded streets, marketplaces, and residential areas while maintaining situational awareness of both potential threats and the civilian population. The risk of civilian casualties imposes restrictions on the use of heavy vehicles, such as tanks or armoured personnel carriers, in narrow streets or alleys, as these vehicles can cause unintended harm to non-combatants. This necessity for caution often leads to the allocation of valuable military resources to ensure the safety of civilians during operations [36, 47].

Military forces must consider evacuation routes and safe zones for civilians when planning their operations. This planning is crucial, as the presence of civilians can complicate tactical manoeuvres and necessitate adjustments in strategy. For instance, panicked civilians fleeing combat zones can inadvertently block troop movements or disrupt tactical plans, creating additional challenges for military personnel [2]. Conversely, civilians may also provide cover for enemy forces, further complicating efforts to secure an area. Such dynamics require military forces to adapt their strategies in real-time, often under intense pressure and in rapidly evolving situations [36].

The unpredictability of civilian behaviour adds another layer of complexity to military operations in urban settings. The interaction between military personnel and civilians can lead to unforeseen consequences, necessitating a flexible approach to operations. For example, the integration of smart city technologies can enhance situational awareness by providing real-time data on civilian movements and behaviours, which can be crucial for military planning [37]. However, the reliance on technology also introduces challenges related to data sharing and trust between military and civilian entities, which can impede operational effectiveness [117].

Intelligence Gathering and Information Operations

Civilians play a complex and dual role in urban warfare, acting as both potential risks and valuable sources of intelligence. Military forces often rely on local populations to gather critical information regarding enemy movements, safe zones, and hidden threats. This reliance is particularly pronounced in urban environments where the civilian population is dense and intertwined with military operations. However, establishing trust with civilians can be a formidable challenge, especially in contexts where military forces are perceived as occupiers or antagonists. As highlighted by John-Hopkins [51], asymmetric conflicts often lead to a situation where the civilian population becomes a target, complicating the dynamics of trust and cooperation between military forces and civilians. The need for tactical decisions that prioritize building rapport with civilians through humanitarian aid, clear communication, and visible efforts to ensure their safety is paramount [51].

Moreover, the potential for enemy forces to manipulate civilian populations adds another layer of complexity to urban warfare. Civilians can be used to disseminate disinformation, create propaganda, or undermine the legitimacy of military operations. This necessitates that military forces engage in effective information operations to counter false narratives while maintaining transparency and credibility. The work of Ljungkvist [17] emphasizes the importance of understanding the urban nature of contemporary warfare, where civilian perceptions can significantly influence operational outcomes. Effective management of civilian perceptions is crucial, as the success or failure of these efforts can directly impact the broader operational environment [118].

In addition to the challenges of building trust and managing perceptions, military operations must also consider the humanitarian implications of their actions. The systematic targeting of civilian infrastructure, as discussed by Sowers et al. [119], illustrates how the destruction of essential services can exacerbate civilian suffering and lead to further alienation from military forces. This highlights the importance of integrating humanitarian considerations into military planning and operations, ensuring that civilian safety and well-being are prioritized alongside military objectives [23, 120]. Ultimately, the dual role of civilians in urban warfare necessitates a nuanced approach that balances military objectives with the imperative to protect and engage local populations effectively.

Psychological and Morale Impacts

The presence of civilians in conflict zones significantly impacts the psychological and morale aspects of tactical decision-making for military personnel. Soldiers operating in urban environments often confront ethical dilemmas, particularly when civilian casualties occur, even unintentionally. The psychological burden of witnessing civilian suffering can lead to moral injury, adversely affecting troop morale and unit cohesion. Research indicates that the emotional stress associated with civilian casualties can lead to a decline in operational effectiveness, as soldiers grapple with the implications of their actions on non-combatants [121].

In modern warfare, the distinction between combatants and civilians has become increasingly blurred, especially in urban settings where civilian populations are dense. This complexity necessitates that tactical leaders incorporate ethical considerations into their operational planning. The need for comprehensive training and support systems is paramount to equip soldiers with the tools necessary to navigate these challenges effectively. For instance, training programs focused on Tactical Combat Casualty Care (TCCC) have been developed to enhance decision-making under stress, emphasizing the importance of situational awareness and ethical considerations in combat scenarios [122]. Such training can help mitigate the psychological toll on soldiers by preparing them for the realities of combat in civilian-populated areas.

Moreover, the implications of civilian casualties extend beyond immediate psychological impacts; they can also affect broader strategic outcomes. Research has shown that military operations resulting in civilian casualties can lead to increased hostility from local populations, further complicating the operational environment and undermining the legitimacy of military actions [121]. Therefore, tactical leaders must consider not only the immediate effects of their decisions on troop morale but also the long-term consequences on community relations and operational success.

To address these multifaceted challenges, military organizations are increasingly recognizing the need for robust support systems that include psychological counselling and ethical training. This holistic approach aims to foster resilience among soldiers, enabling them to maintain morale and unit cohesion even in the face of difficult ethical dilemmas [121, 122]. Ultimately, the integration of civilian considerations into tactical decision-making processes is essential for maintaining operational effectiveness and ensuring the well-being of military personnel.

Strategic Implications of Civilian Harm

Tactical decisions made in urban environments significantly influence strategic outcomes, particularly in military operations where civilian safety is a paramount concern. The implications of civilian casualties extend beyond immediate operational goals, affecting public support both domestically and internationally. High civilian casualties can lead to a loss of legitimacy for military operations, as evidenced by research indicating that civilian harm often fuels resentment among local populations, potentially driving them toward insurgent forces [121, 123]. This phenomenon is particularly pronounced in counterinsurgency and stabilization missions, where the perception of the military as a protector is crucial for gaining local support [124].

The strategic consequences of civilian casualties are multifaceted. For instance, studies have shown that civilian casualties can lead to increased insurgent violence in the long term, as affected communities may seek revenge or feel compelled to support insurgent groups as a reaction to military actions perceived as unjust [121, 123]. Conversely, successful efforts to minimize civilian harm can enhance the legitimacy of military operations and foster long-term stability. This is supported by findings that demonstrate a correlation between reduced civilian casualties and increased local support for military operations, suggesting that protecting civilians is not only a moral imperative but also a strategic necessity [124].

Moreover, the management of civilian casualties and the military's response to them can shape public opinion and influence political dynamics. Research indicates that democracies, in particular, face significant pressure to minimize civilian casualties, as public support for military action can wane in the face of high civilian death tolls [125, 126]. This sensitivity to civilian casualties can lead to more cautious military strategies, which aim to balance operational effectiveness with the need to maintain public support [127]. Ultimately, the tactical decisions made in urban warfare settings must consider the broader strategic implications of civilian harm, as these decisions can either undermine or bolster the legitimacy of military operations and influence the trajectory of conflict.

The Role of Intelligence and Reconnaissance in Urban Settings

Intelligence and reconnaissance are fundamental components of military operations, and their importance is amplified in urban settings due to the complexities and unpredictability of such environments. Cities present unique challenges, including dense populations, layered terrain, and concealed threats, all of which necessitate comprehensive and adaptable intelligence and reconnaissance efforts. These capabilities are critical for understanding the urban battlefield, identifying threats, protecting civilians, and ensuring the success of military objectives.

Understanding the urban terrain is crucial for military operations, particularly in urban combat scenarios where the environment presents unique challenges. Urban areas are characterized by complex features such as multi-story buildings, narrow streets, and underground networks, which significantly complicate reconnaissance and intelligence efforts. Unlike open terrain, urban environments lack predictable patterns, making it difficult to anticipate enemy movements or establish clear lines of engagement. Detailed intelligence is essential to map the city's infrastructure, identify chokepoints, and locate strategic assets like power grids, communication hubs, and transportation networks [128, 129].

The three-dimensional nature of urban combat necessitates a comprehensive understanding of the terrain, including threats from rooftops, underground tunnels, and the interiors of buildings. Advanced technologies such as unmanned aerial vehicles (UAVs) are increasingly employed for reconnaissance in urban environments. These UAVs can autonomously navigate complex terrains, providing real-time information about the layout and condition of the urban landscape [128]. The use of UAVs enhances situational awareness and allows for the identification of potential ambush sites, critical for effective operational planning [130].

Furthermore, technologies such as Light Detection and Ranging (LiDAR) play a pivotal role in urban terrain analysis. LiDAR systems are utilized to create detailed digital terrain models that can differentiate between various urban features, including buildings and vegetation [131]. This capability is essential for understanding the urban landscape and facilitating effective reconnaissance operations. Additionally, through-the-wall imaging (TWI) technologies enable the detection and imaging of targets behind obstacles, which is particularly valuable in urban settings where visibility is often obstructed [129, 132]. These technologies collectively contribute to a more nuanced understanding of urban terrains, allowing military commanders to visualize and strategize effectively in complex environments.

Understanding urban terrain involves integrating various technologies and methodologies to overcome the inherent complexities of urban environments. The combination of UAV reconnaissance, LiDAR mapping, and TWI technologies provides military forces with the necessary tools to navigate and operate effectively within urban landscapes, ultimately enhancing situational awareness and operational success [133].

In urban warfare, the challenge of identifying enemy movements and positions is exacerbated by the complex interplay between military operations and civilian populations. Adversaries often exploit urban environments, utilizing buildings for fortification and underground networks for movement, which complicates reconnaissance efforts. Intelligence gathering becomes crucial in these settings, where traditional military strategies must adapt to the realities of urban conflict. Reconnaissance teams, drones, and advanced surveillance systems are employed to gather actionable intelligence that informs military decision-making. For instance, the integration of technology in urban warfare has been shown to enhance situational awareness and operational effectiveness, allowing for better tracking of enemy movements and positions [51, 134].

Human intelligence (HUMINT) plays a pivotal role in urban warfare, as local civilians can provide valuable insights into enemy activities, safe zones, and potential threats. However, establishing trust within civilian populations is fraught with challenges, particularly in contested areas where loyalties may be divided. The effectiveness of HUMINT is contingent upon the ability of military forces to engage with local communities and build rapport, which can be difficult in environments marked by violence and suspicion [135, 136]. Moreover, the reliance on civilian information necessitates a careful approach to minimize the risk of civilian casualties, as the consequences of misidentifying threats can be dire [137, 138].

In addition to HUMINT, signals intelligence (SIGINT) and electronic surveillance are critical components of modern urban warfare strategies. These methods allow for the interception of communications, which can reveal enemy plans and facilitate the disruption of their coordination efforts. The ability to monitor and analyse electronic communications provides military forces with a significant advantage in understanding enemy tactics and movements [139, 140]. However, the ethical implications of such surveillance practices must also be considered, particularly regarding the potential for civilian privacy violations and the broader impact on community trust [141, 142].

The dynamics of urban warfare necessitate a multifaceted approach to intelligence gathering that combines technological advancements with human insights. The integration of drones and other surveillance technologies has transformed the landscape of urban conflict, enabling real-time monitoring of enemy movements while also raising concerns about the potential for collateral damage [143, 144]. As military operations increasingly occur within civilian environments, the imperative to protect non-combatants while effectively countering enemy actions remains a critical challenge for armed forces engaged in urban warfare [145, 146].

The protection of civilians in urban warfare is a critical concern, particularly considering the complexities introduced by densely populated environments. International humanitarian law emphasizes the principle of distinction, which mandates that military forces differentiate between combatants and non-combatants. This principle is especially challenging to uphold in urban settings where civilian populations are often intermingled with combatants, necessitating accurate and timely intelligence to inform military operations and humanitarian efforts.

Accurate intelligence plays a pivotal role in identifying civilian concentrations and planning military operations that minimize harm to non-combatants. As highlighted by Hultman et al., the presence of military observers alone does not suffice for civilian protection and can paradoxically lead to increased civilian casualties due to the moral-hazard problem associated with humanitarian interventions [147]. This underscores the necessity for robust intelligence frameworks that can effectively inform military strategies while adhering to international law.

Moreover, reconnaissance technologies such as thermal imaging and drone surveillance are instrumental in identifying civilian evacuation routes and makeshift shelters. These technologies enable military forces to coordinate humanitarian aid effectively and adapt their tactics in real-time to changing civilian dynamics, such as the movement of displaced populations or the emergence of new threats. Williams [148] notes that UN peacekeepers are increasingly mandated to protect civilians from imminent physical violence, which necessitates a comprehensive understanding of the operational environment, including civilian agency and resilience. This aligns with the findings of Baines and Paddon [145], who emphasize the importance of recognizing civilian agency in humanitarian protection efforts.

The integration of civilian protection monitoring, as discussed by Krause and Kamler [149], can also enhance the effectiveness of protection strategies in protracted conflicts, such as those seen in Myanmar. Their research indicates that civilian monitoring can positively impact the protection of civilians, suggesting that a collaborative approach involving both military and civilian actors can yield better outcomes in safeguarding non-combatants.

Furthermore, the literature on self-protection strategies adopted by civilians in conflict zones reveals that when formal protection mechanisms fail, civilians often resort to various strategies to safeguard themselves from violence. Jose and Medie [150] highlight that these self-protection efforts are critical yet often underestimated in the broader discourse on civilian protection. This perspective is echoed by Suarez [151], who discusses the strategies employed by civilians in the Eastern Democratic Republic of the Congo, illustrating the need for international actors to understand and support these grassroots efforts.

The urban environment poses significant challenges for intelligence gathering and reconnaissance, primarily due to the complexities introduced by dense infrastructure and the presence of civilians. Traditional surveillance methods, such as satellite imagery and line-of-sight observation, are often obstructed by buildings and urban layouts, which can severely limit their effectiveness [152]. Additionally, signal interference from urban structures can disrupt communications, complicating the operations of electronic intelligence systems [153]. This disruption necessitates the development of advanced methodologies that can adapt to the unique challenges presented by urban environments.

The presence of civilians further complicates intelligence gathering efforts. Distinguishing between combatants and non-combatants requires granular, real-time information, which is often difficult to obtain in densely populated areas [154]. Adversaries may exploit these complexities by employing tactics such as using human shields or disseminating misinformation, thereby misleading intelligence efforts [155]. To counter these challenges, a multi-layered approach is essential. This includes integrating ground reconnaissance, aerial surveillance, electronic monitoring, and human intelligence to create a comprehensive intelligence framework capable of navigating the urban landscape effectively [156].

Recent advancements in artificial intelligence (AI) and machine learning have shown promise in enhancing intelligence gathering capabilities in urban settings. AI technologies can analyse vast amounts of data from various sources, providing actionable insights that traditional methods may overlook [157]. For instance, AI can facilitate the integration of heterogeneous data sources, improving situational awareness and decision-making processes [158]. Furthermore, the application of edge intelligence allows for real-time data processing, which is crucial in dynamic urban environments where conditions can change rapidly [159]. By leveraging these technologies, intelligence agencies can improve their operational effectiveness and resilience against adversarial tactics.

The integration of advanced technologies into urban intelligence has significantly enhanced reconnaissance capabilities, particularly through the use of unmanned aerial vehicles (UAVs), artificial intelligence (AI), and ground-based robotic systems. These technologies collectively improve situational awareness and operational efficiency in urban environments.

UAVs, or drones, have emerged as critical tools for urban reconnaissance due to their ability to provide real-time visual and thermal imagery. They can navigate complex urban landscapes, including narrow streets and rooftops, offering a level of detail that traditional surveillance methods cannot match. Drones are increasingly utilized for various applications such as monitoring, mapping, and infrastructure inspection, which are essential for urban management and security [160-162]. The European Drones Outlook Study highlights the rapid growth of civilian drone applications, emphasizing their role in enhancing urban functionalities [160]. Furthermore, UAVs can operate in hazardous environments, making them invaluable for search and rescue operations, as they are not hindered by obstacles like rubble or debris [163].

Artificial intelligence and machine learning further augment urban intelligence by enabling sophisticated data analysis and pattern recognition. AI-driven systems can process vast amounts of data from diverse sources, including surveillance footage and social media, to create a comprehensive operational picture. This capability allows for the identification of activities and the prediction of future movements, enhancing decision-making processes in urban scenarios [164, 165]. The integration of AI in reconnaissance systems facilitates the automation of data collection and analysis, thereby improving response times and operational effectiveness [164].

In addition to aerial systems, ground-based robotic technologies play a crucial role in urban intelligence. These systems are particularly useful in hazardous environments where human reconnaissance is too risky. For example, ground reconnaissance robots can gather information from confined spaces, such as buildings and underground tunnels, thereby improving situational awareness and reducing risks to personnel [166, 167]. The development of autonomous ground vehicles (AGVs) has been pivotal in performing reconnaissance and rescue operations in challenging terrains, further demonstrating the versatility of robotic systems in urban settings [168].

Moreover, the collaboration between aerial and ground-based systems enhances operational capabilities. The integration of UAVs with ground robots allows for coordinated reconnaissance missions, where aerial systems can provide overhead surveillance while ground robots navigate through confined spaces [169]. This synergy not only expands the operational range of these technologies but also improves their effectiveness in various reconnaissance, surveillance, and inspection tasks [169].

Intelligence and reconnaissance play a pivotal role in shaping strategic and tactical decision-making in urban warfare. The integration of detailed intelligence into mission planning is essential for identifying objectives, allocating resources, selecting routes, and determining engagement strategies. This is particularly evident in the context of military operations, where the effectiveness of reconnaissance-firing systems (RFS) has significantly increased over the years, enhancing target acquisition and engagement capabilities from 10-30% to 50-90% in various conflicts [170]. The use of artificial intelligence (AI) in military intelligence further accelerates decision-making processes, allowing for improved situational awareness and the effective utilization of vast amounts of data [171].

During operations, intelligence is crucial for maintaining situational awareness, which is necessary for executing complex manoeuvres and coordinating between units. The ability to adapt to unforeseen challenges is enhanced by real-time intelligence updates, which inform commanders about changing conditions on the battlefield. For instance, the implementation of UAV reconnaissance has been shown to significantly improve the effectiveness of battalion force protection operations, allowing for timely adjustments to tactical plans based on the evolving operational environment [172]. Moreover, AI-driven mission planning models have been developed to assist in air strike operations, demonstrating the potential of technology to reduce the cognitive burden on commanders and improve operational efficiency [173].

Post-operation analysis is equally important, as reconnaissance efforts help assess mission success and identify areas for improvement. This feedback loop is vital for preparing for future engagements, as it allows military planners to refine their strategies based on lessons learned from previous operations. The integration of AI in post-operation assessments can enhance the analysis of reconnaissance data, providing deeper insights into operational effectiveness and informing future tactical decisions [171, 174].

The significance of intelligence in urban warfare is underscored by historical and contemporary conflicts, where the integration of advanced technologies and strategic intelligence has proven essential for operational success. A notable example is the Battle of Mosul (2016–2017), where intelligence and reconnaissance were crucial for identifying ISIS positions and mapping the city's complex infrastructure. This facilitated the coordination of airstrikes, thereby minimizing civilian casualties and enhancing operational effectiveness [175, 176]. The urban environment, characterized by its dense population and intricate layouts, necessitates sophisticated intelligence frameworks to navigate the challenges posed by urban combat scenarios.

In the ongoing conflict in Ukraine, intelligence efforts have similarly played a pivotal role. The ability to monitor enemy movements and coordinate defences has been critical in cities like Kyiv and Mariupol, where urban landscapes complicate traditional military operations. The integration of artificial intelligence (AI) and smart city technologies has emerged as a transformative factor in urban warfare, enabling more effective decision-making and resource allocation [41, 177]. For instance, the development of intelligent urban computing systems has been highlighted as a means to enhance situational awareness and operational planning in urban environments [178].

Moreover, the evolution of warfare towards more intelligent operations is reflected in the increasing reliance on AI and smart technologies. Research indicates that urban combat operations are increasingly supported by smart city frameworks, which provide essential data for equipment support and operational command [175, 176]. This shift towards intelligence-driven urban warfare not only improves military effectiveness but also emphasizes the importance of protecting civilian lives amidst conflict [176]. The application of augmented reality and other advanced technologies further enhances situational awareness, allowing military forces to operate more effectively in contested urban areas [4].

In urban warfare, intelligence and reconnaissance are indispensable tools for navigating the complexities of densely populated and geographically intricate environments. They enable military forces to understand the terrain, identify threats, protect civilians, and adapt to rapidly changing conditions. While urban settings pose significant challenges to intelligence gathering, advances in technology and multi-layered approaches continue to enhance capabilities. As urbanization increases globally, the role of intelligence and reconnaissance in urban warfare will remain critical to achieving both tactical and strategic success.

Chapter 2

OFFENSIVE STRATEGIES IN URBAN WARFARE

Preparation for Urban Offensive Operations

U rban offensive operations are military actions designed to seize control of a city or urban area from an adversary. These operations are among the most challenging and resource-intensive forms of combat, requiring forces to navigate dense terrain, minimize civilian casualties, and adapt to a complex and unpredictable environment. Offensive operations in urban warfare are typically carried out in phases and involve a combination of tactics, coordination, and specialized equipment. Urban offensive operations present unique challenges and characteristics that distinguish them from traditional military engagements. Key characteristics include Close-Quarters Combat (CQC), the multidimensional nature of urban environments, the presence of civilians, and the high resource demands associated with these operations.

Close-Quarters Combat (CQC) is a defining feature of urban offensives, where engagements often occur in confined spaces such as buildings, stairwells, and narrow streets. This necessitates highly trained infantry capable of executing room-clearing techniques and engaging in hand-to-hand combat. The evolution of CQC techniques, originally developed by Special Operations Forces, has been disseminated to regular infantry units due to the nature of urban operations in Iraq and Afghanistan, highlighting the need for specialized training in these environments [3]. Furthermore, soldiers must maintain a high level of physical fitness and mental preparedness to effectively navigate the stressors associated with CQC, which can lead to overstimulation and require adaptive responses to rapidly changing combat scenarios [7, 179].

The multidimensional battlefield of urban environments complicates military operations. Urban areas are inherently three-dimensional, with threats potentially arising from ground level, rooftops, and subterranean structures such as tunnels and sewers. This complexity necessitates robust situational awareness and reconnaissance capabilities to identify and mitigate threats from all levels [2]. The intricate layout of urban spaces, including the existence of underground facilities and the layout of street networks, poses significant challenges for military planners [2]. The integration of advanced technologies, such as augmented reality systems, can enhance situational awareness and operational effectiveness in these complex environments [4].

The presence of civilians in urban settings adds another layer of complexity to offensive operations. Commanders must navigate the delicate balance between achieving military objectives and minimizing harm to civilian populations, adhering to international humanitarian law [180]. The involvement of civilians in military operations has increased, necessitating a comprehensive understanding of their roles and the potential impacts of military actions on civilian life [181, 182]. Effective civil-military coordination is essential to address the challenges posed by civilian presence, ensuring that military operations do not compromise humanitarian principles [183].

Finally, high resource demand is a hallmark of urban offensives. These operations often require significant manpower, ammunition, and logistical support, consuming more resources than traditional battles due to prolonged engagements and the need for precision-guided munitions [2, 31]. The destruction or obstruction of supply routes in urban environments further complicates logistics, necessitating innovative solutions to sustain operations [31]. The integration of military and civilian resources, particularly in humanitarian contexts, can enhance operational effectiveness and resource management [184].

Circumstances Calling for an Urban Offensive Operation

Urban offensive operations are undertaken when controlling, securing, or neutralizing urban areas becomes essential to achieving broader military objectives. These operations are driven by the unique strategic importance of cities as centres of political power, economic activity, logistical infrastructure, and symbolic value. The circumstances prompting such operations are varied and often multifaceted.

Strategic Importance of Cities: Urban offensive operations are frequently directed at cities that hold political, economic, or symbolic significance. Political and administrative centres, such as capitals, often become primary targets due to their role in governance and decision-making. Capturing such cities disrupts adversarial control and weakens their ability to sustain conflict. For example, Baghdad was a key objective during the 2003 invasion of Iraq, as it symbolized the seat of power.

Economic and logistical hubs, such as cities with ports, railways, or industrial facilities, are also critical targets. Controlling these resources can cripple an adversary's supply chains and operational capabilities, as evidenced by the Battle of Mariupol during the Ukraine conflict, where the port city's strategic importance made it a focal point. Additionally, cities often hold symbolic value, with their capture serving as a morale boost for the attackers and a psychological blow to the enemy. The battle for Kyiv in 2022 highlighted this dynamic, as its strategic and symbolic importance made it a key target.

Presence of Enemy Forces: Urban areas are often strongholds for enemy forces, especially insurgents or militias, who exploit the complexity of the environment for concealment, recruitment, and operational staging. Offensive operations are necessary to eliminate these safe havens, as demonstrated during the Battle of Fallujah in 2004, which aimed to remove entrenched insurgents disrupting broader operations in Iraq.

Guerrilla forces frequently retreat into cities to leverage the urban terrain's complexity. This tactic forces adversaries to engage in urban offensives, such as the Battle of Mosul (2016–2017), where ISIS militants used the city as both a defensive stronghold and a base for operations. Urban offensives are also necessary when adversaries fortify cities to slow down advancing forces or deny access to critical locations.

Threats to Civilian Populations: Urban offensive operations are often launched to protect civilian populations from violence, coercion, or terrorism. For instance, during the Bosnian War, offensive actions in Sarajevo aimed to end the prolonged siege and protect civilians from indiscriminate attacks. In cases where civilians are used as human shields or where humanitarian aid is blocked, offensive operations may be required to secure safe zones and provide assistance.

Preventing Enemy Resupply or Retreat: Urban operations are crucial in disrupting enemy logistics. Capturing cities that serve as transportation hubs can sever supply lines, as seen in General Ulysses S. Grant's campaign during the American Civil War to capture Vicksburg and disrupt Confederate supplies. Similarly, targeting urban areas that provide escape routes can trap adversaries, forcing their surrender or destruction.

Countering Threats to Strategic Stability: Cities often house critical infrastructure, such as power plants, water supplies, and communication networks. Urban offensive operations may be necessary to secure these assets and maintain strategic stability. Additionally, insurgencies destabilizing a region by controlling urban areas may require offensive action to restore government authority, as seen in operations in Grozny during the Chechen Wars.

Retaking Occupied Urban Areas: When adversaries capture a city, retaking it becomes an operational imperative. The liberation of such cities is critical for restoring national sovereignty, public morale, and territorial integrity. The recapture of Kherson during the Ukraine conflict exemplifies the importance of reclaiming strategic urban areas.

Preventing Further Enemy Advances: Urban areas lying along strategic corridors often become pre-emptive targets to block enemy advances. Securing these cities can create buffer zones that protect critical regions and prevent enemy encroachment into more vulnerable areas. Historical examples include urban offensives aimed at halting advancing armies during World War II.

Post-Conflict Stabilization Efforts: After a conflict, offensive operations may be necessary to remove residual enemy forces, eliminate sleeper cells, or secure urban areas experiencing unrest. Stabilizing these areas ensures long-term security and facilitates reconstruction and governance, laying the groundwork for lasting peace.

Urban offensive operations are tailored to address the unique challenges of urban warfare, whether the objective is to neutralize threats, protect civilians, secure vital infrastructure, or reclaim occupied areas. Such operations demand careful planning, precise execution, and a balance between military necessity and humanitarian considerations. By addressing the strategic, symbolic, and logistical importance of cities, urban offensive operations play a pivotal role in achieving broader military and political objectives.

Preparation for an Urban Offensive

The key steps involved in preparing for urban offensive operations include:

1. Conducting Comprehensive Intelligence Gathering

Intelligence is the cornerstone of urban operations. Preparing for urban offensives begins with gathering detailed information about the area of operation, including:

- **City Layout and Infrastructure**: Mapping streets, buildings, underground networks, and other key infrastructure to understand the terrain and identify chokepoints or vulnerabilities.

- **Enemy Forces**: Identifying the strength, positions, fortifications, and tactics of adversaries through reconnaissance and surveillance tools like drones, satellite imagery, and human intelligence (HUMINT).

- **Civilian Presence**: Understanding population density, civilian movement patterns, and potential safe zones to minimize harm during operations.

- **Critical Infrastructure**: Assessing vital utilities such as water, electricity, and communication networks to determine their impact on both military operations and civilian life.

This intelligence should be continually updated to adapt to the dynamic nature of urban warfare.

Figure 13: Human intelligence collectors talk to role players during an intelligence-gathering exercise at Panther Strike 2014 in Camp Williams, Utah, June 20. During the exercise, more than 100 military intelligence Soldiers donned local attire and occupied simulated Afghan villages. Panther Strike is an annual training event that brings together military intelligence Soldiers from across the U.S. and partner nations for a large-scale, dynamic, full-spectrum intelligence exercise. (U.S. Army Photo National Guard photo/Spc. Brianne M. Roudebush/Released). California National Guard, CC BY 2.0, via Flickr.

Figure 14: Unmanned aerial systems help in intelligence gathering. Defense Visual Information Distribution Service, Public Domain, via Getarchive.

2. Developing a Detailed Operational Plan

An effective plan for urban offensive operations must account for the complexities of urban environments and align with strategic objectives. Key elements include:

- **Mission Objectives**: Clearly defining the goals, whether they involve capturing territory, neutralizing enemy forces, or securing critical infrastructure.

- **Phased Approach**: Dividing the operation into distinct phases, such as isolation, reconnaissance, shaping, assault, clearing, and stabilization.

- **Rules of Engagement (ROEs)**: Establishing guidelines to minimize collateral damage and comply with international humanitarian law.

- **Exit Strategy**: Planning for the transition from combat operations to stabilization efforts, ensuring long-term security and governance.

Commanders must also prepare contingency plans to address unforeseen challenges, such as enemy counterattacks or humanitarian crises.

3. Training and Preparing Troops

Urban warfare demands specialized training to address the unique challenges of close-quarters combat, multi-level threats, and the ethical considerations of civilian protection. Training should focus on:

- **Room and Building Clearing**: Teaching soldiers to systematically clear structures, engage in close combat, and identify potential traps or ambushes.

- **Combined Arms Operations**: Integrating infantry, armour, artillery, and air support for coordinated assaults and efficient use of resources.

- **Subterranean Operations**: Preparing troops to navigate and clear underground tunnels, sewers, and basements where adversaries may take refuge.

- **Civil-Military Interaction**: Training forces to engage with civilians, manage evacuations, and handle humanitarian situations sensitively.

Simulated urban environments, such as mock cities or virtual reality training systems, provide valuable hands-on experience for soldiers.

Figure 15: Givati infantry brigade soldiers and Armored Brigade cadets participate in an urban warfare training exercise. Israel Defense Forces, Public Domain, via Flickr.

4. Equipping and Allocating Resources

Urban offensive operations require specialized equipment and ample resources due to their resource-intensive nature. Key considerations include:

- **Weapons and Ammunition**: Deploying precision-guided munitions (PGMs) to minimize collateral damage and ensure accuracy in densely populated areas.

- **Vehicles**: Using mine-resistant vehicles, light armored vehicles, and unmanned ground vehicles (UGVs) to navigate narrow streets and provide troop support.

- **Personal Protective Gear**: Equipping soldiers with body armour, night vision devices, and communication systems to enhance survivability and coordination.

- **Logistical Support**: Establishing robust supply chains to provide food, water, medical supplies, and ammunition, given the high consumption rates in urban combat.

Figure 16: US Army Soldiers display chemical protective equipment during Orient Shield 14. Defense Visual Information Distribution Service, Public Domain, via Picryl.

5. Integrating Technology and Modern Warfare Tools

Technology plays a vital role in urban offensive operations. Advanced tools enhance situational awareness, improve precision, and reduce risks. Preparation should involve:

- **Drones and UAVs:** Using drones for real-time surveillance, target identification, and reconnaissance.

- **Artificial Intelligence (AI):** Leveraging AI to analyse data, predict enemy movements, and optimize tactical decisions.

- **Robotics:** Deploying robotic systems to clear buildings, detect explosives, or provide logistical support in hazardous environments.

- **Electronic Warfare (EW):** Using EW tools to disrupt enemy communications and gain an advantage in the electromagnetic spectrum.

6. Engaging with Civilian Considerations

Urban offensive operations often unfold amidst civilian populations, requiring forces to prioritize humanitarian considerations. Preparation should include:

- **Civilian Evacuation Plans:** Identifying safe zones, establishing evacuation routes, and coordinating with humanitarian organizations.

- **Minimizing Collateral Damage:** Developing strategies to protect critical infrastructure and ensure civilian safety during operations.

- **Information Campaigns:** Communicating with the local population to inform them of evacuation procedures, military objectives, and safety measures.

7. Building Coordination and Interoperability

Urban operations involve multiple units, agencies, and sometimes international coalitions. Effective coordination is essential to avoid confusion and maximize efficiency. Preparation should focus on:

- **Joint Operations Training:** Ensuring that infantry, armoured units, air support, and engineers can work seamlessly together.

- **Interagency Collaboration:** Establishing partnerships with humanitarian organizations, local authorities, and law enforcement to address non-combat aspects of the operation.

- **Command and Control Systems**: Using advanced communication networks to maintain situational awareness and provide real-time updates to all involved parties.

8. Mental and Physical Readiness

The psychological toll of urban warfare can be immense. Preparing troops mentally and physically is crucial for maintaining morale and resilience. Measures include:

- **Stress Management Training**: Teaching soldiers techniques to manage stress, fatigue, and combat-related trauma.

- **Team Cohesion Exercises**: Building trust and collaboration within units to enhance performance under pressure.

- **Health and Wellness Support**: Providing access to medical and psychological care before, during, and after operations.

Urban Offensive Case Studies

Urban offensive operations have become a focal point in contemporary military strategy, particularly in densely populated areas where the complexities of urban warfare present unique challenges. This synthesis examines three significant urban offensives: the Battle of Mosul (2016-2017), the Battle of Fallujah (2004), and the ongoing urban conflicts in Ukraine, drawing on relevant literature to highlight the operational dynamics and implications of these cases.

Figure 17: Iraqi Army soldiers. South of Mosul, Northern Iraq, Western Asia. 23 November, 2016. Mstyslav Chernov, CC BY-SA 4.0, via Wikimedia Commons.

The Battle of Mosul, a critical operation against ISIS, showcased the intense nature of urban warfare characterized by block-by-block fighting. The Iraqi military, supported by coalition forces, faced significant challenges in minimizing civilian casualties, particularly due to ISIS's tactics of using human shields. The operation involved extensive precision airstrikes, which were essential for targeting insurgent positions while attempting to limit collateral damage. This dual focus on offensive capability and civilian protection is echoed in the literature, which emphasizes the need for advanced reconnaissance and precision targeting in urban settings to mitigate risks to non-combatants [2, 34]. The complexities of urban terrain necessitate a nuanced understanding of the operational environment, as highlighted by the challenges faced during the Mosul offensive, where the dense urban landscape significantly influenced tactical decisions [17].

Figure 18: A young boy watches as Head Quarters Troup 11th Armored Calvary Regiment (Personal Security Detachment) rolls by in a stryker in Mosul, Iraq, Oct. 11, 2005. The 11th ACR PSD is in Iraq in support of Operation Iraqi Freedom. .(U.S. Army photo By SPC. Jory C. Randall) (Released). The U.S. National Archives, Public Domain, via Picryl.

Similarly, the Battle of Fallujah illustrated the effectiveness of combined arms tactics in urban warfare. U.S. forces employed a range of military assets to engage entrenched insurgents, demonstrating the importance of coordination among ground troops, air support, and intelligence operations. The operation underscored the difficulties of urban combat, where the presence of civilians and the risk of collateral damage complicate military objectives. The literature points to the evolution of military doctrine in urban environments, emphasizing the need for adaptive strategies that account for the unique challenges posed by urban landscapes [185]. The lessons learned from Fallujah continue to inform contemporary military operations, highlighting the critical nature of urban warfare in modern conflict scenarios.

Figure 19: U.S. Marine Corps Marine armed with a M1014 Joint Service Combat Shotgun, provides security as Iraqi civilians in Fallujah, Iraq, return to the city in working parties on December 11, 2004, so that they can help clean up the city before they are allowed to move back into their homes. Multi-National Forces are currently conducting humanitarian aid missions to help the citizen of Fallujah after a battle with the insurgents destroyed much of the city. (U.S. Marine Corps PHOTO by CPL. Theresa M. Medina) (Released). The U.S. National Archives, Public Domain, via Picryl.

Figure 20: US Army (USA) Soldiers assigned to 3rd Infantry Division (Mechanized) remove Iraqi PMN Anti-Personnel (AP) mines found near Fallujah, Iraq, during Operation IRAQI FREEDOM. The U.S. National Archives, Public Domain, via Picryl.

In the context of the ongoing Ukraine conflict, urban offensives in cities like Kyiv and Mariupol have further illustrated the strategic significance of urban areas in military operations. The defence of Kyiv has been marked by remarkable resilience among defenders, reflecting a broader trend where urban populations actively participate in their defence. Conversely, the siege of Mariupol has revealed the resource-intensive nature of urban offensives, where sustained military engagement leads to significant humanitarian crises [17]. The literature emphasizes the need for comprehensive strategies that incorporate not only military objectives but also the socio-political dynamics of urban populations, as these factors can significantly influence the outcome of urban offensives [186].

The case studies of Mosul, Fallujah, and the ongoing urban conflicts in Ukraine illustrate the complexities and challenges of urban offensive operations. Each case highlights the necessity for adaptive military strategies that prioritize precision, coordination, and an understanding of the urban environment's socio-political context. As urban warfare continues to evolve, the lessons learned from these operations will be crucial for shaping future military doctrines and strategies.

Key Offensive Strategies

Urban offensive operations are complex military endeavours that involve multiple phases, each critical to achieving strategic objectives in urban environments. The phases of such operations can be delineated as follows:

1. Isolation: The initial phase of urban offensive operations focuses on isolating the urban area to prevent enemy reinforcements and supplies from entering or exiting. This is achieved through the establishment of a perimeter and control over key entry and exit points. Historical examples, such as the Battle of Sadr City in 2008, illustrate this tactic where coalition forces implemented strategies to effectively isolate insurgents within the district [5]. This phase is essential for limiting the operational capabilities of the adversary and is supported by literature on urban warfare, which emphasizes the importance of spatial control in military strategy [187].

2. Reconnaissance and Intelligence Gathering: Prior to launching a direct assault, forces must conduct thorough reconnaissance to gather intelligence on the urban layout, enemy positions, and civilian concentrations. This phase often employs advanced technologies such as drones and electronic surveillance systems, which enhance situational awareness and inform tactical decisions [187]. The integration of heterogeneous wireless sensor networks has been shown to significantly improve situational awareness in urban military operations, allowing for real-time intelligence gathering [188]. Such intelligence is crucial for planning effective assaults and minimizing civilian casualties.

3. Shaping the Battlefield: Shaping operations are designed to weaken enemy defences and create favourable conditions for the main assault. This may involve airstrikes, artillery bombardments, and targeted raids aimed at disrupting command and control structures. The literature highlights that effective shaping operations can significantly reduce enemy resistance and enhance the effectiveness of subsequent assaults [187]. The use of advanced technologies, such as the Internet of Things, can also facilitate more precise battlefield support, thereby improving operational outcomes [189].

4. Main Assault: The main assault phase involves the systematic clearing of enemy-held areas, utilizing combined arms tactics that integrate infantry, armour, and air support. Troops typically advance block by block, facing challenges such as narrow streets and debris that complicate mechanized movements. The coordination of various military assets is essential for overcoming enemy resistance, and the literature emphasizes the importance of tactical flexibility in urban combat scenarios [187].

5. Clearing and Securing: After achieving major objectives, forces focus on clearing remaining enemy positions and securing the urban area. This phase includes intensive searches of buildings and underground facilities to neutralize any lingering threats. Establishing safe zones for civilians and restoring order are critical components of this phase, as they help prevent insurgent resurgence and ensure the safety of non-combatants. The literature underscores the necessity of integrating humanitarian considerations into military operations to foster long-term stability [187].

6. Transition to Stabilization: The final phase involves transitioning from combat operations to stabilization efforts. Military forces collaborate with local authorities and humanitarian organizations to restore essential services and support civilian recovery. This phase is vital for rebuilding trust within the community and preventing the re-emergence of conflict. The literature indicates that successful stabilization requires a comprehensive approach that addresses both security and socio-economic needs [187].

Offensive Operations in an Urban Environment

Offensive operations in urban environments adapt standard offensive doctrines to meet the unique demands of urban terrain. Urban combat is fundamentally different from operations in open terrain due to its physical, logistical, and tactical complexities. Urban areas impose constraints on movement, visibility, and communications, requiring commanders to modify strategies to maintain their ability to fix and manoeuvre against the enemy. Additionally, the pace of operations is slower, requiring methodical approaches due to restricted mobility and the need for precise coordination. The integration of combined arms, the effective use of fires, manoeuvring, and specialized equipment are critical components of urban offensive operations [190].

Urban environments demand a slower tempo compared to other battlefields. The congested and compartmentalized nature of cities makes rapid manoeuvres difficult, even for mechanized units. Instead, missions are deliberate and require detailed planning and execution. Urban combat necessitates thorough intelligence, precise control, and close coordination to address challenges such as restricted lines of sight, complex terrain, and the civilian population's presence [190].

Commanders must adapt to the unpredictability of urban combat, relying heavily on reconnaissance and real-time intelligence. Both hasty and deliberate attacks are employed in urban operations, depending on the situation, with phases such as reconnaissance, isolation of enemy forces, gaining a foothold, and seizing objectives being constants across both approaches [190].

Types of Offensive Operations

1. Hasty Attacks: Hasty attacks in urban environments are opportunistic operations conducted with limited preparation. They are initiated in response to a movement-to-contact, meeting engagement, or chance encounter with the enemy. They may also occur as a follow-up to a successful defence or in situations where vulnerable enemy forces present an exploitable opportunity [190].

The defining feature of hasty attacks is their reliance on rapid deployment and immediate action. Upon contact with the enemy, the commander deploys forces to suppress the enemy, exploit gaps or vulnerabilities, and report progress to higher command. However, urban terrain complicates hasty attacks due to incomplete intelligence and the difficulties of coordinating movements and firepower in congested areas [190].

In urban areas, hasty attacks often require manoeuvre units to move through, rather than around, friendly forces fixing the enemy. This makes command, control, and communication especially critical to avoid congestion and ensure seamless transitions between units. Commanders may issue on-order or on-call missions, allowing forces to react to contingencies once objectives are secured [190].

2. Deliberate Attacks: Deliberate attacks are fully synchronized operations that utilize all available resources to achieve specific objectives. They are necessary when the enemy has fortified positions, the urban area is large or congested, or the element of surprise has been lost. These operations demand detailed planning, extensive reconnaissance, and thorough rehearsals [190].

Deliberate attacks focus combat power on the enemy's weak points rather than directly engaging their strongest defences. The strategy mirrors techniques used in assaulting strongpoints, where overwhelming force and coordinated actions are concentrated on exploitable vulnerabilities. Success depends on precise execution, leveraging superior intelligence, and the integration of combined arms [190].

Thorough preparation distinguishes deliberate attacks from hasty ones. Commanders plan each phase meticulously, ensuring forces are synchronized and objectives are achievable. Rehearsals and preparations reduce the likelihood of errors, ensuring the attack can adapt to dynamic battlefield conditions [190].

Consolidation and Reorganization

Consolidation: Consolidation occurs immediately after an engagement and is crucial for securing gains and preparing for potential counterattacks. In urban operations, consolidation is particularly important due to the proximity of enemy forces and the risk of infiltration. After clearing a building or area, assault forces must quickly secure the position, covering likely enemy counterattack routes and reinforcing defences. This ensures the enemy cannot retake ground that has been won [190].

For example, after securing a floor of a building, soldiers are assigned to strategic positions to repel counterattacks or infiltration attempts. This phase provides the unit with the stability required to continue the mission and maintain momentum [190].

Reorganization: Reorganization follows consolidation and involves actions that prepare the unit to resume offensive operations. Key activities include redistributing ammunition, tending to casualties, reestablishing communication lines, and reallocating forces based on the evolving tactical situation. Reorganization ensures the unit remains combat-effective and ready to respond to new threats or exploit emerging opportunities [190].

Isolate and Encircle: Cutting off enemy forces within city blocks

The tactic of isolating and encircling enemy forces in urban warfare is a critical component of offensive military operations aimed at neutralizing adversaries while minimizing collateral damage and preserving critical infrastructure. This method is designed to sever the enemy's access to reinforcements, supplies, and escape routes by systematically enclosing them within specific urban areas, such as city blocks or neighbourhoods. The successful execution of this tactic necessitates meticulous planning, coordination, and execution, leveraging combined arms tactics and intelligence to achieve its objectives.

The primary goal of isolating and encircling enemy forces is to weaken their ability to resist, thereby forcing them into a state of strategic disadvantage. By cutting off external support and limiting mobility, the enemy becomes vulnerable to direct assault, psychological pressure, or eventual surrender. This tactic effectively prevents the enemy from reinforcing their positions or regrouping elsewhere in the urban terrain, ensuring that the offensive operation can proceed without interference [191]. Furthermore, isolation operations are designed to minimize harm to civilians and infrastructure. Encirclement allows for more controlled engagements, which reduces the likelihood of indiscriminate destruction and civilian casualties, aligning with international humanitarian principles [192, 193].

Key phases of isolation and encirclement [192, 193]:

A. Reconnaissance and Intelligence Gathering: Isolation begins with detailed reconnaissance and intelligence efforts to map the city's layout, identify enemy positions, and understand their logistical networks. Surveillance tools, such as drones, satellite imagery, and human intelligence (HUMINT), are used to pinpoint critical routes, chokepoints, and areas of enemy concentration.

B. Establishing a Perimeter: The next phase involves creating a physical or operational perimeter around the targeted urban area. This may include deploying forces to key intersections, bridges, or access roads to block ingress and egress points. The perimeter serves to contain enemy forces within a defined space, preventing their escape or external reinforcements.

C. Cutting Supply Lines: Once the perimeter is established, the operation focuses on severing the enemy's supply chains. This includes targeting logistical hubs, supply convoys, and communication lines. Without access to essential resources such as food, ammunition, and medical supplies, the enemy's operational capability deteriorates over time.

D. Consolidation and Engagement: With the enemy effectively isolated, forces can engage in a systematic clearing operation within the encircled area. The encirclement ensures that combat is localized and that the enemy cannot retreat or regroup. Careful coordination is required to avoid friendly fire or gaps in the perimeter that the enemy could exploit.

Tactical considerations in urban encirclement [192, 193]:

A. Coordinated Movements: Urban terrain complicates the execution of encirclement operations due to its dense infrastructure and limited mobility. Forces must coordinate their movements carefully to avoid congestion, ensure seamless coverage of the perimeter, and maintain lines of communication.

B. Combined Arms Integration: Encirclement operations rely on the effective integration of infantry, armour, artillery, and air support. Mechanized units can secure major thoroughfares, while infantry forces navigate narrow streets and buildings to establish control over smaller zones. Artillery and air support provide firepower to suppress enemy resistance or destroy fortified positions.

C. Managing Civilian Presence: The presence of civilians within the encircled area poses significant challenges. Commanders must prioritize their safety, ensuring evacuation routes are available and that operations minimize harm to non-combatants. Civil-military cooperation is essential to provide humanitarian assistance where needed.

D. Counteracting Enemy Tactics: Encirclement often prompts adversaries to adopt defensive measures, such as fortifying positions, using human shields, or attempting breakout manoeuvres. Forces must anticipate and counter these tactics through flexible planning and constant surveillance.

Encirclement exerts psychological pressure on the encircled enemy forces. Being cut off from reinforcements and resources can demoralize troops, leading to decreased combat effectiveness and increased likelihood of surrender. However, prolonged encirclement can also lead to desperation, prompting the enemy to launch high-risk breakout attempts or retaliate with extreme measures. Commanders must balance the benefits of attrition with the risks of prolonged engagement [192, 193].

The concept of urban encirclement has been a significant tactical manoeuvre in various historical and contemporary conflicts. This synthesis examines three notable case studies: the Battle of Stalingrad, the Battle of Mosul, and the Siege of Mariupol, highlighting the strategies employed and their implications.

A. Battle of Stalingrad (1942–1943): The Battle of Stalingrad stands as a pivotal moment in World War II, where the Soviet Red Army executed Operation Uranus, a counteroffensive that successfully encircled the German Sixth Army. This operation involved a meticulous strategy of cutting off supply lines and surrounding enemy forces within the city, ultimately leading to a significant surrender that marked a turning point in the war. The effectiveness of this encirclement can be attributed to the Soviets' ability to exploit the vulnerabilities of the German command structure, which was characterized by rigid adherence to orders and a failure to adapt to the rapidly changing battlefield conditions [194]. The psychological and physiological stresses experienced by soldiers during urban combat, as documented in studies, further illustrate the intense environment of Stalingrad, where combatants faced extreme conditions that affected their decision-making and operational effectiveness [195, 196].

B. Battle of Mosul (2016–2017): In the campaign to reclaim Mosul from ISIS, coalition forces implemented encirclement tactics that involved controlling key access routes and systematically clearing neighbourhoods. This strategy aimed to isolate insurgents, thereby diminishing their capacity to resist while simultaneously minimizing civilian casualties. The urban landscape of Mosul presented unique challenges, as the presence of civilians complicated military operations. The coalition's approach focused on precision strikes and ground manoeuvres that sought to limit collateral damage, reflecting a growing awareness of the ethical implications of urban warfare [4, 42]. The psychological impact on both combatants and civilians during this operation has been documented, emphasizing the heightened anxiety and stress levels associated with urban combat scenarios [6].

C. Siege of Mariupol (2022): The ongoing siege of Mariupol in the context of the Ukraine conflict exemplifies the complexities and ethical dilemmas associated with urban encirclement. Russian forces encircled the city, effectively cutting off supply lines and exerting immense pressure on Ukrainian defenders. This tactic not only aimed to achieve military objectives but also resulted in significant humanitarian crises, with widespread civilian suffering reported [197]. The siege has raised critical questions regarding the morality of encirclement strategies in urban warfare, particularly when civilian populations are heavily impacted. The psychological toll on defenders and civilians alike has been profound, highlighting the need for a re-evaluation of military tactics in urban settings to better account for humanitarian considerations [4, 42].

While isolation and encirclement are effective strategies in urban warfare, they present several significant challenges and risks that must be carefully managed to ensure operational success.

One of the primary challenges is the logistical demand of maintaining a perimeter and supporting the forces executing the encirclement. Urban environments, with their dense infrastructure and restricted mobility, complicate supply chains and require robust logistical networks to provide continuous support. Ensuring the availability of ammunition, food, water, medical supplies, and communication equipment is critical for sustaining operations, especially during prolonged engagements.

The presence of civilians in urban areas adds another layer of complexity. Civilians not only require protection but also create additional considerations for manoeuvring and engagement. Commanders must carefully plan operations to avoid unnecessary harm to non-combatants and mitigate the risk of humanitarian crises. This often involves coordinating evacuation efforts, establishing safe zones, and working closely with humanitarian organizations, which can divert resources and attention from the primary military objectives.

Enemy breakout attempts pose a constant threat during encirclement operations. Encircled forces, recognizing their disadvantage, may attempt high-risk manoeuvres to breach the perimeter and escape. Such breakouts, if successful, can disrupt the encirclement, allow the enemy to regroup, and prolong the conflict. Preventing breakouts requires vigilant surveillance, rapid response capabilities, and a well-coordinated defensive strategy across the perimeter.

Finally, prolonged engagements are a significant risk if the encircled enemy refuses to surrender. Urban encirclements can become drawn-out battles of attrition, straining the morale and resources of the attacking forces. Prolonged operations may also increase the likelihood of civilian casualties and infrastructure damage, further complicating post-conflict stabilization efforts.

Addressing these challenges requires meticulous planning, real-time adaptability, and a balanced approach that combines tactical precision with humanitarian considerations. By anticipating and mitigating these risks, commanders can enhance the effectiveness of isolation and encirclement strategies while minimizing their potential downsides.

Penetrate and Divide: Using small, mobile units to split enemy formations

The tactic of penetrating and dividing enemy formations is a critical strategy in urban warfare, aimed at disrupting the cohesion of enemy forces, isolating smaller groups, and creating opportunities for targeted engagements. This approach capitalizes on the agility and adaptability of small, mobile units to exploit vulnerabilities in enemy defences, thereby enhancing combat effectiveness in complex urban environments. The fragmentation of enemy formations not only reduces their combat effectiveness but also undermines their command and control structures, providing a significant tactical advantage to the attacking forces.

Penetrating enemy lines necessitates the identification and exploitation of weak points in their defences. This requires meticulous reconnaissance and precise execution to minimize exposure to enemy fire and avoid encirclement. Once penetration is achieved, the objective shifts to dividing the enemy into smaller, less coordinated groups. Fragmented forces are easier to manage, as their reduced size limits their ability to concentrate firepower or coordinate movements effectively. The disruption of communication and logistical networks further weakens the operational capability of the enemy. Small, mobile units are particularly suited for this strategy due to their speed and manoeuvrability, which allow them to navigate urban terrains effectively while maintaining the element of surprise.

The success of penetration and division hinges on detailed reconnaissance to identify the enemy's weak points, defensive positions, and movement patterns. Intelligence gathering through various means, including drones and human intelligence (HUMINT), is vital for understanding the urban battlefield's layout and locating opportunities for penetration. This intelligence enables attacking forces to plan their movements strategically, ensuring that they can exploit vulnerabilities effectively.

Small, mobile units are essential for executing this tactic, as their agility allows them to quickly exploit gaps in enemy defences. These units typically consist of infantry supported by light vehicles or unmanned systems, which enhance their operational flexibility. Their speed enables them to outmanoeuvre the enemy, avoid prolonged engagements, and rapidly shift focus to new objectives, thereby maintaining the initiative in combat scenarios.

Once enemy formations are divided, isolated groups can be systematically engaged and neutralized. Small units rely on close-quarters combat tactics, precision firepower, and superior coordination to overpower fragmented enemy forces. By concentrating their efforts on isolated groups, these units can conserve resources while maintaining pressure on the enemy, thereby maximizing their effectiveness in combat.

Dividing enemy forces disrupts their command and control structures, complicating the ability of commanders to relay orders or coordinate reinforcements. This disarray creates opportunities for attacking forces to gain and maintain the initiative, further enhancing their operational effectiveness. The integration of advanced technologies, such as unmanned aerial vehicles (UAVs) for reconnaissance, can significantly improve situational awareness and operational planning in urban warfare.

The primary advantage of the penetrate and divide strategy is its ability to diminish the enemy's effectiveness by splitting their formations. When enemy units are isolated, their capacity to coordinate defences and concentrate firepower is significantly weakened. This fragmentation allows attacking forces to engage isolated units more efficiently, leading to quicker victories and reduced casualties. Historical analyses of military engagements illustrate that successful penetration tactics often result in disarray among enemy ranks, making it difficult for them to mount a cohesive defence [198]. Furthermore, the concept of combat effectiveness emphasizes that smaller, agile units can exploit weaknesses in enemy formations, leading to a more favourable outcome for the attacking force [198].

Urban environments present unique challenges and opportunities for military operations. The narrow streets and compartmentalized structures of cities can be effectively utilized by smaller units employing the penetrate and divide strategy. These units can navigate areas that larger forces find difficult to access, using the terrain to their advantage while minimizing exposure to enemy fire [198]. The urban landscape allows for ambush tactics and surprise attacks, which can further disrupt enemy operations and morale [198]. Historical case studies have shown that urban warfare often benefits from tactics that leverage the complexities of the environment, allowing for more effective engagements against larger, less mobile forces [198].

The use of small, agile units in the penetrate and divide strategy enhances force protection by reducing exposure to enemy firepower. These units can avoid prolonged engagements, relying instead on mobility and speed to maintain offensive momentum while minimizing casualties [198]. The ability to manoeuvre quickly and adapt to changing battlefield conditions is crucial in modern warfare, where the risk of heavy losses from concentrated enemy fire is significant. By employing tactics that prioritize mobility, forces can maintain pressure on the enemy while safeguarding their own personnel [198].

The psychological effects of dividing enemy forces cannot be underestimated. When an enemy is fragmented, it creates confusion and fear, undermining their morale. Isolated units may feel abandoned and are more likely to surrender or retreat, further compounding the effectiveness of the penetrate and divide strategy [198]. Historical examples demonstrate that psychological warfare plays a critical role in military success, as the perception of vulnerability can lead to real tactical advantages on the battlefield [198].

Despite its advantages, the penetrate and divide strategy presents several challenges that must be carefully managed.

Risk of Overextension: One significant risk is the potential for overextension of small units operating independently. If these units become isolated, they may fall prey to enemy traps or counterattacks. Effective communication and coordination with supporting forces are essential to mitigate this risk [198]. Historical military operations have shown that maintaining a cohesive command structure is vital for the success of such tactics [198].

Navigating Civilian Presence: The presence of civilians in conflict zones complicates the execution of penetration and division operations. Commanders must balance military objectives with the imperative to minimize civilian harm. This necessitates careful planning, including evacuation strategies and strict rules of engagement to protect non-combatants [198]. The ethical implications of military operations in populated areas require a nuanced approach to ensure compliance with international humanitarian law [198].

Enemy Adaptation: Adversaries may adapt to the penetrate and divide strategy by establishing secondary defensive lines or deploying reserves to counteract penetrating forces. This necessitates that attacking units remain flexible and prepared for dynamic responses from the enemy [198]. Historical accounts indicate that successful military strategies often involve anticipating and countering enemy adaptations [198].

Intelligence Requirements Finally, effective penetration relies heavily on accurate and timely intelligence. Incomplete or faulty intelligence can lead to poorly executed operations, resulting in significant losses or mission failure. Historical military campaigns have underscored the importance of intelligence in planning and executing successful military operations [198].

To maximize the effectiveness of penetration and division in urban warfare, commanders should:

- Prioritize reconnaissance to identify enemy weaknesses and avoid heavily fortified areas.

- Use combined arms integration to support small units with artillery, airpower, and mechanized assets.

- Maintain flexibility to adapt to changing battlefield conditions and enemy responses.

- Develop clear communication networks to ensure coordination among dispersed units.

- Implement contingency plans to address potential risks, such as overextension or counterattacks.

Overwhelming Assault: Coordinated attacks with heavy weapon support

The concept of overwhelming assault in urban warfare is characterized by a coordinated, high-intensity attack that leverages a combination of infantry, mechanized units, and heavy weapon support. This tactic is designed to exert such overwhelming pressure on enemy forces that they are unable to mount an effective defence. The primary goal is to saturate the battlefield with firepower, utilizing tanks, artillery, air support, and precision-guided munitions to disrupt enemy positions and break their will to fight. This approach is particularly crucial in urban environments, where dense infrastructure and limited manoeuvrability pose significant challenges to offensive operations.

Heavy weapon support is a critical component of the overwhelming assault. Artillery barrages are employed to suppress enemy positions and destroy fortifications, while tanks provide direct fire support and serve as mobile cover for advancing infantry. Air support, including attack helicopters and close air support (CAS) aircraft, delivers precision strikes against enemy strongholds, enhancing the effectiveness of ground operations. The use of precision-guided munitions (PGMs) is particularly vital in urban warfare, as these weapons allow for targeted strikes with minimal collateral damage, thereby protecting civilian populations and infrastructure.

Figure 21: An M-198 155mm Howitzer of the US Marines firing at Fallujah, Iraq, during the Second Battle of Fallujah. United States Department of Defense From the Department of Defense image search website - Defense Visual Information Center, Public Domain, via Picryl.

The success of an overwhelming assault hinges on the coordination between infantry and heavy weapon support. Infantry units must synchronize their movements with mechanized forces to clear buildings, secure objectives, and neutralize remaining resistance. In urban settings, where heavy weaponry alone cannot effectively clear complex terrain, the role of infantry becomes indispensable. Furthermore, creating breaches in enemy defences is essential for the assault's success. This is accomplished through the use of explosives, engineering vehicles, and anti-fortification munitions, which open pathways for advancing forces to penetrate deeper into enemy territory.

Additionally, the tactic often involves isolating the target area to prevent enemy reinforcements. This is achieved through artillery fire on access routes, airstrikes on supply lines, and establishing perimeter defences to contain the enemy. The overwhelming assault tactic is thus a multifaceted approach that combines various elements of military power to achieve rapid domination over fortified positions, making it a vital strategy in contemporary urban warfare scenarios.

The concept of overwhelming assaults in military strategy presents several advantages and challenges that are critical to understand in the context of modern warfare.

The concentrated use of firepower and synchronized attacks enables forces to achieve rapid domination over enemy positions. This tactic significantly reduces the duration of engagements, thereby limiting the opportunities for counterattacks by the enemy. Research indicates that overwhelming firepower can create a decisive advantage, allowing forces to secure objectives quickly and with minimal resistance, which is crucial in fast-paced combat scenarios [199, 200]. The ability to dominate rapidly not only affects the immediate battlefield but also has strategic implications for the overall campaign, as it can disrupt enemy plans and morale [201].

Overwhelming assaults are particularly effective against heavily fortified positions that resist traditional tactics. The application of heavy firepower and precision strikes can neutralize entrenched defences, allowing attacking forces to breach barriers that would otherwise be impenetrable [199, 200]. This capability is essential in modern warfare, where adversaries often employ sophisticated fortifications. Historical examples demonstrate that overwhelming firepower can dismantle these defences, leading to significant tactical advantages [199].

The psychological effects of overwhelming assaults cannot be understated. The sheer intensity of such attacks can demoralize enemy forces, leading to panic, disarray, and even surrender. Studies have shown that the inability to respond effectively to superior firepower can create a sense of hopelessness among adversaries, which can be exploited to achieve strategic objectives [199, 200]. The psychological component of warfare is critical, as it can lead to quicker victories and reduced casualties in the long run.

By employing heavy weapons to suppress enemy positions, overwhelming assaults can significantly reduce the risk to infantry and mechanized units during the assault. This tactic allows for controlled breaches and precise targeting, which minimizes casualties among advancing troops [199, 200]. The use of overwhelming firepower serves as a protective measure, enabling forces to advance with greater confidence and safety, thus enhancing operational effectiveness [201].

Despite the advantages, overwhelming assaults also present significant challenges that must be managed effectively.

Collateral Damage: The use of heavy weaponry, particularly in urban environments, raises concerns about collateral damage to civilian infrastructure and non-combatants. The risk of unintended consequences necessitates precision targeting and strict rules of engagement to mitigate these risks [199, 200]. Failure to address collateral damage can lead to long-term strategic setbacks, including loss of local support and increased insurgency [201].

High Resource Consumption: Overwhelming assaults require substantial resources, including ammunition, fuel, and logistical support. Prolonged operations can strain supply lines and limit the availability of critical assets, making sustainability a key concern [199, 200]. The high resource consumption associated with these tactics can lead to operational challenges, particularly in prolonged conflicts where resource allocation becomes critical [201].

Dense urban terrain poses significant challenges for the effectiveness of overwhelming assaults. The mobility of mechanized units can be severely restricted by narrow streets, high-rise buildings, and underground networks, necessitating careful planning and adaptation [199, 200]. Commanders must be adept at navigating these challenges to maintain the effectiveness of their assaults in complex environments [201].

Enemy Countermeasures** Adversaries may employ various tactics, such as booby traps, ambushes, or the use of human shields, to counter the effects of overwhelming assaults. Commanders must anticipate and neutralize these threats to maintain momentum and effectiveness in their operations [199, 200]. The dynamic nature of warfare requires constant adaptation and innovation to overcome these countermeasures [201].

To maximize the effectiveness of an overwhelming assault, commanders should:

- **Coordinate Across Units**: Synchronizing the actions of infantry, mechanized forces, and heavy weapon support is critical for maintaining momentum and avoiding friendly fire.

- **Prioritize Intelligence**: Detailed reconnaissance ensures that firepower is directed at critical enemy positions and minimizes unnecessary destruction.

- **Adapt to Terrain**: Urban environments require flexible tactics, including the integration of engineers to clear obstacles and infantry to secure complex terrain.

- **Manage Logistics**: Ensuring the availability of resources, including ammunition and fuel, is essential for sustaining the intensity of the assault.

Importance of Mobility and Positioning in Urban Combat

Urban combat presents unique challenges that require careful attention to mobility and positioning. The dense and compartmentalized nature of urban terrain, coupled with the presence of civilians and the potential for ambushes, necessitates a dynamic and flexible approach to manoeuvring forces and establishing advantageous positions. Effective mobility and positioning are critical for maintaining initiative, mitigating risks, and achieving tactical and strategic objectives in urban warfare.

Urban terrain significantly restricts the movement of mechanized units and vehicles, compelling reliance on dismounted infantry to traverse confined spaces like alleys, staircases, and underground tunnels. Soldiers must be proficient in close-quarters combat and adaptable to the physical constraints posed by the environment. Engineers play a pivotal role in enhancing mobility by clearing obstacles, demolishing barriers, and creating access points for advancing forces. The importance of adaptability in urban combat is underscored by the need for soldiers to respond quickly to the dynamic nature of urban warfare, which often involves navigating through unpredictable and hazardous environments [202, 203].

Moreover, the integration of technology can facilitate mobility in these restricted spaces. For instance, unmanned ground vehicles (UGVs) and drones can scout ahead, identify obstacles, and provide logistical support, thereby reducing risks for personnel [204]. The ability to leverage such technologies is crucial in urban combat scenarios where traditional movement options are limited, highlighting the need for continuous innovation in military tactics and equipment [205].

The urban battlefield is characterized by its dynamic nature, where threats and opportunities can shift rapidly. Mobility is essential for forces to adapt to these changing conditions, whether it involves reinforcing vulnerable positions, exploiting enemy weaknesses, or retreating to regroup. Rapid movement enables forces to evade ambushes and avoid encirclement in hostile areas, which is a critical aspect of urban warfare [203, 206]. The ability to manoeuvre effectively in such environments is not only a matter of physical movement but also involves strategic planning and real-time decision-making to respond to evolving threats [205].

Additionally, the psychological aspects of combat stress must be considered, as they can impact soldiers' performance and decision-making abilities in high-pressure urban environments. Understanding the nature of combat stress and its effects on soldiers is vital for developing effective training and support systems that enhance operational effectiveness in urban combat scenarios [202, 206].

Modern technology plays a transformative role in enhancing mobility in urban combat. The deployment of unmanned systems, such as UGVs and drones, allows for advanced reconnaissance and obstacle clearance, which can significantly improve operational efficiency [204]. These technologies enable forces to maintain situational awareness and coordinate movements even in congested urban environments, thereby facilitating more effective engagement with adversaries.

Furthermore, advancements in communication systems allow for real-time coordination among units, which is essential for executing complex manoeuvres in urban settings. The integration of precision-guided munitions also enhances the ability to strike targets with minimal collateral damage, thereby supporting mobility by allowing forces to engage threats without compromising their positions [203, 205]. The continuous evolution of these technologies underscores the importance of adapting military strategies to incorporate innovative solutions that enhance mobility in urban combat.

Positioning in urban combat involves selecting and securing locations that provide a tactical advantage over the enemy. Proper positioning allows forces to dominate key areas, control movement, and establish defensive or offensive capabilities.

Securing high ground, such as rooftops, upper floors, or elevated structures, is critical in urban combat. From these vantage points, forces can observe enemy movements, direct firepower effectively, and provide overwatch for advancing units. Control of high ground also denies the enemy similar advantages, reducing their ability to target friendly forces or coordinate attacks.

Urban terrain naturally creates chokepoints, such as intersections, bridges, and narrow passages, which can be exploited for tactical advantage. Controlling these chokepoints allows forces to regulate the movement of enemy troops, disrupt supply lines, and create kill zones for ambushes. However, chokepoints must be defended vigilantly, as they are often prime targets for counterattacks.

The urban environment offers abundant opportunities for cover and concealment, such as walls, debris, and buildings. Effective positioning leverages these features to protect forces from enemy fire while enabling them to strike from concealed locations. Proper use of cover minimizes casualties and enhances the survivability of troops in hostile areas.

Figure 22: Four US Marine Corps (USMC) Marines, 1ST Battalion (BN), 8th Marine Regiment (1/8), remain at the ready on a street corner during combat operations with insurgents during a Security and Stabilization Operation (SASO) conducted as part of Operation AL FAJR, which is an offensive operation to eradicate enemy insurgents in Fallujah, Al Anbar Province, Iraq, during Operation IRAQI FREEDOM. The U.S. National Archives, Public Domain, via National Archives and Defense Visual Information Distribution Service.

Strategic points, such as command centres, communication hubs, or supply depots, must be secured to maintain operational effectiveness. Positioning forces at these locations ensures uninterrupted operations and denies the enemy access to critical resources.

The interplay of mobility and positioning in urban combat is a critical aspect of military strategy, particularly in complex environments where traditional tactics may be less effective. Mobility allows forces to reach advantageous positions quickly, while effective positioning enhances their ability to manoeuvre and maintain control over the battlefield. This interdependence is crucial for maintaining initiative, exploiting weaknesses, and defending effectively.

Maintaining Initiative: The ability to move rapidly and establish strong positions enables military forces to dictate the tempo of engagements. Forces that can manoeuvre quickly can keep the enemy off balance, forcing them to react rather than act. This dynamic is essential in urban combat, where the terrain can significantly influence operational tempo. Studies have shown that rapid movement and strategic positioning are vital for maintaining operational effectiveness in urban environments [207, 208].

Exploiting Weaknesses: Mobility is key to exploiting gaps in enemy defences. Forces that can quickly reposition themselves can apply pressure to vulnerable points in the enemy's lines. Effective positioning ensures that these forces can sustain their attacks and capitalize on any weaknesses. Research indicates that the integration of advanced technologies, such as drones and robotics, enhances the ability to scout and identify these vulnerabilities, thereby improving the overall effectiveness of military operations [209, 210].

Defending Effectively: In defensive operations, positioned forces can hold critical areas, while mobile units can reinforce or counterattack as necessary. The ability to maintain control over key terrain is essential for successful defence in urban settings, where the risk of ambushes and surprise attacks is heightened. Effective communication and coordination among units are crucial for ensuring that mobile forces can respond quickly to threats, as urban environments often disrupt traditional communication networks [208, 211].

Mobility and positioning are essential components of urban combat, but they come with significant challenges that complicate operations and demand innovative strategies. The dense, unpredictable, and constrained nature of urban environments exacerbates the inherent difficulties of manoeuvring and establishing tactical advantages.

One of the primary challenges is the restricted movement caused by the physical constraints of urban terrain. Narrow streets, rubble from damaged structures, and the compartmentalized layout of cities limit the ability of vehicles and large units to navigate effectively. These restrictions often force forces to rely on dismounted infantry, which slows operations and increases vulnerability. To overcome these challenges, commanders must implement alternative strategies, such as using engineers to clear paths or leveraging unmanned systems for reconnaissance and movement in confined areas.

Another critical issue is the vulnerability to ambushes that arises from the proximity of enemy forces and the abundance of concealed positions. Urban environments provide adversaries with countless opportunities to hide and launch surprise attacks, putting advancing troops at constant risk. The labyrinthine nature of cities requires forces to maintain high situational awareness and thoroughly secure routes to mitigate the threat of ambushes. This often involves deploying scouts, using drones for overhead surveillance, and implementing strict discipline during movement.

The presence of civilians further complicates mobility and positioning in urban combat. Civilians may occupy areas of tactical importance, creating ethical and operational dilemmas for military forces. Commanders must ensure that movements do not endanger non-combatants or disrupt evacuation efforts, requiring careful planning and adherence to strict rules of engagement. This consideration often necessitates close cooperation with humanitarian organizations and local authorities to balance military objectives with civilian safety.

Finally, communication and coordination are significant challenges in urban combat. The dense infrastructure of cities often disrupts traditional communication networks, making it difficult to relay orders, share intelligence, and synchronize unit movements. This lack of cohesion can lead to disjointed operations and increased risks for troops. Advanced communication technologies, such as encrypted radios and decentralized command structures, can help mitigate these issues, ensuring that forces remain connected and cohesive despite the complexities of the urban environment.

Overcoming these challenges requires adaptability, innovation, and meticulous planning. By addressing the constraints of restricted movement, mitigating ambush risks, prioritizing civilian safety, and enhancing communication, military forces can improve their mobility and positioning in urban combat, ultimately increasing their chances of operational success.

The Role of Intelligence in Mobility

Intelligence is a critical enabler of mobility in combat operations, especially in urban warfare, where the terrain is complex, dynamic, and unpredictable. Effective mobility depends on timely, accurate, and actionable intelligence that allows forces to navigate efficiently, avoid threats, and exploit opportunities. In urban environments, intelligence shapes how forces plan routes, coordinate movements, and adapt to evolving situations.

Intelligence provides detailed information about the physical layout of the urban environment, including streets, alleys, underground networks, and key infrastructure. This knowledge helps commanders identify navigable routes, chokepoints, and potential obstacles, enabling smoother movement of troops and vehicles. Advanced intelligence tools, such as satellite imagery, drones, and ground reconnaissance, are essential for creating accurate maps and identifying areas of concern, such as rubble-filled streets or impassable zones.

For example, in urban operations like those in Mosul or Fallujah, intelligence on building layouts and underground tunnels allowed forces to plan movements while minimizing delays and vulnerabilities. Without this information, units would risk becoming lost, trapped, or exposed to ambushes in unfamiliar terrain.

Mobility in urban combat is inherently dangerous due to the high risk of ambushes and traps set by enemy forces. Intelligence plays a pivotal role in identifying these threats before forces move through contested areas. Human intelligence (HUMINT), signals intelligence (SIGINT), and imagery intelligence (IMINT) can reveal the presence of hidden enemy positions, IEDs (improvised explosive devices), or booby-trapped buildings.

For example, drone reconnaissance and thermal imaging can detect enemy activity behind walls or on rooftops, enabling forces to avoid those areas or prepare countermeasures. This pre-emptive awareness allows troops to maintain operational momentum and minimize casualties.

Timely intelligence enables commanders to make informed decisions about when and where to move forces. By understanding enemy dispositions, movements, and intentions, commanders can adjust plans dynamically to avoid risks and seize opportunities. For instance, intelligence on enemy troop concentrations may lead to the identification of weak points in their defences, enabling a force to bypass heavily fortified areas and exploit gaps.

Additionally, real-time intelligence during operations allows for mid-mission adjustments. For example, if an unexpected threat emerges, such as a collapsed building blocking a primary route, intelligence can help identify alternative paths or assist engineers in clearing the obstacle efficiently.

Intelligence supports the coordination of multiple units, ensuring that their movements are synchronized to avoid congestion, miscommunication, or friendly fire. In urban environments, where units often operate in close proximity and visibility is limited, intelligence ensures that movements are deconflicted and aligned with overall objectives.

For instance, intelligence may highlight optimal times for movement based on enemy patrol schedules or civilian activity patterns. This information helps ensure that units move without compromising their positions or drawing unnecessary attention.

Mobility relies not only on the movement of troops but also on the transportation of supplies, including ammunition, fuel, and medical equipment. Intelligence ensures that supply routes remain secure and accessible, identifying threats such as enemy ambushes, roadblocks, or damaged infrastructure.

By analysing the urban environment and enemy activity, intelligence can help establish reliable supply corridors and anticipate logistical challenges. This proactive approach ensures that mobility is sustained over the course of an operation.

In urban combat, the presence of civilians complicates mobility. Intelligence helps forces understand civilian movement patterns, identify safe zones, and avoid populated areas to minimize collateral damage. For example, during humanitarian operations, intelligence can identify routes that are clear of civilian traffic, allowing forces to move efficiently without disrupting evacuation efforts or endangering non-combatants.

Technological advancements have enhanced the role of intelligence in mobility. Drones provide real-time reconnaissance of urban terrain, while AI-driven analytics process large volumes of data to identify patterns and threats. Ground-based sensors and surveillance equipment further augment the ability to gather actionable intelligence that directly supports movement planning and execution.

The role of intelligence in mobility extends beyond tactical benefits to broader operational and strategic impacts. Effective intelligence ensures that forces maintain momentum, avoid unnecessary engagements, and achieve objectives more efficiently. Conversely, a lack of intelligence can result in delays, increased casualties, and mission failure.

Enhancing Mobility and Positioning

Enhancing mobility and positioning in urban combat requires a combination of preparation, specialized support, advanced technology, and integrated strategies. These elements work together to overcome the unique challenges posed by dense urban environments and ensure forces can navigate and dominate the battlefield effectively.

Effective mobility and positioning begin with well-trained forces capable of handling the complexities of urban terrain. Training programs must emphasize close-quarters combat tactics, urban navigation, and adaptability to dynamic scenarios. Soldiers must be proficient in moving through confined spaces, such as alleys, staircases, and underground tunnels, while maintaining operational cohesion. Additionally, preparation includes developing contingency plans for unexpected obstacles or threats, ensuring that units can adapt quickly to the evolving nature of urban combat.

Combat engineers play a vital role in enhancing mobility by addressing physical barriers that hinder movement. Their expertise in clearing rubble, constructing temporary bridges, and demolishing barriers ensures that forces can maintain momentum and access critical areas. Engineers also create breaches in enemy defences, enabling infantry and mechanized units to penetrate fortified positions. Their contributions are indispensable in maintaining the flow of operations, particularly in environments where natural and man-made obstacles are prevalent.

Modern technology significantly enhances mobility and positioning in urban warfare. Drones provide real-time reconnaissance, allowing forces to scout ahead and identify threats or obstacles. Robotic systems can clear paths, deliver supplies, and even engage targets in hazardous areas, reducing risks to personnel. Advanced navigation systems, including GPS and AI-driven analytics, help units navigate complex urban layouts with greater precision and efficiency. These technological tools not only improve situational awareness but also enable forces to make informed decisions in real time.

The integration of infantry, mechanized units, artillery, and air support is crucial for optimizing mobility and positioning. Infantry units are essential for securing close quarters and engaging in house-to-house combat, while mechanized units provide mobility and firepower to support advancing troops. Artillery offers long-range suppression of enemy positions, and air support delivers precision strikes and surveillance. Together, these elements create a cohesive and adaptable force capable of navigating and dominating urban terrain. Coordinating these assets effectively ensures that mobility and positioning are maintained throughout the operation.

By focusing on training, leveraging engineering expertise, utilizing advanced technology, and integrating combined arms, forces can enhance their mobility and positioning to achieve success in urban combat. These strategies not only improve operational efficiency but also ensure the safety and effectiveness of troops operating in challenging and unpredictable environments.

Integration of Air and Artillery Support in Urban Offenses

The role of air and artillery support in urban offenses is critical for achieving tactical advantages on the battlefield. This support is particularly effective in suppressing enemy defences, clearing and breaching obstacles, providing overwatch and precision fire, and isolating the battlefield.

Figure 23: U.S. Special Forces Green Beret Soldiers, assigned to 7th Special Forces Group (Airborne), Operational Detachment-A, fire an anti-armour mortar system during Integrated Training Exercise 2-16 at Marine Corps Air Ground Combat Center, Twentynine Palms, Calif., Feb. 8, 2016. MCAGCC conducts relevant live-fire combined arms, urban operations, and joint/coalition level integration training that promote operational forces' readiness. (U.S. Air Force photo by Tech Sgt. Efren Lopez/Released). Defense Visual Information Distribution Service, Public Domain, via National Archives and Defense Visual Information Distribution Service.

Air and artillery support are essential for suppressing enemy defences in urban environments. The overwhelming firepower provided by these assets can neutralize fortified positions, sniper nests, and ambush points, thus facilitating safer movement for ground forces. For instance, precision airstrikes can effectively destroy key enemy strongholds, while artillery barrages can disrupt enemy coordination and morale, as noted in the analysis of the Russia-Ukraine conflict where artillery fire was pivotal in achieving decisive effects on the enemy [212]. The use of artillery in urban warfare has been emphasized as a means to apply attrition tactics, which are crucial for weakening enemy capabilities before ground assaults [212].

Artillery and air support are also vital for clearing obstacles and breaching heavily defended positions. High-explosive artillery rounds can demolish barricades or create openings in walls, allowing infantry to penetrate enemy defences with reduced risk. Airstrikes targeting fortified structures can render them ineffective, thus enabling ground forces to advance with minimal resistance. The effectiveness of self-propelled artillery in modern warfare highlights its role in targeting hidden and ground-based threats, which is essential for urban combat scenarios [213]. Moreover, the integration of advanced targeting systems in artillery enhances their capability to engage enemy positions accurately, further facilitating ground operations [212].

Figure 24: U.K. Royal Marines with X-Ray Company, Fire Support Group, 45 Commando conduct a security patrol with Jackals or Mobility Weapon-Mounted Installation Kits (MWMIK) during Integrated Training Exercise (ITX) 2-19 at Galloway lake training area on Marine Corps Air Ground Combat Center (MCAGCC), Twentynine Palms, Calif., Jan. 18, 2019. The Royal Marines took advantage of live fire training areas at MCAGCC while supporting ITX as an adversarial force for urban warfare training. (U.S. Marine Corps Photo by Lance Cpl. William Chockey).

Air assets, including drones and attack helicopters, play a crucial role in providing overwatch for advancing ground troops. They offer real-time surveillance and can deliver precision-guided munitions (PGMs) to eliminate threats near friendly forces. The use of advanced artillery systems equipped with GPS-guided shells allows for precise engagement of enemy positions while minimizing collateral damage. This capability is particularly important in urban settings where civilian populations may be present [212]. The combination of air and artillery support enhances situational awareness and operational effectiveness, as demonstrated in various military operations where coordinated strikes have led to significant tactical advantages [17].

Air and artillery support can effectively isolate sections of the urban battlefield by targeting routes used for enemy reinforcements or retreats. Airstrikes can destroy critical infrastructure such as bridges and roads, while artillery fire can establish fire zones that restrict enemy movement. This strategy not only confines the conflict to a designated area but also disrupts the enemy's logistical capabilities, as evidenced by the strategic use of artillery in recent conflicts [17, 212]. The ability to control movement within urban environments is crucial for maintaining operational superiority and achieving mission objectives.

Challenges of Integrating Air and Artillery Support in Urban Offenses

The dense infrastructure and civilian population in urban areas increase the risk of collateral damage. Artillery shells and air-delivered munitions must be precisely targeted to avoid harming non-combatants or destroying essential infrastructure, such as hospitals or water supplies. Precision-guided munitions and strict rules of engagement are essential for mitigating these risks.

Urban environments often obstruct the line of sight for artillery observers and air support, complicating targeting and coordination. High-rise buildings, narrow streets, and underground structures make it difficult to acquire and engage targets effectively.

The complexity of urban operations requires seamless communication between ground forces, air units, and artillery crews. Disrupted communication networks, common in urban combat, can lead to delays or misfires, endangering both civilians and friendly forces.

Adversaries in urban environments may employ countermeasures such as anti-aircraft weapons, electronic jamming, or using civilians as human shields. These tactics complicate the effective use of air and artillery support and require adaptive strategies to overcome.

Accurate and timely intelligence is paramount in modern warfare, particularly in urban environments where the dynamics can shift rapidly. The deployment of drones, reconnaissance aircraft, and satellite imagery provides critical real-time data on enemy positions, civilian movements, and terrain features. This capability enhances situational awareness and allows for informed decision-making by commanders on the ground. For instance, the integration of advanced technologies in urban warfare, such as augmented reality systems, can facilitate real-time information dissemination, thereby improving operational effectiveness and situational awareness [4]. Furthermore, while intelligent video surveillance technologies are critical in various applications, their specific role in military operations requires further exploration and is not directly supported by the cited reference [214].

Urban warfare necessitates a decentralized command structure that empowers unit commanders to make swift decisions in response to evolving battlefield conditions. This approach enhances responsiveness and flexibility, allowing for immediate requests for air and artillery support as situations develop. The importance of decentralized decision-making is underscored in military strategies that emphasize the need for rapid adaptation to changing circumstances, which is essential for maintaining operational momentum.

The adoption of precision-guided munitions (PGMs) is a critical strategy for minimizing collateral damage while maximizing combat effectiveness. PGMs, such as laser-guided bombs and GPS-guided artillery shells, enable forces to engage specific targets with high accuracy, which is particularly important in urban settings where civilian populations may be at risk. The effectiveness of PGMs in urban warfare scenarios has been well-documented, as they allow for precise strikes against fortified positions or enemy snipers without endangering nearby civilians.

Comprehensive fire plans are essential for ensuring that air and artillery support is effectively synchronized with ground operations. The development of pre-planned fire missions, combined with the ability to adapt to real-time updates, allows for coordinated attacks that can suppress enemy forces while protecting advancing troops. The integration of dynamic targeting processes into fire planning enhances the ability to respond to changing battlefield conditions.

The integration of advanced targeting systems, such as Forward Air Controllers (FACs) and artillery observers equipped with laser designators, significantly enhances the accuracy of fire missions. These systems enable ground forces to mark targets for air and artillery strikes, ensuring that engagements are precise even in densely populated urban areas. The effectiveness of these systems is amplified by the use of real-time data and intelligence, which allows for rapid adjustments to targeting as the situation evolves.

Ethical and Humanitarian Considerations

The integration of air and artillery support in urban offenses necessitates strict adherence to international humanitarian law (IHL) to ensure the protection of civilians and minimize unnecessary harm. Commanders are obligated to distinguish between combatants and non-combatants, a principle foundational to IHL. This principle is articulated in various legal frameworks, including the Geneva Conventions, which emphasize the protection of those not participating in hostilities, such as civilians and medical personnel [215, 216]. The obligation to verify targets through reliable intelligence is critical in urban warfare, where the proximity of civilians to military objectives can complicate operational decisions [217, 218].

Moreover, the rules of engagement must be strictly followed to mitigate risks to civilian populations. This includes employing proportionality and necessity in military operations, which are essential components of IHL [113, 219]. The complexity of urban environments often leads to increased civilian casualties, as evidenced by recent conflicts where the transition from traditional warfare to urban combat has resulted in significant civilian harm [137, 146]. Therefore, commanders must exercise caution and ensure that military actions are justified and that every feasible precaution is taken to avoid civilian casualties [220].

Coordination with humanitarian organizations is also paramount when civilians are at risk. Humanitarian actors, such as the International Committee of the Red Cross (ICRC), play a crucial role in providing assistance and protection in conflict zones. Their involvement can help ensure that military operations are conducted in a manner that respects humanitarian principles and addresses the needs of affected populations [221, 222]. This collaboration is vital in urban settings, where the dynamics of conflict can rapidly change, and the need for humanitarian assistance can become acute [223].

Offensive Fire Support in Urban Warfare

Urban fire support plans integrate various assets and techniques to overcome the challenges of the environment. These include [190]:

A. Use of Mortars: Mortars are the most responsive fire support asset at the small-unit level. Their portability and high angle of fire make them ideal for urban combat. Mortars are employed for tasks such as:

- **Isolating enemy forces** by targeting reinforcements and retreat routes.

- **Clearing rooftops** of snipers or other threats.

- **Providing obscuration** with smoke rounds to conceal friendly movements.

- **Delivering illumination** during night or low-visibility operations.

B. Aerial Observers and UAVs: The use of aerial observers and unmanned aerial vehicles (UAVs) is essential for overcoming the restricted observation in urban terrain. These assets provide real-time intelligence, improve target acquisition, and assist in the delivery of deep fires. UAVs can also identify enemy strongholds and monitor their movements, ensuring that fire missions are accurate and effective.

C. Designating and Marking Targets: In urban environments, clear procedures for target designation and marking are critical. Smoke rounds, laser designators, and infrared strobes can be used to mark targets for artillery or air support. This ensures that fire missions are accurately directed and reduces the risk of striking friendly forces or civilians.

D. Shifting Fires and Coordinating Movements:

As ground forces advance, fire support must be carefully synchronized to shift fires ahead of friendly troops. The forward line of own troops (FLOT) must be clearly designated and communicated to all units. Close coordination between fire support teams and manoeuvre elements ensures that advancing units are not endangered by supporting fires.

Ammunition Selection and Collateral Damage Considerations

Ammunition selection is a critical aspect of urban fire support. High-explosive rounds are effective for breaching structures and neutralizing enemy positions, but their use must be balanced against the risk of collateral damage. Precision-guided munitions (PGMs) are increasingly used to achieve pinpoint accuracy, minimizing harm to civilians and infrastructure [190].

The potential for rubble effects—where destroyed buildings create obstacles for friendly forces—must also be factored into fire planning. Excessive use of high-explosive ordnance can impede mobility and slow the pace of the operation [190].

Communication and Control

Effective communication is paramount in urban fire support. The complexity of the environment and the proximity of friendly forces demand seamless coordination between fire support teams, manoeuvre units, and higher headquarters. Reliable communication systems and standardized procedures ensure that fire missions are executed safely and effectively [190].

Rules of Engagement (ROE) must be clearly understood and strictly enforced to ensure compliance with international humanitarian law. Commanders and fire support teams must balance the tactical necessity of fire missions with the ethical responsibility to protect civilian lives and infrastructure [190].

Integration with Ground Combat Elements

Most fire support in urban offensives is planned and executed at the Ground Combat Element (GCE) level or higher, reflecting the need for centralized coordination in complex operations. However, at the small-unit level, fire support assets such as mortars remain invaluable for responsive and localized missions. The integration of fire support with ground maneuver ensures that attacking forces can maintain momentum, overcome enemy resistance, and secure objectives [190].

Tactics: Clearing, Securing, and Advancing Through City Sectors

Room and Building Clearing

In urban warfare, troops must systematically clear buildings to eliminate enemy presence and secure areas. Techniques such as breaching doors using explosives or mechanical means, deploying flashbang grenades to disorient adversaries, and cautiously advancing through stairwells and hallways are critical. The tactical use of explosive breaching is highlighted as a method to gain entry into fortified structures, which involves careful placement of explosives and maintaining a safe distance during detonation [224]. This method is essential in urban settings where traditional entry methods may be obstructed or too dangerous.

Room clearing is a critical component of urban warfare, particularly during the "Seizing the Objective" phase of an operation. It requires a systematic, well-rehearsed approach to neutralize threats, secure the area, and maintain operational momentum. The eight steps of room clearing, as applied by the US Marine Corps, provide a sequential framework for achieving this goal: dominate, eliminate, control through verbiage, search the dead, search the room, search the living, transition, and mark. Each step plays a vital role in ensuring the safety of personnel and the effectiveness of the operation [190].

1. Dominate: The first step in room clearing is to establish dominance over the space. This can be achieved through forceful entry and the immediate establishment of control. Throwing grenades into the room before entry is a common technique, as it disorients or eliminates threats. The skip-and-bounce technique is preferred, ensuring the grenade is difficult for the enemy to retrieve and throw back. Alternatively, cooking off grenades—holding the grenade for a few seconds before throwing—is reserved for specific combat situations. Marines coordinate grenade usage with visual and verbal alerts, such as the "FRAG OUT" command, to maintain situational awareness while preserving the element of surprise when possible [190].

Once the grenade detonates, assault team members enter the room, moving to dominate positions inside. They shoot targets on the move and maintain a tactical advantage over the enemy. Effective room domination establishes the foundation for the remaining steps [190].

2. Eliminate: After establishing dominance, threats inside the room must be eliminated swiftly. Marines employ rigorously rehearsed techniques to neutralize enemies. One method is the cross-buttonhook combination, where two shooters clear the room from opposite sides. Shooter One enters first, engaging immediate threats while moving across the threshold. Shooter Two follows, using a buttonhook motion to cover the area not visible to Shooter One. Both shooters clear their respective zones using the pieing technique, which involves systematically scanning and engaging the room by sectors [190].

Other methods, such as the cross method, position Marines on opposite sides of the entryway, where they alternately enter the room and clear their assigned zones. If the door is closed, the team employs split positions, with one Marine opening or breaching the door while the other prepares to enter and engage. These techniques ensure all threats are neutralized efficiently, allowing the team to secure the room [190].

3. Control Through Verbiage: In the chaos of room clearing, clear and consistent communication is essential to maintain control. A common, rehearsed language enables the team to coordinate actions effectively. For example, Marines announce the status of the room with commands like "Left/Right side clear," "Checking overhead," "Overhead clear," and "Room all clear." These verbal cues ensure all team members are aware of the situation, reducing the risk of missed threats and enabling smooth transitions to the next steps [190].

4. Search the Dead, Room, and Living: Once the room is cleared and declared secure, the team conducts thorough searches to ensure no hidden threats remain and to gather intelligence. Searches are prioritized as follows [190]:

- **Search the Dead**: Check any enemy killed in action (KIA) for weapons, explosives, or intelligence materials.

- **Search the Living**: Detain and secure any enemy prisoners of war (EPWs), detainees, or non-combatants for further processing.

- **Search the Room**: Conduct a detailed sweep of the area to locate hidden weapons, supplies, or enemy combatants.

Searches must be fast yet thorough, maintaining the operation's tempo while ensuring the area is completely secure. Any detainees or non-combatants are escorted to designated collection points.

5. Transition and Redistribute: Transitioning between rooms is critical to maintaining momentum in urban operations. Marines must have a clear, rehearsed plan for reorganizing and redeploying. After clearing a room, the team redistributes ammunition, attends to any casualties, and prepares to form a two- or four-man stack for the next entry. Effective transitioning ensures that the team remains combat-ready while minimizing downtime [190].

6. Mark: Before leaving a cleared room, the team marks it to signal to friendly forces that the area is secure. Clear visual indicators, such as spray paint, chalk, or chem-lights, are used to mark the room. This step prevents redundant efforts and ensures other units can move through the area without unnecessary delays or confusion [190].

Figure 25: Marines practice clearing rooms in the Urban Assault Course during a deployment for training exercise at Fort Pickett, Va., Dec. 2, 2016. The course strengthened their ability to operate from buddy pairs up to squads, while maintaining communication at all levels. The Marines are with Weapons Company, 3rd Battalion, 8th Marine Regiment. (U.S. Marine Corps photo by Sgt. Clemente C. Garcia). Defense Visual Information Distribution Service, Public Domain, via National Archives and Defense Visual Information Distribution Service.

Room clearing is a methodical process that requires precise execution and teamwork. Each of the eight steps—dominate, eliminate, control through verbiage, search the dead, search the room, search the living, transition, and mark—serves a distinct purpose in ensuring the safety and effectiveness of the operation. Through rigorous training, proper communication, and adherence to these steps, Marines can clear rooms efficiently, neutralize threats, and maintain the momentum necessary to achieve mission success in urban combat [190].

Tactical Control Measures (TCMs)

Tactical Control Measures (TCMs) are essential tools for commanders to ensure effective coordination, communication, and execution of operations in the complex and fragmented environment of urban combat. These measures enable decentralized execution while maintaining overall mission cohesion, especially in scenarios where command and control face significant challenges due to the urban terrain [190].

1. The Importance of TCMs in Urban Operations: Urban warfare is characterized by its dense terrain, limited visibility, and the compartmentalized nature of battlespaces. In this environment, command and control from higher headquarters are often disrupted or delayed, making it imperative for subordinate leaders to take greater initiative. A clear understanding of the commander's intent, including two levels up, ensures that units can operate autonomously yet remain aligned with broader operational goals [190].

The use of detailed graphic control measures is critical for mission success and the avoidance of fratricide. Graphic measures provide a shared visual framework that improves situational awareness and ensures that units can navigate, advance, and secure objectives systematically. These measures also serve as tools for reporting progress, coordinating movements, and maintaining synchronization across multiple units operating in close proximity [190].

2. Phase Lines:

Phase lines are essential TCMs used to report progress or control the advance of attacking units. In urban terrain, phase lines are typically oriented around identifiable features such as principal streets, rivers, or railroad lines. They provide a structured way for commanders to monitor the pace of operations and synchronize unit movements. For example [190]:

- A unit might be tasked with clearing all buildings up to a specific phase line before halting to consolidate or await further orders.

- Phase lines also serve as reference points for reporting, allowing units to quickly communicate their positions relative to the designated line.

By clearly defining phase lines, commanders ensure that units remain coordinated and avoid overlapping efforts, which reduces confusion and the risk of fratricide.

3. Boundaries: Boundaries define zones of action for individual units, ensuring that responsibilities are clearly delineated. In urban environments, boundaries are often aligned with streets or blocks, with both sides of a street included within the same unit's zone. This arrangement prevents gaps in coverage and ensures that units operating on opposite sides of a street remain synchronized [190].

Clear boundaries also reduce the likelihood of friendly fire incidents, as units have a well-defined understanding of where their operational zones begin and end. This clarity is particularly important in the densely populated and visually obstructed urban environment, where the risk of misidentification is high [190].

4. Checkpoints and Contact Points: Checkpoints and contact points are critical for maintaining control and ensuring coordination during urban operations. Checkpoints serve as predefined locations for reporting positions or controlling movement, while contact points are designated locations where units physically meet or establish communication.

Examples of suitable locations for checkpoints and contact points include [190]:

- Street corners
- Railway crossings
- Bridges
- Prominent buildings

By using these identifiable features, commanders can ensure that units remain on track and maintain cohesion, even in the chaotic and fast-changing urban environment. These measures also facilitate resupply, casualty evacuation, and the reinforcement of units as needed [190].

5. Attack Position and Line of Departure (LD): The attack position and Line of Departure (LD) are key control measures for initiating offensive operations in urban combat [190].

- The attack position is often located behind or inside the last large building before crossing into open or contested terrain. This provides a secure staging area for final preparations, such as distributing ammunition, briefing troops, or synchronizing movements.

- The Line of Departure is positioned on the near side of an open area, such as a street or railway line, and serves as the starting point for the assault. The LD ensures that units begin their attack in a coordinated manner, minimizing the risk of premature engagement or disorganized advances.

By designating these positions, commanders establish clear starting points for offensive operations, ensuring that units are prepared and aligned before moving into contested areas.

6. Objectives and Building Designations: In urban operations, objectives are often assigned at the building level. For example [190]:

- Intermediate objectives may include specific buildings or clusters of buildings within a block.

- Final objectives are typically located at the far edge of the built-up area or key pieces of terrain, such as major intersections, command centres, or infrastructure hubs.

To enhance clarity and precision, buildings are assigned alpha-numeric designations. Specific entry points, windows, or doorways may also be labelled to improve situational awareness and facilitate communication during the assault. This level of detail ensures that units can identify their objectives unambiguously, even in the chaotic environment of urban combat.

7. The Challenge of Clearing Buildings: Clearing buildings is a time-intensive process but is often necessary to secure the zone of action and prevent attacks from the rear or flank. A single building may serve as the objective for a rifle squad, platoon, or even a company, depending on its size and strategic importance [190].

In situations where speed is a priority, commanders may direct units to bypass certain buildings. However, this approach comes with risks, as bypassed buildings can harbor enemy forces that may counterattack or interfere with operations. Careful consideration must be given to the trade-offs between speed and security [190].

Combined Arms Integration

Successful urban offensives depend on the effective integration of various military branches, including infantry, armour, artillery, and air support. Tanks and armoured vehicles provide necessary firepower and protection, enabling infantry units to clear and secure urban areas [17]. The coordination among these forces is crucial, as it allows for a comprehensive approach to combat that leverages the strengths of each unit type. The importance of this integration is underscored in contemporary military strategies, particularly in light of recent urban conflicts that have demonstrated the necessity of combined arms operations [35].

Precision Targeting

The use of precision-guided munitions (PGMs) is vital in urban warfare to minimize collateral damage while effectively targeting enemy strongholds and fortified positions. PGMs allow for accurate strikes against specific targets, such as sniper nests or command centres, without endangering civilian lives or infrastructure [21]. This capability is particularly important in densely populated urban areas where the risk of civilian casualties is high, and military operations must be conducted with a high degree of precision and care [23].

Sniper and Counter-Sniper Operations

Skilled snipers play a critical role in urban offensives by providing overwatch for advancing troops and eliminating enemy sharpshooters. The integration of mental skills training in sniper courses has been shown to enhance performance under the stressful conditions typical of urban combat [225]. Effective counter-sniper operations involve not only the identification and elimination of enemy snipers but also the use of advanced techniques for detection and engagement in complex urban environments [226].

Figure 26: A Marine sniper sights through the scope of his 7.62mm M-40A1 sniping rifle from a sand bag rest in the Marine compound near the Beirut International Airport. The Marines have been deployed as part of a multinational peacekeeping force following confrontation between Israeli forces and the Palestine Liberation Organization (PLO). The U.S. National Archives, Public Domain, via National Archives and Defense Visual Information Distribution Service.

Underground Operations

Urban environments often include extensive subterranean networks, such as tunnels and sewers, which adversaries may use for movement and supply. Specialized units are deployed to clear and secure these underground areas, which pose unique challenges due to limited visibility and confined spaces. The need for tactical proficiency in navigating these environments is critical, as they can serve as both routes for enemy movement and hiding spots for ambushes.

Understanding the layout and potential threats within these underground spaces is essential for successful urban operations.

Breaching in Urban Combat

Breaching is a fundamental aspect of urban combat, enabling forces to overcome physical barriers and gain access to urban objectives. The slow, deliberate nature of urban warfare necessitates effective breaching techniques, as even unprepared urban areas can present significant obstacles to movement. In well-defended urban environments, breaching becomes critical to maintaining operational momentum and ensuring the success of assault operations.

Breaching is typically conducted by the assault element of a unit, although in certain scenarios, a separate breaching element may be designated. This flexibility allows commanders to adapt to the scale and complexity of the breaching task. The primary purpose of breaching is to remove or bypass impediments, such as walls, doors, or barricades, to allow the assault element access to the objective.

Urban environments, with their dense construction and defensive fortifications, often include barriers designed to slow or halt attacking forces. Without the ability to breach these barriers, an assault risks becoming bogged down, losing momentum, and exposing forces to enemy counterattacks. Breaching ensures that attacking units can maintain their pace, surprise the enemy, and secure critical positions within urban terrain.

Breaching can be accomplished through explosive, ballistic, thermal, or mechanical methods. Each method is suited to specific situations and types of barriers, and their use depends on factors such as available resources, time constraints, and the tactical environment.

1. Explosive Breaching: Explosive breaching is one of the most effective and rapid methods for creating entry points in urban combat. It involves the use of explosives to destroy doors, walls, or other barriers. This method is often employed when speed is critical, or when barriers are too strong to be removed by other means. However, explosive breaching carries significant risks, including potential harm to nearby personnel, damage to surrounding infrastructure, and noise that can alert the enemy to an impending assault.

Explosives must be handled by trained personnel, such as combat engineers or members of the assault element with specialized training. Proper placement and calculation of charges are essential to minimize unintended damage and maximize the breach's effectiveness.

2. Ballistic Breaching: Ballistic breaching uses direct fire weapons to neutralize barriers. This method is particularly useful for destroying locks, hinges, or doors. Shotguns, rifles, and other firearms equipped with breaching rounds are commonly used for this purpose. Ballistic breaching is relatively quick and requires minimal equipment, making it a practical choice in situations where explosives are unavailable or inappropriate.

However, ballistic breaching may not be effective against reinforced barriers, and it generates significant noise that can compromise operational stealth.

3. Thermal Breaching: Thermal breaching employs tools such as cutting torches to penetrate metal barriers, including door hinges, grates, or fortified entrances. This method is precise and minimizes collateral damage, making it ideal for situations requiring discretion or where nearby infrastructure must remain intact.

The primary drawback of thermal breaching is the time required to complete the task, as well as the need for specialized equipment and training. Additionally, the use of thermal tools may produce noise and sparks, potentially alerting the enemy.

4. Mechanical Breaching: Mechanical breaching relies on tools such as crowbars, axes, saws, and sledgehammers to physically remove or destroy barriers. It is a straightforward and resource-efficient method, often employed when other options are unavailable. Mechanical breaching is particularly effective against non-reinforced doors, windows, or barricades.

The main limitation of mechanical breaching is its slower pace and the physical effort required, which can fatigue personnel and expose them to enemy fire during prolonged breaching operations.

Assault Entry in Urban Combat

Assault entry is a critical aspect of urban warfare, enabling infantry squads to gain access to structures while adapting to the challenges of confined environments, unpredictable enemy positions, and complex terrain. Effective assault entry methods must consider tactical, logistical, and situational factors. Six primary methods of assault entry—Top-Level Entry, Bottom-Level Entry, Vehicle-Elevated Entry, Assisted Lifted Entry, Ladder-Assisted Entry, and Helicopter-Borne Entry—each have unique advantages and disadvantages. The selection of an entry method depends on mission objectives, structural conditions, and the enemy's defensive posture [190].

Top-level entry involves initiating the assault from the roof of a structure and moving downward. This method is ideal for achieving surprise, as enemy defences are often oriented toward lower levels. By starting at the top, squads can disrupt enemy positions, take advantage of gravity, and use the roof as a platform for deploying grenades or explosives [190].

Top-level entry is particularly useful in densely built areas with interconnected rooftops, such as townhouse-style residences. These environments often lack access for vehicles, requiring infantry squads to rely on their own mobility and ingenuity. However, this method also has drawbacks. Once inside the structure, withdrawing from a top-level entry is challenging, especially under heavy enemy fire. This limits the squad leader's flexibility to disengage and regroup. Additionally, securing access to rooftops can be time-intensive, particularly in structures with varying heights.

Bottom-level entry, also known as clearing from the ground up, provides greater control and flexibility. Starting at the lowest level of a structure allows squads to systematically clear floors while maintaining secure egress points. This method is less resource-intensive, requiring fewer personnel, and is particularly advantageous when dealing with casualties, as injured Marines can be evacuated more easily.

However, bottom-level entry also presents challenges. Moving upward through a structure forces squads to contend with enemy defences on higher floors, where adversaries can use gravity and narrow choke points, such as stairwells, to their advantage. This method requires careful planning to neutralize these vulnerabilities while maintaining the squad's momentum [190].

Figure 27: A Marine Corps M998 High-Mobility Multipurpose Wheeled Vehicle (HMMWV) passes through the city with a load of supplies during Operation Provide Comfort, a multinational effort to aid Kurdish refugees in southern Turkey and northern Iraq. The U.S. National Archives, Public Domain, via Picryl.

Vehicle-elevated entry enables squads to access upper levels of buildings directly, bypassing heavily defended lower floors. Using vehicles, such as armoured personnel carriers or utility vehicles, Marines can ascend to second- or third-story windows or balconies, forcing enemies to retreat downward into streets where supporting forces await [190].

While effective, this method requires precise coordination between infantry squads and vehicle crews. Vehicles must be positioned securely to avoid exposing the assault team to enemy fire. Additionally, vehicle availability and manoeuvrability in narrow urban streets may limit the feasibility of this approach.

Assisted lifted entry is a straightforward method in which one Marine assists another by physically lifting them into a higher position, such as a window or balcony. This technique is simple and requires minimal equipment, making it suitable for situations where time is critical or resources are limited.

Despite its simplicity, this method has significant limitations. The Marine performing the lift remains exposed in the street, increasing their vulnerability to enemy fire. Assisted lifted entry also requires coordination and physical effort, which can slow down the assault and reduce the unit's combat readiness during the entry process [190].

Ladder-assisted entry allows squads to access upper levels using ladders. This method combines the advantages of vehicle-elevated and assisted lifted entries, providing flexibility in environments where vehicles cannot reach and reducing the physical strain on personnel.

However, ladder-assisted entry introduces logistical challenges. Transporting and deploying ladders requires planning and may slow the pace of operations. Additionally, ladders can expose assault teams during the climb, requiring cover from supporting elements to mitigate risk.

Helicopter-borne entry is a highly desirable but complex assault method. Using helicopters, squads can land directly on rooftops, initiating a top-down assault that maximizes surprise and momentum. This approach is ideal for high-priority targets or heavily fortified structures where traditional entry methods are impractical [190].

However, helicopter-borne entry involves significant coordination and resources. Helicopters must navigate urban terrain, often under hostile fire, to insert troops safely. The operation requires precise timing, secure landing zones, and contingency plans in case of mechanical failures or enemy interference. Despite these challenges, helicopter-borne entry remains an effective option for elite units in specialized operations [190].

Rotation Plan in Urban Operations

A well-structured rotation plan is essential for maintaining operational tempo, efficiency, and adaptability during urban combat. By enabling units to rotate between key roles—Assault, Support, and Security—it ensures sustained effectiveness over extended engagements. Additionally, a rotation plan provides flexibility to address unforeseen circumstances, such as casualties or shifting tactical requirements, helping units maintain momentum and avoid overexertion.

Urban operations involve tasks that are both physically and mentally demanding, such as assaulting enemy positions, providing suppressive fire, and securing perimeters. A rotation plan ensures that no single unit is overly fatigued or exposed to prolonged risk, distributing responsibilities across multiple elements to maintain combat readiness.

Rotation also supports dynamic mission requirements. For instance, after clearing a building, an Assault unit can rotate into a less intensive Security role, while another unit moves forward to take over the assault. This adaptability allows commanders to respond effectively to factors like objective size, enemy resistance, or unexpected casualties, maintaining operational flexibility and effectiveness.

A rotation plan typically organizes units into three core roles:

1. **Assault Unit**: Responsible for seizing objectives, clearing buildings, and neutralizing threats.

2. **Support Unit**: Provides suppressive fire, logistical support, and other resources to assist the assault.

3. **Security Unit**: Secures the perimeter, prevents reinforcements, and ensures the safety of rear elements.

The rotation plan usually anticipates one to two rotations during an operation but remains flexible to adapt to the situation. For example:

- After the Assault unit secures a structure, it transitions to the Security role, maintaining the newly captured position and preparing for follow-on tasks.

- The Support unit transitions into the Assault role, advancing to seize the next objective.

- The Security unit rotates into the Support role, providing suppressive fire and logistical support.

This cycle ensures that each unit experiences a balanced mix of high-intensity and lower-intensity tasks, allowing for sustained combat effectiveness throughout the mission.

Key considerations for a rotation plan:

Planning Rotations: Commanders must anticipate when and where rotations will occur based on factors such as mission tempo, structure size, and resistance intensity. Preplanning rotations at logical points—such as after clearing a significant building or securing a position—helps maintain cohesion and operational flow. Flexibility remains critical, allowing adjustments for casualties, unexpected threats, or other developments.

Criteria for Initiating Rotations: Clear criteria for rotations ensure smooth execution. These may include:

- Completion of a specific objective (e.g., clearing a building).

- Resource depletion or personnel exhaustion in a unit.

- Casualties requiring redistribution of responsibilities.

- Tactical changes, such as reinforcements or shifts in mission focus.

Simplicity and Clarity: While rotations add operational flexibility, excessive complexity can lead to confusion and inefficiency. A simple, well-communicated rotation plan ensures that all units understand their roles and responsibilities, minimizing the risk of miscommunication and operational delays.

A well-implemented rotation plan offers several critical benefits in urban combat operations. By rotating roles, units can maintain a steady operational tempo, ensuring consistent progress toward mission objectives without overburdening any single element. This structured approach enhances adaptability, allowing commanders to respond swiftly to challenges such as casualties, resource depletion, or shifting tactical scenarios without disrupting the overall mission. Additionally, rotations prevent burnout by distributing high-intensity and lower-intensity tasks evenly among units. This balance fosters improved morale, sustained focus, and operational efficiency throughout the engagement, ensuring that personnel remain effective and mission-ready.

Effective rotations require precise coordination to avoid gaps in coverage or exposure to enemy fire. Commanders must clearly communicate the rotation sequence and designate specific points for transitions.

Unforeseen casualties can disrupt rotation plans. Commanders should include contingency measures, such as reserve units to fill critical gaps or adjustments to the planned sequence.

The urban environment, with its narrow streets, dense structures, and concealed enemy positions, complicates rotations. Thoroughly securing positions before initiating a rotation helps mitigate the risk of counterattacks.

Subsequent rotations should be planned to ensure continuity. For instance, after completing its role as Security, the former Assault unit may transition into the Support role for the next phase of operations. Aligning follow-on rotations with the operational timeline ensures that all units remain adequately rested and resupplied before assuming demanding roles again.

Chapter 3

DEFENSIVE STRATEGIES IN URBAN WARFARE

Defensive Strategies versus Offensive Strategies in Urban Warfare

Defensive and offensive strategies in urban warfare serve distinct purposes and employ fundamentally different approaches due to the unique demands of urban environments. While both aim to achieve military objectives within densely populated and infrastructurally complex areas, their execution is shaped by the advantages and challenges of the terrain, the role of civilians, and the specific goals of each operation.

1. Purpose and Objectives: Defensive strategies focus on holding and protecting key positions, denying the enemy access to critical areas, and delaying their advance. They are inherently reactive, designed to sustain control over territory, preserve resources, and buy time for broader strategic goals, such as reinforcements or negotiations. For instance, defenders aim to protect infrastructure, civilians, or symbolic landmarks that are vital for operational or political reasons.

In contrast, offensive strategies aim to seize control of urban areas, eliminate threats, or gain strategic advantages by attacking enemy positions. Offenses are proactive and designed to penetrate enemy defences, disrupt their operations, and secure key objectives, often at the cost of greater resource expenditure and risk.

2. Use of Terrain: Defensive strategies exploit the inherent advantages of urban terrain to slow or neutralize enemy advances. Defenders use the environment's natural compartmentalization—narrow streets, high buildings, and underground tunnels—to create choke points, ambush zones, and fortified positions. By turning the terrain into a force multiplier, defenders can maximize their impact while minimizing exposure.

Offensive strategies, on the other hand, must overcome these terrain-based challenges. Attackers need to navigate through constrained spaces, neutralize defensive strongholds, and advance in an environment that limits visibility and manoeuvrability. Offensive forces often rely on combined arms tactics, such as air and artillery support, to break through urban defences, which can lead to significant collateral damage if not carefully managed.

3. Role of Civilians: Civilians play a critical role in shaping urban warfare strategies, and their presence is often a greater challenge for offensive operations than defensive ones. Defensive strategies typically involve protecting civilians and infrastructure while leveraging the complexity of civilian-dense areas for concealment and support. Defenders may use civilians for logistical aid or as part of their strategic advantage, though ethical considerations are paramount to avoid violations of international law.

Offensive strategies, however, must account for the risks of collateral damage and civilian casualties, which can undermine political and moral objectives. Attackers face the added burden of distinguishing combatants from non-combatants in densely populated areas, complicating targeting decisions and slowing their advance.

4. Resource Allocation and Logistics: Defensive strategies are generally more resource-efficient, as defenders can establish fortified positions, stockpile supplies, and operate within a limited area. The fixed nature of defence reduces logistical demands, allowing forces to sustain themselves over longer periods with fewer resources. However, prolonged defence can lead to attrition if resupply routes are cut off.

Offensive strategies require significantly more resources, including ammunition, fuel, and medical supplies, due to the dynamic nature of advancing into hostile urban terrain. Attackers must also invest heavily in reconnaissance, specialized equipment, and force protection to mitigate the risks posed by fortified defences and hidden threats.

5. Tactical Focus and Execution: In defensive strategies, the focus is on delaying, disrupting, and denying the enemy's objectives. This involves setting up obstacles, ambushes, and fortified positions to make enemy advances costly and time-consuming. Defensive operations are typically slower and rely heavily on the defender's ability to predict enemy movements and adapt to evolving threats.

Offensive strategies emphasize speed, coordination, and momentum. Attackers must break through defensive lines, clear buildings, and secure objectives while maintaining pressure on the enemy. Offensive operations often involve a higher degree of synchronization between units, as well as greater reliance on intelligence and precision fire support to neutralize enemy strongholds.

6. Risk and Adaptability: Defensive strategies inherently involve less exposure to risk, as defenders can use prepared positions and knowledge of the terrain to mitigate the attacker's advantages. However, static defences can become vulnerable if the enemy successfully isolates or overwhelms them, requiring defenders to remain adaptable and prepared for counterattacks.

Offensive strategies carry higher levels of risk, as attackers are often exposed to ambushes, sniper fire, and other defensive countermeasures. The dynamic nature of offensive operations demands constant adaptability and resilience to maintain momentum while countering unexpected challenges.

7. Psychological and Moral Impacts: Defensive strategies often rely on the psychological advantage of holding familiar ground, which can boost morale among defending forces. However, prolonged defence under siege conditions can lead to fatigue and diminished morale if reinforcements or relief are delayed.

Offensive strategies place significant psychological pressure on attackers, who must contend with the uncertainty of urban combat, including concealed threats and potential ambushes. Successful offensive operations, however, can demoralize defenders and civilians aligned with the enemy, shifting the momentum of the conflict.

Defensive and offensive strategies in urban warfare differ fundamentally in their objectives, use of terrain, and operational demands. Defensive strategies capitalize on the advantages of the urban environment to hold ground and exhaust attackers, while offensive strategies aim to overcome these challenges through coordination, firepower, and precision. Both approaches require careful planning, adaptability, and a deep understanding of the urban battlefield to achieve their respective goals while minimizing the risks and costs associated with fighting in densely populated and structurally complex areas.

Principles of Defence in Urban Settings: Delay, Disrupt, and Deny

Defensive Strategies in Urban Warfare: When and Why They Are Required

Defensive strategies in urban warfare are crucial when forces need to hold territory, deny access to critical areas, protect civilians or infrastructure, or gain time for broader strategic goals. The dense populations, complex structures, and limited mobility of urban terrain naturally favour defenders, making these strategies indispensable in a range of tactical and strategic scenarios. Below are key circumstances and reasons for employing defensive strategies in urban warfare.

Urban areas often house critical infrastructure, such as government buildings, communication hubs, industrial facilities, transportation networks, and utilities. Defensive strategies are employed to prevent these assets from falling into enemy hands, as their loss can have severe tactical, operational, or even strategic consequences. For instance, defending bridges, airports, or power plants ensures the continuity of logistics and support for ongoing operations. The loss of strategic infrastructure can cripple supply chains and operational capabilities, while its protection sustains the defender's ability to operate effectively and imposes significant obstacles on attackers.

Cities frequently hold symbolic, logistical, or political importance, making them key targets during conflicts. Defensive operations are often required to deny the enemy control of these areas, which might otherwise serve as staging grounds for future offensives or as sources of resources and legitimacy. Losing a major city or urban area can undermine morale and national authority, while defending these locations prevents the enemy from establishing strongholds or consolidating power.

In certain situations, defensive operations are less about permanently holding a position and more about delaying the enemy to buy time for reinforcements, evacuation, or repositioning of friendly forces. Urban terrain is particularly well-suited for slowing enemy advances due to its compartmentalized and maze-like environment. Delaying the enemy disrupts their operational tempo, depletes their resources, and creates opportunities to reposition forces or secure civilian evacuations, contributing to the defender's broader strategic goals.

Urban warfare often involves densely populated areas, placing civilians at high risk. Defensive strategies are vital for protecting civilian populations, maintaining control over humanitarian corridors, and securing safe zones. The presence of civilians can have profound ethical and strategic implications, with civilian casualties potentially undermining the defender's legitimacy. Effective protection of civilians helps maintain international support and minimize collateral damage, making it a key consideration in urban defence.

When facing numerically or technologically superior forces, defensive strategies allow smaller or less-equipped units to leverage the complexities of urban terrain. Cities provide natural cover and concealment, enabling defenders to mitigate disadvantages in manpower or firepower. Urban environments also facilitate guerrilla tactics, sniping, and ambushes, effectively neutralizing some of the attacker's strengths and creating opportunities to level the battlefield.

Defensive strategies are also essential when friendly forces risk being surrounded or cut off from reinforcements and supplies. By holding key positions within urban areas, defenders can maintain control of critical routes and prevent encirclement. Preserving lines of communication and supply is vital for sustaining operations, and preventing encirclement ensures defenders retain mobility and the ability to withdraw or counterattack if necessary.

In many cases, defending urban areas serves broader strategic or political objectives, such as maintaining sovereignty, protecting national morale, or creating conditions for peace negotiations. Capital cities and other major urban centres often hold immense symbolic importance and may need to be defended at all costs to prevent political collapse. Holding symbolic cities bolsters national resistance and morale, while defending them can support political negotiations or create stalemates favourable to the defending force.

Finally, urban areas often become bases of operation for insurgents or guerrilla forces due to their population density and concealment opportunities. Defensive strategies are crucial for neutralizing these threats and restoring control over contested areas. Eliminating insurgent strongholds prevents prolonged instability and disrupts the enemy's ability to recruit, resupply, or stage further attacks.

Guerrilla forces have emerged as a significant component of modern urban warfare, characterized by their unconventional tactics and adaptability to complex urban environments. These irregular combatants, often composed of insurgents, militias, or freedom fighters, utilize asymmetrical warfare strategies to counter stronger conventional military forces. Their operational effectiveness in urban settings can be attributed to several defining characteristics, including their reliance on mobility, local knowledge, and the ability to blend into civilian populations.

Guerrilla forces are distinguished by their use of asymmetrical tactics, which include hit-and-run attacks, ambushes, and sabotage. These methods are designed to disrupt and harass larger, more organized military units over time, rather than engaging in direct confrontations [227]. Their operational structure is typically decentralized, allowing for greater flexibility and rapid relocation, which is crucial in urban environments where conventional forces may struggle to maintain control [228]. Additionally, guerrilla fighters possess intimate knowledge of the urban terrain and cultural dynamics, enabling them to exploit vulnerabilities in their adversaries' operations [25].

The reliance on civilian populations is another hallmark of guerrilla warfare. By embedding themselves within communities, guerrilla forces can conceal their movements and operations, complicating the efforts of conventional forces to distinguish between combatants and non-combatants [45]. This tactic not only provides logistical support but also creates ethical dilemmas for conventional forces, who must navigate the challenges of minimizing civilian casualties while pursuing military objectives [51]. Furthermore, guerrilla forces often operate with limited resources, relying on improvised weapons and locally sourced materials, which enhances their resilience and adaptability in urban combat scenarios [229].

The urban landscape significantly amplifies the effectiveness of guerrilla tactics. Dense infrastructure, narrow alleyways, and interconnected buildings provide natural concealment and cover, allowing guerrilla fighters to evade detection and launch surprise attacks [19]. The psychological impact of persistent guerrilla operations can also be profound, as conventional forces face the constant threat of ambushes and harassment, leading to increased stress and uncertainty in their operational environment [21].

Moreover, guerrilla forces excel at prolonging engagements, turning urban battles into drawn-out conflicts that can drain the resources and morale of conventional forces. By leveraging the compartmentalized nature of cities, they can disrupt supply lines and operational tempo, creating a war of attrition that is difficult for conventional armies to sustain [230]. This strategic advantage is particularly evident in case studies such as the Battle of Mosul, where ISIS utilized guerrilla tactics to effectively resist coalition forces in a densely populated urban setting, employing tunnel networks and improvised explosive devices [25].

The presence of guerrilla forces in urban warfare presents significant challenges to conventional military operations. One of the primary difficulties is the complexity of counterinsurgency operations, where conventional forces must balance combat effectiveness with the imperative to protect civilian lives. This constraint is often exploited by guerrilla fighters, who can operate with relative impunity in civilian areas [45]. Additionally, the high resource costs associated with countering guerrilla tactics—such as extensive intelligence gathering and specialized equipment—can strain the logistical and financial capabilities of conventional forces [228].

The unpredictable nature of guerrilla warfare further complicates military operations, as these forces can adapt quickly to changing circumstances and strike unexpectedly, forcing conventional forces to remain in a constant state of readiness [229]. This dynamic creates a significant operational burden, stretching manpower and resources thin, and complicating the overall strategy for urban warfare [45].

Delay, Disrupt, and Deny

Defending in urban settings requires a tailored approach that leverages the complexities of the urban terrain to counteract the attacker's advantages. The principles of delay, disrupt, and deny form the foundation of effective defensive strategies, enabling defenders to hold ground, weaken the enemy, and maintain operational control despite resource constraints or numerical disadvantages. These principles are interconnected and applied in a complementary manner to maximize the defender's effectiveness in urban warfare.

Delay

The principle of delay in military strategy is crucial for slowing down an enemy's advance, allowing for necessary defensive preparations and reinforcements. Urban environments, characterized by dense infrastructure and confined spaces, inherently favour delay tactics. The complexities of urban warfare, including three-dimensional terrain, significantly hinder rapid movement for attacking forces, making delay tactics particularly effective in these settings [25].

Key tactics for implementing delay include the use of obstacles and barriers, sniper and ambush positions, and deceptive defences. Physical barriers such as barricades and rubble can effectively impede enemy movement, creating choke points in narrow streets or alleyways [25]. For instance, during the Battle of Stalingrad, Soviet forces utilized the city's ruins to create a labyrinth of fortified positions, which forced German troops to engage in costly urban combat, significantly slowing their advance [194]. This tactic exemplifies how urban environments can be transformed into defensive strongholds that leverage the terrain to disrupt enemy operations.

Sniper and ambush positions are another critical component of delay tactics. By positioning snipers and small units in concealed or elevated locations, defenders can engage advancing forces, compelling them to slow down and take cover. This not only disrupts the enemy's operational tempo but also instils a sense of caution and fear, further delaying their progress. The psychological impact of such tactics can be profound, as they create uncertainty and hesitation among attacking forces [194].

Deceptive defences, such as false strongpoints or abandoned positions, serve to mislead attackers, causing them to waste time and resources securing areas of no strategic value. This tactic can be particularly effective in urban warfare, where the complexity of the environment can obscure the true intentions and capabilities of the defending forces [25]. The combination of these tactics can create a formidable defence that not only delays the enemy but also sets the stage for potential counterattacks or strategic withdrawals to more defensible positions.

The historical context of urban warfare further illustrates the effectiveness of delay tactics. The Battle of Stalingrad serves as a prime example, where Soviet forces turned the city into a fortified stronghold, utilizing the urban landscape to their advantage. The German advance was significantly slowed, allowing Soviet forces to regroup and launch counteroffensives, ultimately leading to a pivotal victory [194]. This historical precedent underscores the importance of delay tactics in shaping the outcomes of urban conflicts.

Disrupt

Disruption tactics in urban warfare are designed to undermine an attacker's cohesion, organization, and operational effectiveness. Urban environments inherently present challenges for attackers due to their complex terrain, which limits visibility, communication, and manoeuvrability. This complexity can significantly hinder coordination among attacking units, making urban settings particularly advantageous for defenders employing disruption strategies [49, 119].

One key tactic for disruption is the interdiction of supply lines. By cutting off or interfering with the logistics and supply routes within urban areas, defenders can create critical resource shortages for attackers, particularly in terms of ammunition, food, and medical supplies. This tactic has been observed in various conflicts where controlling urban logistics has proven essential for maintaining operational superiority [231, 232]. The effectiveness of such strategies is amplified in urban settings, where the dense infrastructure can be leveraged to obstruct and complicate supply chain routes [233].

Counterattacks also play a vital role in disruption tactics. Small, mobile counterattacks targeting weak points or rear elements of an attacking force can compel attackers to divert resources and attention away from their primary objectives. This tactic not only disrupts the attackers' plans but also creates confusion, forcing them to adapt under pressure, which increases the likelihood of operational errors [234, 235]. Historical examples, such as the Battle of Mosul, illustrate how effective counterattacks can be in urban warfare, where defenders utilized mobility and local knowledge to exploit vulnerabilities in the attacking forces [51].

Furthermore, communication jamming is a critical tactic in urban warfare, particularly in the context of electronic warfare. Urban environments provide numerous opportunities for defenders to disrupt the attackers' communication networks, complicating coordination and command structures. By jamming radios or other communication devices, defenders can create significant operational challenges for attackers, further exacerbating the difficulties posed by the urban terrain [236, 237]. The combination of these tactics—interdicting supply lines, executing counterattacks, and employing communication jamming—can effectively disrupt an attacker's operational capabilities and cohesion, leading to a strategic advantage for defenders in urban combat scenarios.

Deny

The principle of denial in military strategy emphasizes the importance of preventing adversaries from achieving their objectives. This is particularly relevant in urban warfare, where defenders can leverage the inherent compartmentalization of the environment to control critical areas and restrict enemy movement. Effective denial tactics can significantly alter the dynamics of a conflict, as evidenced by historical examples and theoretical frameworks.

One key tactic in denial operations is the establishment of strongpoints. Defenders can create fortified positions in strategically significant locations, such as intersections or high-rise buildings, which complicates the attackers' efforts to bypass or capture these areas. The First Battle of Grozny (1994-1995) serves as a notable example, where Chechen defenders utilized fortified buildings and small, mobile units to contest critical terrain, effectively denying Russian forces control over the city [231]. This tactic illustrates how urban environments can be manipulated to enhance defensive capabilities.

Another critical aspect of denial is the use of demolitions and sabotage to destroy essential infrastructure. By targeting bridges, utilities, and other critical resources, defenders can significantly hinder an attacker's operational capabilities. This strategy not only denies physical access but also creates psychological barriers, instilling doubt and fear in the attacking forces.

Securing civilians and resources is also vital in denial operations. Protecting non-combatants ensures that defenders maintain morale and support from the local population, which can be crucial in prolonged conflicts. Historical accounts indicate that successful defence strategies often involve integrating civilian safety into military operations, thereby enhancing the legitimacy and effectiveness of the defenders' efforts.

Decentralizing the defence is another effective tactic that allows small, well-trained teams to hold key positions independently. This strategy was evident during the First Battle of Grozny, where Chechen forces employed decentralized units to contest Russian advances, demonstrating the effectiveness of small, agile groups in urban combat scenarios [238].

Integration of Delay, Disrupt, and Deny

The integration of the principles of delay, disrupt, and deny into urban defence strategies is crucial for enhancing resilience against potential threats. Each principle plays a distinct yet complementary role in a cohesive defensive framework. Delay tactics are primarily focused on buying time, which is essential for slowing down an attacker's momentum as they attempt to penetrate urban areas. This can be achieved through various means, such as creating obstacles or employing defensive measures that hinder rapid movement. For instance, urban planning that incorporates barriers or controlled access points can effectively delay an attacker, allowing defenders to prepare and respond more effectively.

The strategic use of urban geography, such as narrow streets or complex layouts, can also serve to slow down adversaries, thereby providing defenders with critical time to mobilize resources and coordinate their response. Disruption strategies aim to weaken the attacker's coordination and deplete their resources during their advance. This can involve targeted actions that disrupt communication lines or logistical support for the attackers, thereby creating confusion and inefficiency within their ranks.

Urban environments can be leveraged to create conditions that are unfavourable for attackers, such as using the dense urban fabric to complicate their movements and operations. Additionally, the integration of technology, such as surveillance systems and real-time data analytics, can enhance the ability to disrupt attackers effectively. Denying access to critical objectives is the final principle, which focuses on securing key locations to ensure that attackers cannot achieve their goals, even if they make progress elsewhere. This requires a robust understanding of urban infrastructure and the identification of vital assets that must be protected.

Effective denial strategies often involve a combination of physical barriers, security personnel, and community engagement to create a sense of ownership and vigilance among residents. Furthermore, urban design that promotes defensible space can enhance the ability of communities to resist incursions and maintain control over their environments.

The interplay between these principles allows defenders to maximize the advantages of urban terrain while forcing attackers to expend significant resources for minimal gains. By effectively integrating delay, disrupt, and deny strategies, urban defenders can create a multi-layered defence that not only protects critical assets but also enhances overall community resilience against various threats.

Fortifying Defensive Positions in Buildings, Streets, and Alleys

Fortifying defensive positions in urban warfare involves using the inherent characteristics of buildings, streets, and alleys to enhance protection, control movement, and create advantageous positions against attacking forces. The dense infrastructure of urban environments provides defenders with natural opportunities to create strongholds, ambush zones, and chokepoints. Proper fortification of these areas ensures that defensive operations are sustainable and resilient against sustained assaults.

Figure 28: U.S. Marines take a low position and formulate a battle strategy to flush insurgents from their fighting positions during a rooftop gun battle in Ramadi, Iraq. U.S. NAVY, Public Domain, via GetArchive.

Fortifying buildings is a critical aspect of urban defence, particularly in conflict situations where structures can provide essential cover and strategic advantages. To effectively transform buildings into fortified positions, several key strategies must be employed, including securing entry points, optimizing interior layouts, utilizing upper floors and rooftops, implementing booby traps and obstructions, and enhancing camouflage and concealment.

Securing Entry Points: The first line of defence in fortifying a building involves reinforcing entry points such as doors and windows. These vulnerabilities can be addressed through the use of barricades, sandbags, and improvised materials like furniture or debris. Research indicates that effective fortification of entry points can significantly delay or prevent enemy access, thereby enhancing the overall security of the structure [239]. The historical context of fortified settlements shows that similar strategies have been employed across various cultures to protect against invasions and attacks [240].

Interior Defence: The interior layout of a building plays a crucial role in its defensive capabilities. Establishing designated firing positions and fallback routes can optimize the building for defence. Reinforcing walls or intentionally breaching them can create advantageous pathways for rapid movement, allowing defenders to respond quickly to threats [241]. Studies on historical fortified architecture reveal that the spatial organization of interiors has been a fundamental consideration in military architecture, facilitating both defence and tactical manoeuvrability [242].

Upper Floors and Rooftops: Elevated positions, such as rooftops, provide defenders with superior fields of fire and observation capabilities. Placing snipers or machine gunners in these locations can effectively control surrounding areas and cover critical entry points. However, it is essential to reinforce rooftops against indirect fire or aerial attacks, as historical examples demonstrate that such vulnerabilities can lead to significant losses during assaults [243]. The strategic use of elevation in fortifications has been a common practice throughout history, particularly in regions prone to conflict [244].

Booby Traps and Obstructions: To deter attackers, the integration of booby traps, tripwires, and physical obstructions like barbed wire can be highly effective. These measures not only delay enemy movements but also increase the risk of casualties during assaults, thereby enhancing the defensive posture of the building [245]. The archaeological record shows that fortified settlements often included such deterrents, reflecting a long-standing understanding of their importance in defence strategies [239].

Camouflage and Concealment: Finally, modifying the external appearance of fortified buildings to blend with their surroundings is crucial for avoiding detection by enemy forces. Effective camouflage can significantly enhance the survivability of a structure during conflict. Historical analyses of fortified settlements indicate that successful concealment strategies have often been employed to protect vital resources and personnel from enemy reconnaissance [240].

Fortifying streets in urban warfare is an important strategy that enhances the defensive capabilities of military forces. The fortification of streets involves creating barriers, establishing overwatch positions, and preparing ambush points, all of which serve to channel enemy movements and restrict their operational freedom. This response synthesizes relevant literature to support the various aspects of street fortification in urban combat scenarios.

Chokepoints, such as narrow streets and intersections, are vital in urban warfare. Fortifying these areas with barriers, wrecked vehicles, and rubble can significantly slow down advancing enemy forces. Such tactics are essential as they create natural chokepoints that defenders can exploit. For instance, barriers can be strategically placed to impede vehicle movement, while heavy weapons can be positioned at these chokepoints to maximize their impact against enemy forces [180]. The use of wrecked vehicles and debris not only serves as physical obstacles but also enhances the psychological impact on the enemy, making them more cautious in their advance [180].

Overwatch positions are another critical element of street fortification. Buildings that overlook key streets can be occupied by defenders, allowing for effective control of the area. The presence of snipers and machine gunners in these elevated positions can dominate open spaces, creating a hazardous environment for attackers attempting to cross these streets [180]. This strategic occupation of buildings not only provides defenders with a tactical advantage but also allows for better situational awareness and intelligence gathering on enemy movements [180].

The implementation of physical obstacles and blockades is crucial for disrupting enemy vehicle movements. Concrete blocks, steel beams, and abandoned vehicles can be used to create barriers that hinder the mobility of mechanized units. Additionally, the incorporation of anti-tank barriers and mines further enhances the defensive posture against armoured threats [180]. Such measures are essential in urban environments where the mobility of both defenders and attackers is often restricted by the built environment [180].

Creating unobstructed firing lanes is another vital aspect of fortifying streets. By clearing unnecessary obstacles, defenders can establish clear lines of sight and target advancing attackers effectively while maintaining cover. This tactic not only improves the defensive capabilities of the forces but also increases the likelihood of inflicting casualties on the enemy [180]. The establishment of these firing lanes is critical in urban warfare, where the complexity of the environment can often hinder effective engagement [180].

Finally, preparing ambush points along main roads and side streets is a strategic manoeuvre that can lead to significant tactical advantages. Concealed positions allow defenders to launch surprise attacks on advancing enemy forces, inflicting heavy casualties and disrupting their operations [180]. This tactic is particularly effective in urban settings where the terrain can be utilized to conceal defensive positions, making it difficult for attackers to anticipate and counter such threats [180].

Fortifying alleys for defensive purposes presents a strategic advantage in urban warfare, leveraging their inherent characteristics such as narrowness, concealment, and limited accessibility. These features can be effectively utilized to slow or trap enemy forces, while simultaneously providing defenders with opportunities for manoeuvrability and tactical positioning.

Narrow Pathways as Barriers: The narrowness of alleys inherently restricts vehicle movement, making them ideal for the implementation of anti-tank obstacles or traps. Defenders can utilize debris to block entrances or reinforce walls to create impassable barriers, effectively controlling access points. This tactic is supported by military strategies that emphasize the importance of terrain in shaping defensive operations, particularly in urban environments where mobility is constrained [246]. The use of physical barriers in narrow spaces has been documented as a successful method to impede enemy advances and protect defensive positions [246].

Crossfire Opportunities: The confined nature of alleys allows defenders to establish crossfire positions, creating zones where attackers are vulnerable to fire from multiple directions. This tactical advantage is crucial in urban combat scenarios, where the ability to engage the enemy from concealed positions can significantly enhance the effectiveness of defensive operations. Studies on urban warfare tactics highlight the effectiveness of utilizing terrain features, such as alleys, to create overlapping fields of fire and maximize defensive capabilities [246].

Escape Routes: Defenders can also use alleys to establish fallback positions or escape routes. Preplanned pathways ensure that forces can retreat or reposition without exposing themselves to enemy fire. This strategic use of urban terrain allows for greater flexibility in defensive operations, enabling defenders to maintain operational continuity even under pressure. The importance of maintaining escape routes in combat scenarios has been emphasized in military literature, which advocates for the integration of terrain analysis into tactical planning [246].

Vertical Defence: The presence of adjacent buildings provides defenders with opportunities to establish overwatch positions over alleys, allowing for targeted engagements against enemy forces moving below. This vertical dimension of defence complicates the attackers' ability to neutralize fortified positions, as they must contend with threats from multiple elevations. The concept of utilizing vertical space in urban environments is well-documented in military strategy, underscoring its significance in enhancing defensive postures [246].

Concealed Movements: Alleys also facilitate concealed movements for defenders, enabling them to reposition, resupply, or launch counterattacks without detection. Mapping and securing these pathways are critical to preventing enemy infiltration and maintaining operational security. The strategic use of concealment in urban warfare has been extensively studied, with findings indicating that effective use of terrain can significantly impact the outcome of engagements [246].

Integrating defensive positions in urban environments is a multifaceted endeavour that requires careful planning and execution to ensure effective protection against potential threats. A cohesive defensive strategy must consider several key elements, including interlocking fields of fire, layered defence, control of access points, and robust communication and coordination systems.

Interlocking Fields of Fire: The arrangement of defensive positions should facilitate mutual support among them. This concept is critical in ensuring that no single position can be isolated or overwhelmed without assistance from adjacent defenders. Research indicates that effective military strategies often rely on the ability to create overlapping fields of fire, which enhances the overall defensive capability of a unit [74]. This principle is not only applicable in military contexts but can also be observed in urban planning where defensive measures are integrated into the infrastructure to provide comprehensive coverage against threats [247].

Layered Defence: Establishing multiple defensive lines creates depth in the defence, allowing defenders to retreat to pre-prepared positions if attackers breach an outer line. This strategy is supported by historical military practices, where depth in defence has proven effective in maintaining operational integrity even under pressure [74]. The concept of layered defence is also relevant in urban environments, where buildings and streets can be utilized to create a series of defensive layers that can absorb and mitigate attacks [247].

Control of Access Points: Entrances to defensive zones, such as major intersections or alleyways, must be heavily fortified. This includes the establishment of checkpoints and barriers that not only prevent enemy penetration but also serve as early warning systems. Research emphasizes the importance of controlling access points in urban warfare, as they are critical junctures that can significantly influence the outcome of engagements [74]. Effective fortification of these points can deter enemy movements and provide defenders with strategic advantages.

Communication and Coordination: Effective communication among defensive positions is paramount for maintaining situational awareness and coordinating responses to threats. Urban terrain can disrupt traditional communication methods, necessitating the implementation of backup systems such as runners or prearranged signals. Studies have shown that successful military operations often hinge on the ability to communicate effectively under challenging conditions, highlighting the need for redundancy in communication systems [74]. This is particularly relevant in urban settings where physical obstructions can impede direct lines of communication.

Challenges of Fortification: While fortifying defensive positions offers significant advantages, it also presents challenges that must be addressed. One major concern is the exposure of fortified positions to heavy weapons, such as artillery and airstrikes. Historical analyses indicate that static defences can be vulnerable to such threats, necessitating the use of camouflage and the dispersion of critical assets to mitigate risks [74]. Additionally, over-reliance on static positions can constrain mobility, limiting defenders' ability to adapt to dynamic threats. Therefore, flexible plans and mobile reserves are essential components of a successful defensive strategy [74].

Moreover, ethical considerations must guide the construction and use of fortifications, especially in populated areas. Fortifications can disrupt civilian life and damage infrastructure, raising significant humanitarian concerns. It is crucial for military planners to balance the need for security with the potential impact on non-combatants, ensuring that defensive measures do not exacerbate existing vulnerabilities within the civilian population [74].

Utilizing Local Geography and Infrastructure as Defensive Assets

Defensive Operations in Urban Warfare (USA Marines Approach)

Defensive operations in urban warfare are designed to repel enemy attacks, safeguard strategic assets, and create opportunities for counteroffensives. These operations demand meticulous planning, resource allocation, and a nuanced understanding of the urban environment, leveraging its complexities to the defender's advantage [190].

The objectives of defensive operations in urban settings vary widely based on the overarching strategic and tactical context. One primary goal is gaining time, where delaying enemy advances allows for reinforcements, the repositioning of friendly forces, or the completion of broader strategic plans. Denying key terrain is another critical aim, ensuring that vital areas such as bridges, government buildings, and industrial hubs do not fall into enemy hands. Similarly, retaining key terrain is essential to safeguard crucial infrastructure that underpins both operational success and civilian needs [190].

Another important objective is controlling avenues of approach. By directing or limiting enemy movements through predefined chokepoints, defenders can shape the battlefield to their advantage. Defensive operations may also support other missions by acting as an economy-of-force effort, enabling larger offensives or manoeuvres to succeed elsewhere. Finally, urban defence aims to erode enemy forces by inflicting heavy casualties, weakening their capacity to sustain prolonged operations [190].

Effective urban defence hinges on a strategic focus on key terrain features, buildings, and zones that offer significant defensive value. By leveraging these areas, defenders can maintain their ability to manoeuvre and launch counterattacks while disrupting enemy advances and preserving critical assets.

The scheme of manoeuvre in urban defensive operations revolves around the strategic integration of fire support, obstacles, and tactical positioning to disrupt enemy advances and neutralize threats effectively. Indirect fires, such as mortars and artillery, play a crucial role in disrupting enemy formations, forcing them into vulnerable engagement areas where they can be neutralized. At the same time, direct engagement by infantry units targets the enemy at maximum effective range, employing coordinated firepower to maximize impact and reduce the threat of prolonged resistance [190].

Obstacles, including barricades, rubble, and minefields, are meticulously placed to slow or redirect enemy movements, creating opportunities to exploit their disarray. These barriers are designed not only to delay advances but also to funnel enemy forces into preselected zones where they are most exposed to defensive firepower. Forward observers and tactical leaders are integral to this process, guiding the deployment of supporting assets like artillery and air support. By ensuring alignment with the overall defensive strategy, these assets enhance the defenders' ability to maintain control and disrupt enemy operations in the dense and complex urban environment [190].

In urban defensive operations, different types of defence strategies are employed to adapt to the complexities of the environment and achieve tactical and strategic objectives. One approach is the block from a battle position, which involves defending specific structures, such as buildings or complexes. These positions are fortified with all-around defence, supported by obstacles and concentrated firepower. Firing positions are strategically placed to cover key avenues of approach, ensuring mutual support and creating a cohesive defensive perimeter.

Another method is the block in sector, where units are assigned to defend a specific urban sector. The success of this strategy relies heavily on meticulous planning to establish interlocking fields of fire and optimal positioning of forces. By maintaining control over their sector boundaries, units can effectively repel enemy advances and protect critical areas within the urban landscape [190].

The delay in sector strategy focuses on trading space for time, leveraging ambushes, tactical withdrawals, and strategically placed obstacles to slow the enemy's momentum. This approach is particularly useful for inflicting maximum damage on enemy forces while preventing them from achieving their objectives without engaging in a decisive confrontation.

Fighting positions in urban environments are designed with specific considerations to maximize their effectiveness. These positions prioritize protection from direct and indirect fires, often utilizing reinforced structures or constructed overhead shielding. Dispersion and concealment are critical to prevent enemy targeting, while access to covered routes ensures that resupply, reinforcement, or withdrawal can be executed safely. Interlocking fields of fire and escape routes further enhance the defensive capabilities of these positions, allowing defenders to maintain resilience and adaptability in the face of an assault [190].

Urban defensive operations rely on a well-defined distribution of responsibilities among units to ensure cohesion and effectiveness in the face of complex challenges. The main effort constitutes the primary force responsible for repelling enemy advances at critical defensive positions. This group is heavily supported by additional combat assets, such as anti-armour weapons and indirect fire systems, to strengthen their capacity to hold ground and neutralize enemy threats. The main effort focuses its firepower and resources on key areas that are essential to maintaining the overall integrity of the defence [190].

Supporting efforts play a complementary but equally vital role in the defensive strategy. These units are tasked with securing flanks, reinforcing the main effort when necessary, and controlling secondary routes that could serve as potential avenues for enemy movement. By providing additional layers of security and flexibility, the supporting efforts ensure that the defence remains resilient and adaptive. Their role in preventing encirclement and responding to unexpected developments is critical for maintaining the overall stability and effectiveness of the defensive operation. Together, the coordination between the main effort and supporting efforts forms the backbone of a successful urban defence.

Tactical Control Measures (TCMs) are essential for maintaining synchronization and coordination during defensive operations in urban environments. These measures provide a framework for organizing actions, reducing risks, and ensuring clarity across all levels of command. Phase lines, which rely on easily identifiable landmarks such as streets, rivers, or prominent structures, are used to manage the movement of forces, controlling advances or withdrawals with precision. These landmarks act as reference points, ensuring that units remain aligned with the overall defensive strategy [190].

Checkpoints and contact points serve as predefined locations where units can report their positions or establish physical contact with one another. These points are critical for maintaining situational awareness and facilitating seamless communication between units operating in a dense and complex urban setting. Similarly, target reference points (TRPs) are employed to guide fire control and ensure accurate targeting. By designating specific landmarks for this purpose, commanders can direct firepower effectively while minimizing the risk of collateral damage or misdirected strikes [190].

The implementation of clear and effective TCMs is crucial for avoiding confusion and reducing the risk of fratricide. These measures enable coordinated actions across the chain of command, fostering a unified and efficient defensive effort that leverages the advantages of urban terrain while mitigating its inherent challenges [190].

A robust security plan is an indispensable element of urban defensive operations, aimed at preventing enemy infiltration and maintaining situational awareness in a complex environment. Observation posts (OPs) are strategically placed at vantage points to monitor enemy movements and provide early warnings of potential threats. These posts serve as the first line of defence, offering critical intelligence that allows defenders to anticipate and respond to enemy actions before they escalate.

Reinforced patrols play a vital role in navigating the urban terrain, actively identifying vulnerabilities and securing areas that may be susceptible to enemy exploitation. These patrols provide dynamic coverage, ensuring that no gaps exist within the defensive framework and that potential threats are addressed swiftly. By maintaining a proactive presence, patrols enhance the overall security posture and reduce the likelihood of successful enemy incursions.

Perimeter security forms the backbone of the defence, requiring continuous monitoring to detect and prevent breaches. This involves establishing layers of observation, obstacles, and responsive capabilities to ensure that defensive positions remain secure and ready to repel attacks. A well-coordinated security plan integrates these elements, creating a comprehensive system that preserves the integrity of the defence while maintaining a high level of situational awareness [190].

Local Geography and Infrastructure as Defensive Assets

Urban defensive operations rely heavily on the effective use of local geography and infrastructure to strengthen positions and hinder enemy advances. Cities provide a unique environment where natural and man-made features can be leveraged to create robust defensive networks. Understanding and utilizing these assets is essential for maximizing the defender's advantages and exploiting the challenges faced by attackers in urban settings.

The layout of urban terrain offers numerous opportunities to create chokepoints and limit enemy movement. Streets, alleyways, and intersections can be fortified and used as natural barriers to slow the advance of enemy forces. By channelling the enemy into predetermined paths, defenders can set up interlocking fields of fire, ambush points, and strategically placed obstacles. This approach disrupts the attacker's tempo and forces them into predictable patterns of engagement, giving defenders the upper hand.

Buildings and other structures are critical assets in urban defence. Reinforced concrete buildings, multi-story complexes, and industrial facilities provide excellent cover and vantage points for defenders. High-rise buildings, for example, offer an opportunity to establish overwatch positions that dominate surrounding areas, allowing defenders to monitor enemy movements and provide suppressive fire. Basements and underground structures, such as parking garages or subway tunnels, can be utilized for secure storage, troop movements, or surprise counterattacks. Each structure's location, construction, and accessibility must be carefully evaluated to determine its defensive potential.

The existing infrastructure of a city can also play a vital role in impeding enemy forces. Bridges, for example, can be reinforced to hold defensive positions or destroyed to deny enemy access. Utilities such as power grids and water supplies, while essential for civilians, can be strategically controlled to limit their availability to the enemy. Roads and railways, which are often critical for logistics, can be blocked or mined to disrupt enemy supply chains and communication networks.

Effective use of local geography and infrastructure requires thorough reconnaissance and planning. Defenders must understand the city's layout, identify strategic points, and anticipate how the enemy might exploit the terrain. Coordination with engineers and other specialized units ensures that defensive positions are reinforced appropriately and that infrastructure can be modified or neutralized as needed.

The ability to adapt to and use local geography and infrastructure is a defining aspect of successful urban defence. By leveraging the inherent complexities of urban terrain, defenders can create significant challenges for attackers, delay their progress, and impose heavy costs on their operations. This strategic use of the environment transforms the urban landscape into a formidable defensive asset, enabling defenders to maintain control and protect key objectives.

Effective Use of Snipers, Ambush Points, and Traps

In urban warfare, the effective utilization of snipers, ambush points, and traps is critical for defenders aiming to disrupt, delay, and neutralize enemy forces. The urban environment presents unique challenges, such as limited visibility, vertical structures, and numerous concealed positions, which can be exploited by defenders to impose significant psychological and physical costs on attackers. These tactics require meticulous planning, precise execution, and adaptability to the ever-changing dynamics of urban combat.

Figure 29: A sniper from Belgium fires his Barrett M82 .50 caliber rifle during a range session from 700 to 2200 meters away on July 12, 2018 during the International Special Training Centre Desert Sniper Course at Chinchilla Training Area, Spain. The two-week course is designed to teach trained sniper teams the necessary skills to operate in a desert environment. (U.S. Army photo by 1st Lt. Benjamin Haulenbeek). Defense Visual Information Distribution Service, Public Domain, via National Archives and Defense Visual Information Distribution Service.

Snipers are a pivotal component of urban defence, providing precision fire from concealed and elevated positions. Their ability to engage high-value targets, such as enemy officers or technical assets, can significantly disrupt the enemy's chain of command, creating chaos within their ranks. The dense urban landscape allows snipers to utilize multi-story buildings, rooftops, and reinforced windows to establish advantageous firing positions that offer both a wide field of view and adequate cover. To maintain their effectiveness, snipers must frequently relocate to avoid detection and counter-sniper efforts, which adds to the psychological pressure on enemy forces, compelling them to move cautiously and allocate resources to counteract the sniper threat [6, 248].

Ambush points are another essential tactic in urban warfare, designed to exploit the predictable movements of enemy forces through narrow streets, alleyways, or choke points. These ambushes are meticulously planned to maximize surprise and lethality, often employing interlocking fields of fire to create overlapping kill zones. The success of an ambush hinges on precise timing and coordination among defensive units, with many ambushes supported by explosive devices or pre-sighted artillery fire to enhance their destructive potential. Beyond inflicting physical damage, ambushes instil fear and hesitation within enemy ranks, disrupting their momentum and increasing their vulnerability to subsequent attacks [249].

Traps, including booby traps, improvised explosive devices (IEDs), and false pathways, serve to further hinder and demoralize enemy forces. These devices are often concealed within the urban terrain, making them difficult to detect in cluttered and confined settings. Traps can target both personnel and vehicles, and their effectiveness is amplified by the attackers' need to advance cautiously, investing significant time and resources in clearing operations. This not only slows their progress but also provides defenders with opportunities to regroup, resupply, or mount counterattacks [195, 248].

Figure 30: The bomb disposal team of the Afghan Army 215 Corps neutralizes an IED in Sangin, Helmand. Al Jazeera English, CC BY-SA 2.0, via Wikimedia Commons.

Improvised Explosive Devices (IEDs) have become a defining feature of modern warfare, particularly utilized by non-state actors, insurgent groups, and terrorist organizations. These groups, including the Taliban, Al-Qaeda, and ISIS, leverage IEDs as a means to counteract the technological and numerical superiority of conventional military forces. The hallmark of asymmetric warfare, IEDs allow weaker forces to engage in high-impact attacks with relatively low-cost weaponry, thus leveling the battlefield against more powerful adversaries [250, 251]. The strategic deployment of IEDs reflects a broader trend where insurgents and guerrilla fighters adapt their tactics to exploit vulnerabilities in their opponents, often resulting in significant operational challenges for state militaries [252, 253].

IEDs are particularly prevalent in conflicts characterized by insurgency and guerrilla tactics, as evidenced by their extensive use in Iraq, Afghanistan, Syria, and various regions in Africa, such as Somalia and Nigeria. These devices serve multiple tactical purposes: they disrupt military movements and supply chains, instil psychological fear among opposing forces, and exert economic pressure by forcing adversaries to allocate substantial resources to counter-IED measures [251, 254]. The unpredictable nature of IEDs not only lowers morale among military personnel but also affects civilian populations, creating a climate of fear and instability [250, 253]. Furthermore, the low-cost nature of IEDs makes them an attractive option for insurgents, as they can inflict considerable damage without the financial burden associated with conventional weaponry [251, 252].

While non-state actors predominantly employ IEDs, there are instances where state-backed proxies or irregular forces utilize similar tactics to achieve strategic objectives without direct state involvement. This indirect support can manifest in various forms, from providing resources to training insurgents in the use of IEDs [251]. However, conventional state militaries generally refrain from using IEDs due to ethical and legal constraints, as their deployment often violates international humanitarian law, which seeks to protect civilians and minimize indiscriminate harm [250, 255]. States that resort to such tactics risk damaging their international reputation and legitimacy, as adherence to established norms of warfare is a critical aspect of maintaining global standing [251, 255].

IEDs are a prevalent tool in modern asymmetric warfare, primarily utilized by non-state actors and insurgent groups. Their deployment is rarely, if ever, seen in conventional military operations due to the ethical, legal, and reputational implications associated with their use. The strategic advantages offered by IEDs, however, continue to make them a favoured weapon among insurgents and terrorist organizations, shaping the dynamics of contemporary conflicts [250, 251, 253].

Figure 31: Disposing of an IED near the USA's Shindand base in Afghanistan. ISAF Headquarters Public Affairs Office from Kabul, Afghanistan, CC BY 2.0, via Wikimedia Commons.

Traps in modern warfare represent a critical aspect of asymmetric conflict, where less conventional forces utilize various devices and techniques to gain an advantage over more powerful adversaries. These traps can be categorized into several types, each with distinct characteristics and applications.

Improvised explosive devices (IEDs) are among the most prevalent traps in contemporary warfare, particularly in insurgent and guerrilla tactics. These devices are often low-cost yet highly effective, making them a preferred choice for non-state actors. Pressure plate IEDs are activated by the weight of a person or vehicle, while tripwire IEDs detonate when a wire is disturbed, commonly hidden in paths or doorways [256]. Command-triggered devices allow remote detonation, providing flexibility in targeting [257]. Additionally, booby-trapped explosives, which are concealed in everyday items, pose significant risks to unsuspecting individuals [115]. The widespread use of these traps by groups such as the Taliban and ISIS underscores their effectiveness in disrupting conventional military operations [49].

Mechanical traps, which do not rely on explosives, include various devices designed to injure or incapacitate enemies. Spike pits, often camouflaged, can cause severe injuries and are sometimes treated with toxins to increase their lethality [256]. Swinging traps, which release heavy objects when triggered, and spring-loaded weapons, such as spears or spiked boards, are also employed to inflict harm without the use of explosives [49]. These traps exploit the element of surprise and the environment, making them effective in guerrilla warfare scenarios.

Although less common due to ethical and legal constraints, chemical and biological traps have been utilized in warfare to incapacitate or harm opponents. Poisoned objects, such as spikes or food supplies, can deliver harmful substances to unsuspecting targets [257]. Gas-release traps, which disperse chemical agents like tear gas, are designed to create chaos and confusion in enemy ranks [49]. The use of such traps raises significant moral and legal questions, often leading to international condemnation.

Environmental traps leverage natural or constructed terrain to hinder enemy movement or cause harm. Flood traps, created by breaching dams or reservoirs, can inundate areas and obstruct troop movements [256]. Collapsing structures, rigged to fall when triggered, can cause injuries and block strategic pathways [49]. Concealed fire hazards, where flammable materials ignite upon disturbance, also serve to create dangerous environments for advancing forces [257]. These traps are particularly effective in urban warfare, where the landscape can be manipulated to the defender's advantage.

Psychological traps aim to instil fear and confusion rather than cause direct physical harm. Decoy traps, which appear to be active devices but are harmless, can mislead and waste enemy resources [49]. Delayed-activation devices create uncertainty, as they may detonate hours or days after being set, prolonging psychological stress among troops [257]. Information traps manipulate the battlefield environment to create a false sense of security, leading to unexpected attacks when complacency sets in [256].

Traps are predominantly employed by insurgent groups, guerrilla fighters, and terrorist organizations, who utilize these tactics to level the playing field against conventional military forces. Historical examples include the Viet Cong's extensive use of traps during the Vietnam War and various resistance movements in World War II [49]. These groups operate in environments that allow them to exploit local terrain and urban settings effectively, enhancing the impact of their traps [257].

Conventional state militaries typically avoid the use of traps, particularly those that are indiscriminate, due to ethical and legal restrictions imposed by international law, including the Geneva Conventions [49]. Professional militaries prioritize precision and accountability, adhering to rules of engagement that discourage the use of traps that could harm civilians or non-combatants [257]. The potential damage to a military's reputation and the risk of being accused of war crimes further deter the use of such tactics [256].

The Geneva Conventions, established in 1949 and supplemented by additional protocols in 1977 and 2005, are pivotal international treaties that set forth the standards for humanitarian treatment during armed conflicts. These conventions are foundational to international humanitarian law (IHL), aiming to protect individuals who are not participating in hostilities, such as civilians, medical personnel, and aid workers, as well as those who are no longer participating, including wounded soldiers, prisoners of war, and shipwreck survivors [258, 259]. The conventions are grounded in principles of humanity, neutrality, and impartiality, mandating that all parties in a conflict treat the wounded and sick without discrimination and ensure humane treatment of prisoners of war [260].

The four main Geneva Conventions serve distinct purposes: the first convention protects wounded and sick soldiers on land; the second extends similar protections to those at sea; the third focuses on the humane treatment of prisoners of war; and the fourth protects civilians, particularly in occupied territories [258, 259]. Each convention emphasizes the prohibition of torture and inhumane treatment, ensuring that individuals are treated with dignity [261]. The additional protocols further expand these protections, particularly in non-international conflicts, and introduce enhanced protections for vulnerable groups such as women and children, while also reinforcing the principle of distinction between combatants and civilians [260, 262].

The Geneva Conventions of 1949 have achieved nearly universal acceptance, with 196 countries ratifying them. These include all member states of the United Nations, as well as some non-member entities like the Holy See and the State of

Palestine. This makes the Geneva Conventions among the most widely ratified treaties in the world.

However, there are a few notable non-parties to the Additional Protocols of the Geneva Conventions, which were adopted in 1977 and 2005 to strengthen humanitarian protections:

1. **United States**: The U.S. has ratified the four main Geneva Conventions but has not ratified Additional Protocol I (focused on international conflicts) or Additional Protocol II (focused on non-international conflicts). The U.S. has cited concerns about provisions that could expand protections to irregular forces, such as guerrilla fighters, and potential impacts on U.S. military operations.

2. **India**: India has ratified the original Geneva Conventions but has not ratified Additional Protocols I and II due to concerns about how they might apply to internal conflicts, such as those involving insurgent groups.

3. **Pakistan**: Similar to India, Pakistan has not ratified Additional Protocols I and II, citing concerns related to internal insurgencies and potential interpretations of the protocols that could impact its sovereignty.

4. **Turkey**: Turkey has ratified the original Geneva Conventions but has not ratified Additional Protocols I and II, likely due to concerns about internal conflicts with Kurdish groups and the broader implications of the protocols.

5. **Iran**: Iran has ratified the Geneva Conventions but has not ratified Additional Protocols I and II, possibly due to concerns about the protocols' implications for conflicts in the region.

6. **Israel**: Israel has ratified the four Geneva Conventions but has not ratified Additional Protocol I, which includes provisions about military occupation, citing concerns over its implications for Israel's ongoing conflict with Palestine.

7. **Myanmar (Burma)**: While Myanmar has ratified the original Geneva Conventions, it has not ratified the Additional Protocols, likely due to ongoing internal armed conflicts and the implications of applying international humanitarian law in these scenarios.

8. **South Sudan:** Although South Sudan became independent in 2011, its status regarding the ratification of the Geneva Conventions and Additional Protocols remains unclear or incomplete.

9. **Other States:** Some small states or micro-nations (e.g., Andorra) have not formally ratified the Additional Protocols, though this is largely due to administrative reasons rather than substantive opposition.

The Geneva Conventions primarily apply to states, and non-state actors, such as insurgent groups or separatist movements, are not parties to these treaties. However, non-state actors are still bound by customary international humanitarian law, which incorporates many principles of the Geneva Conventions.

While the main Geneva Conventions are universally accepted, the lack of ratification of certain Additional Protocols reflects ongoing geopolitical and military concerns among states. The reluctance to ratify these protocols often stems from concerns over sovereignty, internal conflicts, and the potential impact on military operations and domestic policies.

Despite the near-universal ratification of the Geneva Conventions, violations persist, leading to significant humanitarian crises. Mechanisms for accountability are embedded within the conventions, including provisions for war crimes tribunals and international courts [263, 264]. However, enforcement remains a challenge, particularly in asymmetric conflicts and civil wars, where state and non-state actors may disregard these legal frameworks [259]. The ongoing relevance of the Geneva Conventions is underscored by their adaptability to modern warfare, including the complexities introduced by non-state actors and cyber warfare [265].

The psychological impact of snipers, ambushes, and traps is profound, as they create a constant sense of insecurity and unpredictability for attackers. Urban combat is inherently stressful, characterized by limited situational awareness, and the knowledge that any movement could trigger an ambush or trap exacerbates these challenges. For defenders, these tactics enable smaller forces to exert disproportionate influence, compensating for numerical or technological disadvantages [248, 266].

Figure 32: Cpl. Matthew Dooley with 1st Battalion, 1st Marine Regiment, provides security while conducting a helo insertion to a nighttime urban terrain raid training during Exercise Iron Fist 2014 aboard Camp Pendleton, Calif., Feb. 18, 2014. Iron Fist is an is an amphibious exercise that brings together Marines and sailors from the 15th Marine Expeditionary Unit, other I Marine Expeditionary Force units, and soldiers from the JGSDF, to promote military interoperability and hone individual and small-unit skills through challenging, complex and realistic training. United States Marine Corps, Public Domain, via GetArchive.

To effectively employ these tactics, defenders must integrate them into a cohesive defensive strategy. Coordination between snipers and ambush teams is crucial to cover retreat routes or eliminate escaping forces, while traps can be strategically deployed to funnel attackers into predetermined kill zones. Effective communication and situational awareness are essential to avoid fratricide and ensure judicious application of these methods, particularly in areas with civilian presence [267].

Urban Camouflage and Concealment Tactics

Urban camouflage and concealment tactics are essential in urban warfare, where the dense infrastructure and civilian environments present unique challenges and opportunities for military operations. The effectiveness of these tactics hinges on the ability to obscure personnel, vehicles, and equipment from enemy observation while leveraging the complexities of urban landscapes to gain strategic advantages. Urban environments differ significantly from natural terrains, necessitating specific adaptations in camouflage techniques to account for the distinct visual and environmental characteristics present in cities.

In urban settings, effective camouflage requires a careful selection of colours, patterns, and materials that align with the built environment. Unlike natural terrains that feature organic shapes and earthy tones, urban areas are characterized by geometric patterns and a variety of artificial materials. Research indicates that successful urban camouflage must mimic the straight lines and macro patterns of buildings and other structures, as well as incorporate smaller details that reflect the urban landscape [268, 269]. For instance, in a concrete-dominated environment, muted gray tones may be employed, while areas with vibrant graffiti might require more innovative patterns to avoid detection [268, 269].

Concealment tactics extend beyond visual blending; they also involve utilizing physical obstructions to remain hidden. Urban warfare allows combatants to exploit natural elements such as buildings, walls, and rubble for concealment. Alleyways, courtyards, and abandoned structures provide critical cover, while rooftops and basements offer vertical concealment, complicating enemy targeting efforts [19]. Additionally, the use of obscurants like smoke grenades can temporarily block enemy lines of sight, facilitating troop repositioning or evasion [19].

Deception plays a pivotal role in urban camouflage strategies. The deployment of decoys, such as fake weapon emplacements or false troop positions, can mislead adversaries and divert their attention from actual forces [19]. Furthermore, disguising military vehicles as civilian objects—such as delivery trucks or construction equipment—can significantly reduce their visibility in urban environments [19]. This tactic not only enhances concealment but also complicates the enemy's ability to distinguish between combatants and non-combatants, which raises ethical concerns under international humanitarian law (Massingham, 2023).

The dynamic nature of urban combat necessitates continuous adaptation of camouflage and concealment tactics. Urban environments are subject to rapid change due to destruction and shifting battle lines, requiring forces to constantly reassess their surroundings and modify their strategies accordingly (Coward, 2009). For example, a previously concealed position may become exposed due to the collapse of nearby structures or changes in enemy positioning [19].

Modern technology also influences urban camouflage and concealment. The use of thermal and infrared imaging by contemporary militaries poses challenges to traditional concealment methods, as these technologies can detect heat signatures that may reveal hidden personnel or equipment [19]. To counteract this, forces may utilize thermal blankets or materials designed to disperse heat, as well as advanced camouflage patterns that disrupt heat detection [19]. Additionally, electronic concealment techniques, such as minimizing electromagnetic signatures, are increasingly important in urban warfare scenarios [19].

Ultimately, urban camouflage and concealment tactics are not merely about remaining hidden; they also involve controlling visibility to achieve tactical advantages. A well-concealed force can surprise an enemy, gaining a critical edge in combat, while poor concealment can lead to increased vulnerability to ambushes and airstrikes [19]. The successful execution of these tactics requires a profound understanding of urban terrain, meticulous planning, and effective coordination among military units [19].

Case Studies: Defensive Strategies from Key Historical Urban Conflicts

Urban conflicts throughout history have showcased a variety of defensive strategies that leverage the unique characteristics of urban environments. These strategies have evolved over time, adapting to technological advancements and the complexities of urban warfare. This synthesis examines key historical and recent urban conflicts, highlighting the defensive tactics employed by defenders in cities under siege.

The Siege of Stalingrad (1942–1943) serves as a quintessential example of urban defensive warfare. Soviet forces effectively utilized the city's industrial ruins and rubble as natural fortifications, employing a defence-in-depth strategy that created multiple layers of resistance. This approach forced German troops to engage in brutal street-to-street combat, significantly slowing their advance and inflicting heavy casualties. The Soviets' tactic of "hugging the enemy" minimized the effectiveness of German air and artillery support by engaging them in close quarters, which was critical for their defensive success [270]. Furthermore, the use of snipers and fortified positions contributed to the high attrition rates experienced by the German Sixth Army, ultimately leading to their encirclement and surrender [270].

Figure 33: Soviet soldiers in a trench during the battle of Stalingrad. Unknown author, Public domain, via Wikimedia Commons.

In the Vietnam War, the Battle of Hue (1968) demonstrated the effectiveness of urban terrain in defensive operations. North Vietnamese forces capitalized on the dense urban landscape by fortifying key structures and employing guerrilla tactics, including ambushes and booby traps. Their intimate knowledge of the city allowed them to resist a technologically superior U.S. military for an extended period, showcasing how urban infrastructure can be integrated into defensive strategies.

The conflict highlighted the devastating impact of urban warfare on civilian populations, as the fighting resulted in significant destruction and loss of life.

Figure 34: Returning Fire: Marines A Company, 1st Battalion, 1st Marines [A/1/1] fire from a house window during a search and clear mission in the battle of Hue (official USMC photo by Sergeant Bruce A. Atwell). USMC Archives from Quantico, USA, CC BY 2.0, via Wikimedia Commons.

The Siege of Sarajevo (1992–1996) further illustrates the challenges of urban defence in modern warfare. Defenders in Sarajevo constructed underground tunnels to maintain supply lines and fortified critical buildings to withstand prolonged artillery bombardments. They employed snipers and small-unit tactics to disrupt Serbian advances, demonstrating the importance of local knowledge in urban combat [271]. However, the siege also underscored the humanitarian crises that often accompany urban warfare, with widespread civilian suffering and casualties [271].

The First Battle of Grozny (1994–1995) during the Chechen War showcased decentralized urban defence tactics against a numerically superior Russian force. Chechen fighters utilized the city's narrow streets and extensive underground networks to conduct ambushes and evade detection, effectively turning the urban environment into a formidable defensive stronghold. This asymmetrical approach allowed them to inflict significant casualties on Russian troops, who struggled to adapt to the complexities of urban warfare.

In more recent conflicts, the Battle of Mosul (2016–2017) exemplified the sophisticated defensive strategies employed by ISIS. The group fortified buildings, rigged explosives in key chokepoints, and utilized tunnels for movement and storage. They also integrated modern technology, such as drones for reconnaissance, complicating coalition efforts to retake the city. The use of civilians as human shields presented ethical dilemmas for attacking forces, highlighting the complexities of modern urban warfare.

Figure 35: US Army (USA) Soldiers assigned to the 327th Infantry Regiment, conduct a raid on a suspected Fedayeen rebel's home in Mosul, Iraq, during Operation IRAQI FREEDOM. The U.S. National Archives, Public Domain, via Picryl.

The defence of Mariupol during the ongoing Ukraine conflict (2022) further illustrates contemporary urban defensive strategies. Ukrainian forces utilized industrial areas, such as the Azovstal Steel Plant, as strongholds, engaging in close-quarters combat and employing snipers. Despite facing overwhelming Russian firepower, the defenders prolonged the siege, demonstrating the effectiveness of leveraging urban terrain to delay enemy advances. This conflict reflects the ongoing evolution of urban defence tactics in response to modern warfare challenges.

In the context of the Israel-Gaza conflict, urban defence has become critical for militant groups like Hamas, which have constructed extensive tunnel networks and fortified buildings to resist Israeli advances. The integration of civilians into these defensive strategies complicates military operations and raises significant ethical concerns. Israel's own defensive strategies in urban settings emphasize securing settlements and preparing for counter-infiltration, showcasing the multifaceted nature of urban warfare today.

Figure 36: The IDF uncovers hidden tunnels in the Gaza Strip, used by Hamas. Defense Forces, CC BY-SA 2.0. via Flickr.

The Gaza Strip is characterized by a complex network of underground tunnels, often referred to as the "Gaza metro," which has been developed primarily by Hamas and other Palestinian military organizations. This extensive tunnel system serves multiple strategic purposes, including the storage and concealment of weapons, the movement of fighters, and the execution of offensive operations against Israel. The tunnels also facilitate the transportation of hostages and enable infiltration into both Israel and Egypt, effectively masking militant activities within the densely populated urban environment of Gaza [272, 273].

The infrastructure of these tunnels is substantial, with estimates suggesting that they extend over 500 kilometres throughout the Gaza Strip [274]. This network not only plays a crucial role in military strategy but also reflects the socio-political dynamics of the region, where the tunnels are used to circumvent blockades and restrictions imposed by Israel and Egypt. The sociological implications of these tunnels are significant, as they represent both a means of resistance against perceived oppression and a method of sustaining military operations in a context of ongoing conflict [272, 274].

The operational use of these tunnels has been documented in various military engagements, where they have been employed to launch surprise attacks on Israeli forces and to facilitate the movement of militants during conflicts. For instance, during Israel's military operations, the tunnels have been utilized to evade detection and to regroup forces, showcasing their importance in the tactical landscape of the region [272, 273]. Furthermore, the tunnels have been described as a vital component of Hamas's military infrastructure, allowing for a degree of operational flexibility that is important in asymmetric warfare scenarios [273, 274].

Chapter 4
SMALL UNIT TACTICS AND TECHNIQUES

Infantry Tactics: Fire Teams, Squads, and Platoon Tactics for Clearing Buildings

Fire Teams, Squads and Platoons

Utilising a US Army context, the fire team represents the smallest manoeuvre element within the Army that is controlled by a leader. It serves as the foundational building block for all Army tactical operations. These fire teams come together to form squads, and squads combine to create platoons, which are led by Army lieutenants as part of an infantry rifle company. The success of tactical operations relies on soldiers at every level understanding their roles, responsibilities, and tactical missions, as well as the steps required to accomplish them [275, 276].

A fire team consists of four soldiers: a fire team leader, a rifleman, an automatic rifleman, and a grenadier. Two fire teams, referred to as Alpha and Bravo Teams, combine under the leadership of a squad leader to form a nine-soldier infantry rifle squad. The experience and rank of fire team members range from privates straight out of training to specialists with several years of experience. The team leader, typically a sergeant with three to five years of service, oversees the team's actions and ensures cohesion [275, 276].

Each fire team member is assigned a specific position and weapon, tailored to their role within the team. The rifleman carries the M16 or M4 rifle, equipped with night-vision devices and aiming optics. Their primary role is engaging targets within their weapon's effective range, and they often fulfill secondary responsibilities like navigation, security, or ammunition support. The M4 Carbine, lighter and more manoeuvrable than the M16, allows the rifleman to excel in close-quarters combat, particularly in urban settings [275, 276].

Figure 37: Army National Guard Soldier fires M4 rifle on range at the 47th Winston P. Wilson (WPW) and 27th Armed Forces Skill at Arms Meet (AFSAM) at Robinson Maneuver Training Center, Ark, 2018. Defense Visual Information Distribution Service, Public Domain, via National Archives and Defense Visual Information Distribution Service.

The automatic rifleman wields the M249 Squad Automatic Weapon (SAW), a light machine gun that delivers sustained, accurate automatic fire. With a maximum effective range of 1,000 meters, the M249 provides vital suppressive fire, enabling the team to manoeuvre or hold defensive positions. Its higher rate of fire and heavier ammunition load underscore the automatic rifleman's critical role in both offense and defence [275, 276].

Figure 38: Armed with an FNMI 5.56 mm M249 Squad Automatic Weapon (SAW), US Army (USA) SPECIALIST (SPC) Dunn, a GUNNER with 2nd Brigade Reconnaissance Troop, sits on a hilltop providing perimeter security for the Quick Response Force (QRF) during a Search and Seizure mission, in support of Operation IRAQI FREEDOM. The U.S. National Archives, Public Domain, via Picryl.

The grenadier operates an M203A1 grenade launcher mounted on an M4 rifle. This weapon offers unparalleled versatility, capable of firing a variety of rounds, including high-explosive, smoke, and nonlethal projectiles. The M203 allows the grenadier to engage enemies in concealed positions, such as trenches or buildings, that are otherwise inaccessible to direct fire. This capability makes the grenadier essential in providing indirect fire support, marking targets, and signalling tactical actions during operations [275, 276].

Figure 39: A member of the Canadian 2nd Battalion Royal 22e Régiment fires his C7A2 5.56-mm Automatic Rifle mounted M203A1 grenade launcher during live-fire platoon attacks at range 800 during the biennial Rim of the Pacific (RIMPAC) Exercise, Marine Corps Base Camp Pendleton, July 10. 2018. Defense Visual Information Distribution Service, Public Domain, via National Archives and Defense Visual Information Distribution Service.

The fire team leader, armed with an M4, leads by example, directing the team's movements and managing the rate and placement of their fire. This leader ensures the team adheres to unit standards and provides critical support to the squad leader as needed. The fire team is further divided into buddy teams, pairs of soldiers who support each other during combat, enhancing the team's effectiveness and resilience [275, 276].

The diverse weaponry within a fire team ensures balanced firepower and tactical flexibility. The rifle provides precision engagement, the M249 delivers suppressive fire, and the M203 enables indirect fire against entrenched enemies. This combination allows the fire team to handle a wide range of battlefield scenarios, from engaging enemy troops at a distance to clearing urban environments [275, 276].

The M16A1, chambered in 5.56×45mm NATO, is the lightest rifle in the series, weighing 6.35 pounds without a magazine and sling, and 7.06 pounds with a 30-round magazine. It has the shortest overall rifle length (44.25 inches with a bayonet knife) and uses a barrel rifling twist of 1:12 inches. It fires at a muzzle velocity of 3,250 feet per second and has a cyclic rate of fire of 700–800 rounds per minute. Its maximum effective range is 460 meters for point targets, and it lacks burst or area-target functionality [275, 276].

The M16A2/A3, also chambered in 5.56×45mm NATO, is heavier at 7.78 pounds without a magazine and sling, and 8.79 pounds with a 30-round magazine. It is longer, measuring 44.88 inches with a bayonet knife, and features a tighter barrel rifling twist of 1:7 inches, improving performance with heavier projectiles. The muzzle velocity is slightly reduced to 3,100 feet per second. These models introduce a burst-fire mode of 90 rounds per minute and increase the maximum effective range to 550 meters for point targets and 800 meters for area targets. The A3 variant retains an automatic firing option, while the A2 does not [275, 276].

The M16A4, also chambered in 5.56×45mm NATO, is similar to the A2/A3 in both weight and dimensions, weighing 9.08 pounds with no magazine and 10.09 pounds with a 30-round magazine. It offers identical performance to the A2/A3, with a muzzle velocity of 3,100 feet per second, burst-fire capability, and maximum effective ranges of 550 meters for point targets and 600 meters for area targets [275, 276].

The M4 carbine, also chambered in 5.56×45mm NATO, is the most compact and versatile of the series. It features an adjustable buttstock, reducing its length to 29.75 inches (closed) or 33 inches (open). It weighs 6.49 pounds without a magazine and sling, and 7.50 pounds with a 30-round magazine. Due to its shorter barrel, the muzzle velocity is reduced to 2,970 feet per second. The M4 maintains a similar rate of fire to the M16A2/A3, but its maximum effective range is slightly reduced to 500 meters for point targets and 600 meters for area targets [275, 276].

Overall, the M16A2/A3 and A4 models offer greater range and accuracy, benefiting from improved rifling and calibre performance, while the M4 carbine prioritizes portability and manoeuvrability for close-quarters and urban combat scenarios. The M16A1, though lighter, is outpaced by the more advanced models in terms of versatility and effective engagement distances [275, 276].

The M249 Squad Automatic Weapon (SAW) is a versatile light machine gun chambered for 5.56mm ammunition, typically used with a 4:1 mix of ball and tracer rounds packaged in 200-round drums. Each drum weighs approximately 6.92 pounds. The weapon has a total length of 40.87 inches and weighs 16.41 pounds, with an additional 16 pounds for its M122 tripod mount, including the traversing and elevating (T&E) mechanism [275, 276].

The M249 boasts a maximum range of 3,600 meters and a maximum effective range of 1,000 meters when mounted on a tripod with the T&E mechanism. Using a bipod, the effective range decreases to 800 meters for area targets and 600 meters for point targets. Grazing fire is effective up to 600 meters on uniformly sloping terrain. Suppression can be maintained effectively up to 1,000 meters [275, 276].

In terms of fire rate, the M249 supports a sustained rate of 100 rounds per minute in 6- to 9-round bursts with 4- to 5-second intervals, requiring a barrel change every two minutes. The rapid rate doubles to 200 rounds per minute in similar bursts with 2- to 3-second intervals, also necessitating a barrel change every two minutes. The cyclic rate ranges between 650 and 850 rounds per minute in a continuous burst, with a barrel change required every minute. The weapon's standard ammunition load is 1,000 rounds, carried in five 200-round drums [275, 276].

Mounted on the M122A1 tripod, the weapon achieves a height of 16 inches and offers a controlled traverse of 100 mils, with a normal sector of fire covering 875 mils. It allows for precise elevation and depression adjustments: +200 and -200 mils under control, or up to +445 and -445 mils freely. This adaptability, combined with its firepower and range, makes the M249 a critical asset for infantry squads in both offensive and defensive roles [275, 276].

The M203/M203A1 grenade launcher is a lightweight, single-shot, 40mm weapon designed to be mounted under an M16 rifle or M4 carbine. The launcher measures 30.5 cm (12 inches) in barrel length and adds 1.4 kg (3.0 pounds) when unloaded, increasing to 1.6 kg (3.5 pounds) when loaded. Fully loaded with its rifle counterpart, the system weighs 5.0 kg (11.0 pounds). It features six right-hand rifling lands for projectile stability [275, 276].

The 40mm ammunition weighs approximately 227 grams (8 ounces) per round. The launcher is equipped with a front leaf sight assembly and a rear quadrant sight for aiming. It operates as a single-shot weapon, achieving a muzzle velocity of 76 meters per second (250 feet per second) and withstanding a chamber pressure of up to 206,325 kilopascals (35,000 psi) [275, 276].

The launcher's maximum range is approximately 400 meters (1,312 feet). Its effective range varies by target type: 350 meters (1,148 feet) for fire-team-sized area targets and 150 meters (492 feet) for vehicle or weapon point targets. For safety, the minimum firing range is 130 meters (426 feet) during training and 31 meters (102 feet) in combat scenarios. The grenades arm themselves at a range of approximately 14 to 38 meters (46 to 125 feet) after firing [275, 276].

The M203/M203A1 supports a rate of fire of 5 to 7 rounds per minute, with a minimum combat load of 36 high-explosive rounds. These characteristics make it a versatile weapon for infantry squads, offering indirect fire capabilities to engage targets in dead space, vehicles, or fortified positions [275, 276].

A rifle squad is the foundational combat unit in an infantry platoon, consisting of two fire teams and a squad leader. The squad leader, typically a staff sergeant (E-6) with six to eight years of military experience, has progressed through the ranks, often beginning as a rifleman in a fire team. This leadership role demands tactical acumen, operational oversight, and the ability to lead by example in high-pressure environments [275, 276].

The squad leader carries full responsibility for the actions and performance of their squad, ensuring both operational effectiveness and the well-being of the Soldiers under their command. This includes manoeuvring the squad during missions, directing the rate and distribution of fire, and coordinating their movements to achieve tactical objectives. The squad leader's decisions directly influence the success of the mission, requiring a comprehensive understanding of terrain, enemy behaviour, and mission priorities [275, 276].

In addition to tactical leadership, the squad leader is integral to the training and readiness of their team. They ensure squad members are proficient in both individual and collective tasks, fostering a cohesive unit capable of executing complex operations. This involves hands-on training, regular performance assessments, and maintaining a focus on continuous improvement to meet evolving mission requirements [275, 276].

The squad leader also assumes a logistical role, managing the squad's essential needs. They are responsible for requesting and distributing critical supplies such as ammunition, water, rations, and equipment, ensuring the squad remains operationally capable in diverse conditions. Additionally, they maintain accountability for their Soldiers and equipment, conducting inspections of weapons, clothing, and gear, and directing maintenance to ensure readiness [275, 276].

Within the rifle platoon, composed of three rifle squads and a headquarters element, the squad leader is a critical link between the fire teams and the platoon leader. The squad's manoeuvrability and firepower contribute significantly to the platoon's overall effectiveness, with the squad leader acting as the tactical leader on the ground. By managing their team's performance and maintaining communication with higher command, the squad leader helps align squad-level actions with broader platoon objectives [275, 276].

As part of the infantry rifle platoon, rifle squads execute a wide range of missions, from offensive assaults to defensive operations and reconnaissance tasks. The squad leader's ability to adapt to these varied demands ensures that the team remains flexible and effective in combat scenarios. This leadership position, with its combination of tactical, logistical, and administrative responsibilities, underscores the importance of experience and capability in guiding small-unit operations to success [275, 276].

Clearing Buildings

Clearing buildings in urban warfare is a complex and high-stakes operation that requires a thorough understanding of infantry tactics at various levels, including fire teams, squads, and platoons. Each unit plays a pivotal role in ensuring the success of the mission while minimizing risks to personnel and civilians.

Role of Fire Teams: Fire teams, typically composed of four soldiers, are the foundational units in infantry tactics. Each member of the fire team has a designated role that contributes to the overall effectiveness of the unit during building clearance operations. The team leader orchestrates the entry and movement, ensuring clarity in sectors of fire and responsibilities among team members. The rifleman engages targets with precision, while the automatic rifleman provides suppressive fire to neutralize threats. The grenadier enhances the team's capabilities by using indirect fire to target enemies in cover or in adjacent rooms [277-279].

Fire teams utilize systematic room-clearing techniques, such as the "cross" or "buttonhook" methods, which are designed to ensure comprehensive coverage of the room and mitigate the risk of overlooking potential threats. These methods emphasize speed, surprise, and aggression, which are critical for dominating the space and neutralizing any hostile elements quickly [280-283]. Additionally, the use of buddy pairs within the fire team fosters mutual support and accountability, enhancing the team's operational effectiveness during the clearance process [282, 284].

Role of Squads: A squad, consisting of two fire teams and a squad leader, is responsible for coordinating the actions of its fire teams to clear larger areas, such as floors or small buildings. The squad leader, typically a staff sergeant, plays a crucial role in tactical decision-making based on real-time intelligence. This includes directing fire teams to work in concert, maintaining momentum, and ensuring communication throughout the operation [285-287].

Squad tactics often involve cover-and-move principles, where one fire team advances while the other provides suppressive fire. This method not only protects the advancing team but also prevents enemy reinforcements from disrupting the operation. The squad leader's situational awareness is vital for adapting to dynamic combat conditions, allowing for effective resource allocation and strategic planning [288]. In multi-story buildings, the squad leader may assign specific tasks to fire teams, such as securing stairwells or providing overwatch, ensuring comprehensive coverage of the operational area.

Role of Platoons: Platoons, typically composed of three to four rifle squads and led by a lieutenant, are tasked with clearing larger and more complex structures. The platoon leader is responsible for developing the overarching strategy for building clearance, which includes determining entry points, routes of advance, and support requirements. This often necessitates coordination with indirect fire assets and engineers to manage obstacles and enhance operational effectiveness.

During operations, the platoon leader assigns specific objectives to squads, such as securing stairwells or controlling exterior approaches. This level of organization allows for a more comprehensive approach to building clearance, integrating various units to suppress enemy movements and create safer conditions for those operating within the building. The strategic deployment of machine guns or sniper teams further enhances the platoon's capability to neutralize threats before entry, thereby reducing risks to infantry personnel.

Tactical Principles for Clearing Buildings** Several tactical principles guide the building clearance process. Isolation and security are paramount; units must isolate the building to prevent enemy reinforcements or escapes. Speed and coordination are essential to maintain momentum and avoid delays that could allow the enemy to regroup. The principle of violence of action emphasizes the need for overwhelming firepower during entry, which is critical for quickly neutralizing threats and protecting friendly forces.

The use of grenades, such as fragmentation or flashbangs, is a key tactic for disorienting enemies before entry. Each soldier is assigned a specific sector of fire, ensuring comprehensive coverage and minimizing blind spots. Effective communication and strict discipline among units are vital to synchronize movements and respond to evolving situations, thereby reducing the risk of friendly fire incidents.

Use of Entry Teams, Breaching Techniques, and Room-Clearing Procedures

In urban warfare, entry teams, breaching techniques, and room-clearing procedures are essential for successfully navigating and securing enemy-held buildings. These operations require precision, coordination, and versatility, as forces face the challenge of operating in confined, unpredictable, and often hostile environments. Entry teams combine specialized breaching methods with disciplined clearing techniques to neutralize threats, protect team members, and minimize harm to civilians or infrastructure.

Entry teams are small, specialized units trained to breach and enter structures quickly and effectively. These teams operate under the principle of rapid domination, aiming to secure control of a room or building before the enemy can organize a response. Members of an entry team typically have designated roles, including the point man (leading the breach), the cover man (providing suppressive fire), and rear security (guarding the team's flank or rear).

To achieve their objectives, entry teams employ various entry techniques suited to the tactical situation. Among these techniques, dynamic entry, the cross method, slice the pie, narrow angle, and various breaching techniques are crucial for effective operations.

Dynamic Entry is characterized by its fast and aggressive approach, aiming to quickly overwhelm enemy forces. This technique often involves a "knock and announce" procedure followed by breaching the door using mechanical, ballistic, or explosive methods. Once inside, team members prioritize identifying and neutralizing imminent threats while securing advantageous positions within the room. This approach is particularly effective in scenarios where time is critical, such as hostage rescues or when facing immediate threats [289]. The emphasis on speed and aggression in dynamic entry is supported by tactical studies that highlight its effectiveness in high-stakes environments, where rapid action can significantly alter the outcome of an operation [290].

The Cross Method involves two operators positioned on either side of a doorway, entering the room in a coordinated manner. This technique allows for overlapping fields of fire and ensures that no threats are overlooked as each operator covers their assigned sector. The cross method balances speed with precision, enabling teams to secure rooms methodically while minimizing the risk of friendly fire [291]. Research indicates that such coordinated tactics enhance situational awareness and operational effectiveness during room clearing [292].

Slice the Pie is a more cautious technique that focuses on assessing and clearing a room by exposing only small portions at a time. This method allows team members to limit their visibility to potential threats while methodically scanning the room. It is particularly useful in situations where stealth is paramount, such as when civilians are present or when the enemy is well-entrenched [293]. The effectiveness of this technique is underscored by tactical analyses that advocate for gradual exposure to threats, thereby reducing the risk of ambush [294].

Narrow Angle techniques involve operators using the edges of doorways or windows to gain insights into a room while remaining concealed. This method allows for reconnaissance without fully committing to an entry, which is essential in situations where threats are uncertain or the environment is hostile [295]. Tactical literature emphasizes the importance of reconnaissance in enhancing situational awareness and informing subsequent entry strategies [296].

Breaching Techniques are integral to gaining entry into secured structures. Various methods, including explosive, ballistic, mechanical, hydraulic, and thermal breaching, are employed based on the specific circumstances of the operation.

Explosive breaching is particularly effective against reinforced barriers but requires specialized training to mitigate risks [297]. Ballistic breaching offers a quick solution for standard doors, while mechanical breaching provides a quieter alternative for stealth operations [298]. Hydraulic and thermal breaching techniques are also valuable in scenarios requiring precision and minimal noise, respectively [299]. The choice of breaching method is critical and is often dictated by the operational context and the nature of the barriers encountered [300].

Room-Clearing Procedures following a successful breach are essential for ensuring safety and effectiveness. Initial domination involves team members quickly securing key positions to establish control and minimize exposure. This is followed by threat neutralization, where identified threats are engaged with speed and precision [301]. The importance of clear communication and overlapping fields of fire during this phase cannot be overstated, as they are vital for avoiding friendly fire and ensuring comprehensive threat management [302]. Finally, the search and transition phase allows teams to identify any remaining threats or hazards before moving to the next area, ensuring a systematic approach to securing the environment.

Threat engagement in room-clearing operations is a critical aspect of tactical response, emphasizing the need for precision and the minimization of risks to non-combatants. This process involves several key components, including Positive Identification (PID), the use of overlapping fire, and the escalation of force. PID is essential to ensure that a target is indeed hostile before engagement, particularly in environments where civilians or hostages may be present. The importance of PID is underscored in various tactical frameworks, which highlight the necessity of accurate threat assessment to avoid unnecessary casualties [303, 304].

Operators utilize overlapping fire techniques to maintain interlocking fields of fire, which allows them to cover all angles while minimizing the risk to teammates. This tactic is crucial in dynamic environments where threats can emerge from multiple directions, necessitating a coordinated approach to engagement [304, 305]. Additionally, the escalation of force is a strategic consideration, where operators may opt for non-lethal options such as tasers or flashbangs to subdue threats without resorting to lethal force, thereby further protecting non-combatants [303, 304].

In the context of clearing blind spots, operators must systematically address potential concealment areas created by furniture, doors, or other obstacles. This involves using angles to approach and check under or behind these obstacles, as well as cautiously opening additional doors to avoid ambushes. Tools such as mirrors or cameras are employed to inspect hard-to-reach areas, enhancing situational awareness and safety during operations [303, 304]. The systematic approach to clearing these areas is critical in ensuring that all potential threats are identified and neutralized effectively.

Maintaining security after a room has been cleared is another vital component of threat engagement. This involves establishing perimeter security, where at least one operator guards the entry point to prevent adversaries from regrouping or escaping. Sector reassignment may also occur, allowing operators to rotate and cover new areas or support adjacent teams, thereby enhancing overall operational security [303, 304]. The search and secure phase is equally important, as teams must search the cleared area for weapons, intelligence, or other critical items while ensuring that all personnel, including civilians, are accounted for [304, 305].

As teams prepare to exit and transition to the next objective, team consolidation and communication with command are essential. Members regroup to confirm readiness, and updates on progress and room status are communicated to ensure coordinated efforts moving forward. This transition phase is critical for maintaining operational momentum and ensuring that all team members are aligned on the next steps [303, 304].

Special scenarios, such as hostage situations or the presence of booby traps, require additional considerations. In hostage situations, the priority is to isolate and protect hostages while neutralizing threats, which necessitates a delicate balance of aggression and caution [304, 305]. Operators must also remain vigilant for signs of traps or explosives, particularly in areas with suspicious devices, and utilize night vision devices or infrared aiming systems during low-light operations to maintain effectiveness [303, 304].

Cover and Movement in Confined Spaces

Cover and movement are fundamental tactics in confined spaces, ensuring the safety and effectiveness of a team operating in environments like urban interiors, tunnels, or ships. These operations are inherently challenging due to restricted mobility, limited visibility, and increased proximity to threats. The principles of cover and movement in such environments are designed to minimize vulnerability while maintaining operational momentum.

In confined spaces, cover refers to physical barriers that provide protection from enemy fire or observation. Movement involves the tactical advancement or repositioning of team members to achieve objectives while remaining shielded. Combining these two elements ensures that one team member or element is always in a protected or overwatch position, allowing others to move safely. This method, known as "bounding overwatch," is especially critical in confined spaces, where threats can appear suddenly and from unexpected angles.

One of the most important aspects of cover in confined spaces is understanding the environment and using available structures effectively. Walls, furniture, doorframes, and even debris can serve as cover, shielding operators from hostile fire or observation. However, these barriers must be chosen carefully. Solid materials, such as reinforced concrete or steel, offer better protection than wooden furniture or drywall. Operators must also be wary of secondary dangers, such as ricochets or collapsing structures, which are more likely in confined and enclosed areas.

Movement in confined spaces requires precision and coordination. Operators must move deliberately, maintaining a low profile to reduce their exposure to threats. In narrow corridors or tight rooms, movement is often linear, with operators advancing in single-file or staggered formations. These formations ensure that team members can cover each other while maintaining clear lines of fire. Speed is important, but it must not compromise the thoroughness of clearing procedures or the ability to maintain situational awareness.

Communication is critical during cover and movement operations in confined spaces. Verbal commands and hand signals are used to coordinate actions and maintain awareness of team positions. Operators must ensure that movements are synchronized to avoid gaps in coverage that could expose the team to threats. Effective communication also helps prevent friendly fire incidents, a significant risk in tight and chaotic environments.

In confined spaces, the presence of multiple angles and potential ambush points makes maintaining 360-degree security paramount. Operators must constantly adjust their fields of fire and ensure that all possible threat vectors are covered. This includes checking corners, overhead spaces, and blind spots. Each movement forward requires the team to reassess and secure the environment before proceeding.

The challenges of confined spaces also necessitate the use of specialized equipment to enhance cover and movement. Shields, ballistic helmets, and body armour provide additional protection, while compact weapons allow for easier manoeuvrability.

Body armour used in urban warfare typically consists of three main components: ballistic plates, soft armour, and modular attachments. Together, these components provide comprehensive protection against a wide range of threats, including small arms fire, fragmentation, and blunt force trauma. Modern body armour systems are also designed to accommodate the additional gear required in urban operations, such as communication devices, ammunition pouches, and hydration systems.

The ballistic plates in urban warfare body armour are typically made of advanced materials such as ceramic composites, ultra-high molecular weight polyethylene (UHMWPE), or steel. These plates are inserted into carriers and are designed to stop high-velocity rifle rounds, which are common in urban combat scenarios. Plates are rated according to the National Institute of Justice (NIJ) standards, with Level III and Level IV plates providing protection against rifle threats. Given the prevalence of close-range engagements in urban areas, most soldiers opt for Level IV plates, which can withstand armour-piercing rounds.

Soft armour is an integral part of urban warfare body armour, providing protection against lower-velocity threats such as handgun rounds and fragmentation from grenades or improvised explosive devices (IEDs). Made from materials like Kevlar or Twaron, soft armour is flexible and lightweight, allowing for greater mobility. It is typically worn as part of a vest or incorporated into the carrier that holds the ballistic plates.

Modularity is a key feature of body armour for urban warfare. Soldiers operating in urban environments often need to adapt their gear to specific missions. Modern body armour systems are designed with MOLLE (Modular Lightweight Load-carrying Equipment) webbing, which allows users to attach and reconfigure pouches, holsters, and other equipment. This adaptability is crucial in urban operations, where soldiers may need to carry breaching tools, extra magazines, or medical supplies.

Figure 40: Troops wearing the new Virtus body armour and load carrying system in a mock foot patrol scenario. Defence Imagery, CC BY-SA 2.0, via Flickr.

In urban warfare, body armour must also address the risks associated with explosives and shrapnel. Soldiers frequently encounter IEDs, grenades, and other explosive devices that generate high-velocity fragments. To counter these threats, many body armour systems include neck, groin, and shoulder protectors. These additional panels provide expanded coverage, reducing the risk of injuries from fragments or secondary projectiles caused by collapsing structures.

Mobility is a critical consideration for body armour in urban combat. Soldiers often need to move quickly through confined spaces, climb over obstacles, and engage in close-quarters combat. Excessively heavy or bulky armour can impede movement, reduce situational awareness, and increase fatigue. As a result, modern body armour is designed to strike a balance between protection and weight. Advances in materials science have enabled the development of lightweight yet highly protective armour systems, ensuring that soldiers remain agile and effective.

Thermal regulation is another factor in the design of urban warfare body armour. Soldiers operating in urban environments may face extreme heat or prolonged physical exertion. To address these challenges, many armour systems include ventilation channels or moisture-wicking fabrics to enhance comfort and reduce heat stress. Some advanced systems even integrate cooling technologies to maintain optimal body temperature during extended missions.

Another emerging trend in urban warfare body armour is the integration of technology. Smart armour systems equipped with sensors can monitor a soldier's vital signs, detect impacts, and even provide ballistic data to command centres. These innovations enhance situational awareness and enable commanders to make informed decisions during urban operations.

Tools like mirrors or cameras help clear blind spots, and night vision devices or flashlights enhance visibility in low-light conditions. Smoke or flashbangs can be deployed to disorient adversaries, creating opportunities for safe movement.

Cover and movement in confined spaces demand high levels of training and discipline. Operators must be familiar with their equipment and trained to respond instinctively to threats. Regular drills that simulate confined environments help teams develop the muscle memory and situational awareness needed for real-world operations. Mental resilience is equally important, as confined spaces can be physically and psychologically demanding, with limited room for error.

Using Drones and Robots for Reconnaissance and Combat Support

The integration of drones and robots into modern warfare has transformed military operations, particularly in the realms of reconnaissance, combat support, and logistics. These technologies provide significant advantages, especially in complex environments such as urban warfare, where traditional methods may falter. Drones and robots enhance situational awareness, operational efficiency, and the safety of personnel, marking a pivotal shift in military strategy.

Drones are primarily utilized for reconnaissance, offering capabilities that are crucial in urban settings characterized by limited visibility due to dense structures. Equipped with advanced technologies such as high-resolution cameras and thermal imaging, drones can gather real-time intelligence on enemy positions and movements. This capability allows military commanders to make informed decisions without risking personnel. For example, drones can effectively map sniper locations or detect improvised explosive devices (IEDs) in hostile areas, thereby minimizing exposure to danger for ground troops [306]. The versatility of drones, which can operate at various altitudes and in confined spaces, further enhances their utility in urban combat scenarios [307].

Figure 41: A photo of the Uavtek Nano Drone bug being used by UK military.
Uavtek, CC BY-SA 4.0, via Wikimedia Commons.

Ground robots complement aerial drones by providing ground-based reconnaissance and support. These robots are equipped with cameras and sensors that enable them to navigate hazardous environments, such as tunnels or buildings suspected of containing explosives. Their ability to neutralize threats, inspect dangerous areas, and identify chemical or radiological hazards significantly reduces risks to human operators [308]. In urban combat, robots can perform critical tasks such as breaching operations and scouting in tight spaces, where human entry may be too dangerous [164].

Figure 42: A TALON tracked military robot picks up a downed unmanned aerial system at Al Asad Air Base, Iraq, May 19, 2020. (U.S. Army photo by Spc. Derek Mustard). Defense Visual Information Distribution Service, Public Domain, via National Archives and Defense Visual Information Distribution Service.

Combat support is another area where drones and robots excel. Armed drones can deliver precision strikes against enemy positions, disrupt supply lines, and provide close air support to ground forces with minimal collateral damage. Their ability to loiter over a battlefield for extended periods ensures continuous support for ground troops, enhancing tactical advantages [309]. Similarly, ground robots equipped with weaponry can engage targets and provide cover for advancing soldiers, thereby reducing the risk to human life while maintaining effective firepower [309].

Figure 43: MTGR – Light, Full Featured Tactical Ground Robot. Robobotics, CC BY-SA 4.0, via Wikimedia Commons.

Logistics and supply missions also benefit from the deployment of drones and robots. Drones can swiftly deliver essential supplies, such as ammunition and medical kits, to isolated units under fire, circumventing ground obstacles and enemy lines [308]. Ground robots can transport heavy equipment and evacuate casualties, facilitating operations in urban terrains that may be mined or under surveillance [310]. The integration of advanced technologies, including artificial intelligence (AI), further enhances the capabilities of these systems, enabling autonomous navigation and adaptive decision-making in dynamic environments [311].

Despite the advantages, the deployment of drones and robots in military operations presents challenges. These systems are vulnerable to electronic warfare tactics, such as jamming and hacking, which can disrupt their functionality [312]. Moreover, ethical considerations arise regarding the use of autonomous systems capable of lethal force, necessitating clear policies and oversight to ensure compliance with international laws [313].

The deployment of Lethal Autonomous Weapons Systems (LAWS) raises complex legal and ethical questions under international law, particularly in the context of International Humanitarian Law (IHL) and human rights law. The fundamental principles of IHL, such as distinction, proportionality, and necessity, are critical in assessing the legality of these systems in armed conflict.

The principle of distinction mandates that parties to a conflict must differentiate between combatants and non-combatants, as enshrined in the Geneva Conventions and their Additional Protocols [314]. Autonomous systems must be capable of accurately identifying legitimate military targets to comply with this principle. This requirement poses significant technological challenges, as the systems must be programmed to avoid civilian casualties and damage to civilian infrastructure [315, 316]. The implications of failing to adhere to this principle can lead to violations of international law, as LAWS could potentially engage in indiscriminate attacks if not properly regulated [317, 318].

Proportionality is another crucial aspect of IHL that must be considered in the context of LAWS. This principle stipulates that the harm caused to civilians or civilian property must not be excessive in relation to the anticipated military advantage gained from an attack [319, 320]. Autonomous systems must be equipped with advanced algorithms and real-time decision-making capabilities to assess potential collateral damage effectively. The complexity of urban warfare environments further complicates the application of this principle, as rapid changes on the battlefield can affect the proportionality assessment [321, 322].

The principle of necessity also governs the use of lethal force by autonomous systems. It dictates that lethal force should only be employed when essential to achieve a legitimate military objective, and all feasible precautions must be taken to minimize harm [323, 324]. This necessitates robust decision-making protocols and oversight mechanisms to ensure compliance with IHL [325]. The integration of human oversight in the decision-making process is a point of contention, with some experts advocating for a "human-in-the-loop" approach to maintain accountability in lethal engagements [315, 318].

Accountability for actions taken by autonomous systems is a significant concern under international law. States and military commanders are responsible for ensuring that their operations comply with IHL. However, the deployment of autonomous systems complicates the assignment of accountability, particularly when decisions to use lethal force are made by machines [326, 327]. The call for maintaining human control over critical decisions is echoed in various legal discussions, emphasizing the need for a framework that ensures compliance with international norms [316, 321].

International treaties and conventions, such as the Convention on Certain Conventional Weapons (CCW), also play a role in regulating the use of LAWS. The CCW aims to balance military necessity with humanitarian concerns, addressing weapons that cause unnecessary suffering or are indiscriminate in nature [314]. Although binding regulations specific to LAWS have yet to be established, ongoing discussions within the CCW highlight the international community's efforts to grapple with the legal and ethical implications of these technologies [322, 328].

Moreover, human rights law, particularly the right to life as articulated in the International Covenant on Civil and Political Rights (ICCPR), imposes additional obligations on the deployment of lethal autonomous systems. Any deprivation of life must be lawful and not arbitrary, necessitating that autonomous systems are designed and operated in a manner that respects this right [316, 318]. The Martens Clause further reinforces the need for compliance with humanitarian principles, even in the absence of specific treaties governing new technologies [314, 315].

Coordinating with Armor and Support Units in Urban Spaces

Coordinating with armour and support units in urban spaces is a complex yet critical aspect of urban warfare. The dense, highly compartmentalized nature of urban environments presents unique challenges that require seamless integration between infantry, armoured vehicles, and various support elements to maximize operational effectiveness while minimizing risks. Effective coordination ensures that all units leverage their capabilities in a complementary manner, overcoming the limitations of individual components and adapting to the unique demands of urban combat.

Armoured units, such as tanks and infantry fighting vehicles (IFVs), provide essential firepower, protection, and mobility in urban settings. However, their effectiveness is limited by narrow streets, obstacles, and the high density of potential threats. Infantry units play a crucial role in guiding armour through these environments, providing close protection against threats such as enemy infantry, rocket-propelled grenades (RPGs), and improvised explosive devices (IEDs). Communication and synchronization are essential, as infantry and armour must work in tandem to clear streets, secure intersections, and neutralize fortified positions.

Figure 44: A Russian BMP-3 amphibious infantry fighting vehicle. Vitaliy Ragulin, CC BY-SA 3.0, via Wikimedia Commons.

One key aspect of coordination is ensuring that armoured units are used effectively without exposing them to unnecessary risks. Tanks and IFVs are particularly vulnerable to ambushes and attacks from elevated positions, such as rooftops or upper floors of buildings. Infantry units are tasked with clearing these threats in advance or providing overwatch as the armoured units move. Conversely, armoured vehicles provide suppressive fire and demolish fortified positions, creating safe pathways for infantry to advance. This mutual dependency underscores the need for constant communication between infantry squad leaders and armoured unit commanders.

Support units, including artillery, engineers, and logistics teams, play equally vital roles in urban operations. Artillery units provide indirect fire support, neutralizing enemy positions that are inaccessible or too dangerous for direct engagement. Precision munitions are particularly valuable in urban environments to reduce collateral damage and avoid civilian casualties. Engineers are essential for clearing obstacles, breaching fortifications, and constructing or repairing pathways to ensure mobility for both infantry and armoured units. Logistics teams sustain operations by delivering ammunition, medical supplies, and other essentials, often under challenging and contested conditions.

Communication systems must be robust and adaptable to facilitate effective coordination between infantry, armour, and support units. Urban environments often disrupt radio signals due to building interference, necessitating the use of alternative methods such as wired communications, visual signals, or pre-established communication protocols. Commanders at all levels must prioritize maintaining situational awareness and relaying real-time information to ensure units can adapt to the dynamic nature of urban combat.

Modern military units utilize a diverse array of communication methods tailored to the urban environment. Radio communication remains the backbone of tactical operations, primarily using Very High Frequency (VHF) and Ultra High Frequency (UHF) bands for short- to medium-range communications. These frequencies are particularly effective in urban settings due to their ability to penetrate walls and other obstacles more efficiently than higher frequencies. However, challenges such as multipath interference, caused by reflections off buildings, and jamming threats necessitate enhanced communication systems to ensure reliability [329]. Frequency-hopping and encryption technologies are commonly employed to mitigate these issues. Frequency-hopping radios rapidly switch frequencies in a predetermined sequence, complicating interception and jamming efforts, while encryption ensures that intercepted communications remain secure [329].

In addition to radio systems, satellite communication (SATCOM) is crucial for long-range coordination between units and higher command. SATCOM provides a reliable communication alternative when terrestrial systems are compromised, especially in extended operations or isolated urban zones. Nevertheless, SATCOM devices can be bulkier and susceptible to interference in areas with dense infrastructure or active electronic warfare [25]. Wired communication systems, such as field telephones and fibre-optic cables, are also utilized in urban warfare for secure and interference-free communication, particularly in fixed positions or command centres, despite their limited mobility and vulnerability to physical disruption [180].

Advancements in digital communication have introduced Tactical Data Links (TDLs) and mobile ad hoc networks (MANETs) into urban operations. TDLs facilitate real-time information sharing, including positions and mission updates, across various platforms and units. MANETs enable decentralized networks that adapt dynamically as units move, ensuring continuous connectivity even in the absence of traditional infrastructure [25]. Blue Force Tracking (BFT) systems enhance situational awareness by providing real-time GPS-based tracking of friendly units, integrating with tactical radios and command systems to display unit positions on digital maps, thereby reducing the risk of fratricide and improving coordination [180].

Urban environments introduce significant communication challenges, including signal interference from buildings and electromagnetic congestion from civilian communication systems. The high density of civilian infrastructure, such as cell towers and Wi-Fi networks, exacerbates interference issues [180]. Additionally, the proximity of enemy forces increases the risk of signal interception, jamming, and cyberattacks on digital systems. Limited line-of-sight due to urban terrain also obstructs direct communication, which is critical for certain radio and laser systems [25].

To address these challenges, military forces employ various strategies and technologies. Relay stations, whether portable or vehicle-mounted, extend the range and reliability of communication systems by being positioned on rooftops or high points to overcome line-of-sight issues. Drones and robots equipped with communication relays act as mobile nodes, bridging gaps in the network and providing additional data streams [180]. Backup systems, including hand signals and pre planned brevity codes, are maintained to ensure coordination in case of electronic system failures. Training and protocols are essential to instil communication discipline among troops, minimizing unnecessary transmissions and maintaining operational security [180].

Chapter 5
COMBINED ARMS IN URBAN WARFARE

The Importance of Combined Arms in Achieving Urban Combat Objectives

The significance of combined arms in urban combat is paramount, as it effectively integrates diverse military capabilities to address the unique challenges posed by urban warfare. Urban environments, characterized by dense infrastructure, limited visibility, and the prevalence of close-quarters combat, necessitate a multifaceted approach where no single combat branch can operate independently. The synergy created through combined arms operations—leveraging infantry, armour, artillery, engineers, and aviation—enables military forces to navigate the complexities of urban combat more effectively [330].

Combined arms is a military strategy and operational approach that integrates different branches of armed forces—such as infantry, armour, artillery, engineers, aviation, and naval forces—into a cohesive and coordinated effort to achieve a common objective. The principle of combined arms is to exploit the unique strengths of each unit or capability while mitigating their individual weaknesses through mutual support. By leveraging diverse military assets in a synchronized manner, combined arms operations maximize combat effectiveness and adaptability across various environments, including conventional battlefields, urban areas, and asymmetric warfare scenarios.

Figure 45: U.S. Soldiers, with Alpha Troop, 1st Battalion, 6th Infantry Regiment, 2nd Armored Brigade Combat Team, 1st Armored Division, pose for a photo with a M2 Bradley Infantry Fighting Vehicle and a AH-64 Apache Helicopter in the Central Command (CENTCOM) area of responsibility, Nov. 23, 2020. The soldiers are in Syria to support the Combined Joint Task Force-Operation Inherent Resolve (CJTF-OIR) mission. CJTF remains committed to working by, with, and through our partners to ensure the enduring defeat of Daesh. (U.S. Army photo by Spc. Jensen Guillory). Spc. Jensen Guillory, Public Domain, via Wikipedia.

The core idea behind combined arms is that no single combat arm can achieve optimal success independently in modern warfare. For instance, while infantry units excel in close-quarters combat and securing terrain, they may be vulnerable to enemy armoured vehicles, indirect fire, or airstrikes. Tanks and armoured vehicles provide the firepower and protection needed to support infantry, but they are less effective in confined spaces or without dismounted support to protect them from ambushes or anti-armour threats. Similarly, artillery and air support offer long-range firepower to neutralize distant or fortified targets but require accurate reconnaissance and coordination to avoid harming friendly forces. By combining these elements, commanders create a force that is greater than the sum of its parts, capable of addressing a wide range of battlefield challenges.

Infantry units form the backbone of urban operations, responsible for tasks such as clearing buildings and securing perimeters. However, they face vulnerabilities from ambushes, improvised explosive devices (IEDs), and fortified enemy positions. The integration of armoured units, such as tanks and infantry fighting vehicles (IFVs), enhances the infantry's capabilities by providing essential firepower, mobility, and protection. Tanks can neutralize enemy strongpoints while offering cover for advancing infantry, and IFVs further bolster infantry mobility and protection, allowing for rapid responses to emerging threats [330].

Artillery and indirect fire assets play a critical role in urban combat by suppressing enemy positions and isolating areas of resistance. The use of precision-guided munitions minimizes collateral damage, which is particularly important in urban settings with civilian populations. Additionally, artillery can create breaches in walls or obstacles, facilitating infantry and armoured unit advances [330]. Combat engineers also contribute significantly by breaching fortifications, clearing mines, and ensuring mobility through dense urban terrain. Their role is vital; without engineers, progress through heavily fortified urban environments would be considerably hindered [330].

Aviation assets provide a strategic advantage in urban warfare by offering reconnaissance, fire support, and logistical capabilities. Unmanned aerial vehicles (UAVs) deliver real-time intelligence, enabling commanders to monitor enemy movements and assess the battlefield. Attack helicopters and fixed-wing aircraft can provide close air support, delivering precision strikes against fortified positions. Moreover, transport helicopters are essential for medical evacuation, resupply, and rapid troop deployment in contested urban zones [330].

Effective combined arms operations require seamless coordination among various military branches to achieve mission objectives. For instance, infantry may depend on tanks for fire support while engineers breach barricades under artillery cover. This level of coordination is crucial to prevent friendly fire and ensure that all units work towards a common goal. Commanders must possess a comprehensive understanding of each unit's capabilities and limitations, tailoring their strategies to the specific challenges presented by the urban battlefield [330].

Urban combat also demands adaptability, as the environment can change rapidly due to enemy countermeasures and civilian movements. Combined arms operations provide the flexibility needed to respond to these dynamic conditions. For example, if infantry encounters a heavily fortified enemy position, engineers and armour can be redirected to breach and neutralize it. Similarly, if enemy forces attempt to regroup, aviation assets can quickly engage them to prevent further resistance [330].

The psychological impact of combined arms is another critical factor in urban warfare. The visible presence of multiple combat branches working in concert can demoralize enemy forces, leading to disarray and a decreased willingness to fight. This psychological edge, combined with the physical dominance achieved through integrated operations, enhances the likelihood of mission success while minimizing risks to friendly forces [330].

In urban warfare, combined arms operations employ specific tactical applications to maximize the effectiveness of various military branches working in unison. One key tactic is isolation and encirclement, where artillery, unmanned aerial vehicles (UAVs), and infantry collaborate to cordon off urban areas. This approach prevents enemy forces from receiving reinforcements or escaping, effectively cutting off their supply lines and ability to manoeuvre.

Clearing and securing operations are another critical aspect. Infantry units, supported by armoured vehicles and combat engineers, systematically advance through enemy-held buildings and infrastructure. Engineers address obstacles such as barricades and mines, while armour provides the firepower and protection necessary for infantry to secure each structure. This systematic approach ensures a methodical clearing of hostile positions while minimizing friendly casualties.

To counter the prevalent threat of snipers in urban combat, counter-sniper operations are employed. UAVs provide real-time surveillance to identify enemy sharpshooters, and snipers from the attacking force neutralize these threats. This coordination enables infantry units to advance safely without the constant risk of being pinned down by precision enemy fire.

Close Air Support (CAS) is another vital component, involving precision airstrikes that work in tandem with ground forces to target fortified enemy positions. These strikes are carefully coordinated to avoid collateral damage, ensuring that friendly troops and civilians in the area remain unharmed. By integrating these tactics, combined arms operations ensure a cohesive and effective approach to achieving objectives in the challenging urban combat environment.

Coordination Between Infantry, Armour, Engineers, and Artillery

Coordination in urban warfare involves multiple military branches, including infantry, armour (tanks and other armoured vehicles), engineers, and artillery units. Infantry units are the primary force engaged in clearing and securing urban environments. Armoured units provide firepower, protection, and mobility support. Engineers handle obstacles, such as clearing mines, breaching walls, and constructing barriers. Artillery units deliver indirect fire to suppress or neutralize enemy positions.

Urban warfare is inherently complex and dangerous, requiring the combined strengths of multiple branches to mitigate vulnerabilities. Infantry alone may struggle against fortified enemy positions, armour may be vulnerable without close infantry support, and engineers need protection while conducting their tasks. Artillery provides essential long-range support but requires precise coordination to avoid harming friendly forces or civilians. Without seamless coordination, operations risk delays, higher casualties, and failure to achieve objectives.

The goal of coordination is to achieve operational objectives in complex urban environments. This involves synchronizing the unique capabilities of each unit to overcome the challenges of dense terrain, fortified positions, and close-quarters combat. Infantry clears and holds positions, armour supports their advances and protects them from heavy enemy fire, engineers address physical barriers and prepare the urban landscape for mobility, and artillery delivers long-range fire to weaken enemy defences or disrupt reinforcements.

Coordination occurs in urbanized areas, ranging from small towns to sprawling cities. Urban warfare presents unique challenges, such as narrow streets, dense buildings, and the presence of civilians. Coordination is particularly critical at key locations, such as intersections, fortified buildings, or chokepoints, where different units must work together to secure progress and maintain operational momentum.

Coordination is essential throughout all phases of urban operations. During the planning phase, units work together to establish roles, objectives, and operational plans. During the assault phase, real-time coordination ensures that each unit performs its task effectively, whether it is clearing a building, securing an intersection, or suppressing enemy fire. After an area is secured, ongoing coordination ensures continued support for holding and stabilizing the area.

Coordination is achieved through detailed planning, effective communication, and shared situational awareness. Commanders define clear roles and establish communication channels between infantry, armour, engineers, and artillery units. Real-time communication systems, such as radios and tactical data links, enable rapid adjustments during combat. Joint training exercises before deployment familiarize units with each other's capabilities, limitations, and procedures, enhancing their ability to operate cohesively.

Example - Coordination in Urban Warfare - A Hypothetical Scenario

Situation: A coalition force is tasked with clearing a heavily fortified urban district held by an insurgent group. The district includes narrow streets, dense residential buildings, and known enemy strongpoints such as repurposed schools and warehouses. The insurgents have fortified positions, planted improvised explosive devices (IEDs), and set up sniper nests to control key chokepoints.

Mission: The mission is to clear the district of enemy forces while minimizing civilian casualties and preserving key infrastructure. The operation will involve close coordination between infantry, armour, engineers, and artillery to ensure success.

Step-by-Step Coordination:

1. Planning Phase

- **Intelligence Gathering**: UAVs and reconnaissance teams map enemy positions, IED locations, and potential ambush points. Engineers identify obstacles and potential breach points.

- **Mission Briefing**: Infantry, armour, engineers, and artillery commanders meet to plan a synchronized assault. They define objectives, assign roles, and establish communication protocols.

2. Initial Movement

- **Infantry Advance**: Infantry platoons begin moving through the outskirts to secure entry points into the district. Their task is to identify and neutralize sniper positions and small enemy outposts.

- **Artillery Support**: Artillery units fire precision-guided munitions to suppress known enemy positions, such as fortified strongpoints, while avoiding civilian areas.

- **Armoured Units in Reserve**: Tanks and armoured personnel carriers (APCs) provide overwatch from secure positions, ready to support infantry with direct fire if necessary.

- **Engineers Follow Infantry**: Combat engineers move in tandem with infantry to clear IEDs, establish safe lanes, and prepare breach points.

3. Breaching a Fortified Position

- **Scenario**: A school building has been converted into an insurgent stronghold. The structure is heavily fortified, with machine gun nests, snipers, and an IED-laden perimeter.

- **Engineers Prepare the Breach**: Using mine-clearing vehicles and handheld tools, engineers neutralize IEDs and cut through barricades. They prepare explosive charges to breach walls, creating multiple entry points for infantry.

- **Artillery Fires Smoke Rounds**: To obscure enemy vision, artillery units deploy smoke rounds around the building, providing cover for advancing forces.

- **Infantry Clears the Building**: Infantry teams storm the building using coordinated entry tactics such as dynamic entry and room-clearing procedures. Armoured vehicles provide direct fire support to suppress enemy defenders inside.

4. Urban Advance and Clearing Operations

- **Infantry Supported by Armor**: As infantry advances deeper into the district, tanks and APCs provide overwatch and heavy firepower to neutralize enemy strongpoints. Tanks are positioned at intersections to dominate streets and protect advancing infantry.

- **Artillery Continues Fire Support**: Artillery shifts focus to deeper enemy positions based on real-time updates from infantry. UAVs help adjust fire to ensure accuracy and avoid collateral damage.

- **Engineers Maintain Mobility**: Engineers clear rubble to keep streets passable for armoured vehicles and supply convoys. They also construct temporary bridges or ramps if necessary.

5. Securing the Area

- **Consolidation of Gains**: Once a block is cleared, infantry secures key buildings and sets up defensive positions. Engineers fortify these positions using sandbags and barriers.

- **Artillery Suppression**: Artillery continues to suppress enemy reinforcements attempting to retake lost ground.

- **Armor Holds Critical Chokepoints**: Tanks maintain control of intersections and open areas, preventing insurgents from regrouping.

6. Exit and Handover

- **Transition to Stabilization**: Once the district is cleared, forces focus on stabilizing the area. Infantry patrols ensure no enemy fighters remain, and engineers assist with removing unexploded ordnance (UXO).

- **Civil Affairs Teams Enter**: To win hearts and minds, civil affairs units work alongside infantry to distribute aid and restore essential services.

This example illustrates how infantry, armour, engineers, and artillery coordinate to achieve tactical objectives in urban warfare:

- **Infantry:** The primary force for close-quarters combat and building clearing.

- **Armor:** Provides heavy firepower, protection, and overwatch in urban chokepoints.

- **Engineers:** Ensure mobility, breach fortifications, and neutralize IEDs.

- **Artillery:** Offers suppression, precision strikes, and situational adaptation.

Each element relies on the other for success, and real-time communication ensures flexibility and adaptability during complex operations. This coordination is the foundation of effective combined arms in urban environments.

Recent conflicts have underscored the critical importance of combined arms operations in urban warfare, particularly evident in the Battle of Mosul (2016–2017) and the ongoing Ukraine conflict (2022-). These engagements illustrate the necessity for integrated military strategies that leverage various combat elements to adapt to the complexities of urban environments.

In the Battle of Mosul, coalition forces effectively utilized a combination of infantry, armoured units, engineers, unmanned aerial vehicles (UAVs), and precision airstrikes to systematically dismantle ISIS strongholds within a densely populated urban landscape. This intricate approach was essential for minimizing civilian casualties while maximizing operational effectiveness in a challenging environment where traditional warfare tactics could lead to significant collateral damage [17, 41]. The integration of these diverse military assets allowed for a more nuanced engagement strategy, enabling forces to adapt to the dynamic and unpredictable nature of urban combat [23].

Similarly, the Ukraine conflict has highlighted the importance of combined arms in urban warfare, particularly in battles such as those in Bakhmut. Here, the integration of artillery, drones, infantry, and armoured units has been crucial in maintaining momentum against entrenched defenders. The urban setting necessitates a shift from conventional warfare tactics to more flexible and responsive strategies that can address the unique challenges posed by urban terrains, such as limited visibility and restricted manoeuvrability [49, 331]. The ongoing conflict has demonstrated that successful military operations in urban areas require a coordinated effort among various branches of the armed forces to effectively counter entrenched positions and adapt to rapidly changing battlefield conditions [332].

Moreover, the evolution of warfare in urban settings has prompted military strategists to reconsider traditional doctrines. The Russian invasion of Ukraine serves as a case study in this regard, illustrating how modern conflicts increasingly rely on hybrid warfare tactics that blend conventional military operations with irregular and asymmetric strategies [333, 334]. This shift emphasizes the need for a comprehensive understanding of urban warfare dynamics, where the integration of various combat elements is not merely advantageous but essential for achieving operational success [335].

Role of Armoured Vehicles in Supporting Infantry Operations

Armoured vehicles are important assets in urban warfare, providing protection, mobility, and firepower to support infantry operations. Different nations employ a variety of armoured vehicles specifically designed or adapted for urban combat.

Protection is one of the primary functions of armoured vehicles. They are engineered to withstand enemy fire and explosive threats, thereby ensuring the safety of the troops inside. For instance, rolled homogeneous armour (RHA) is commonly used in military applications due to its excellent energy-absorbing properties, making it ideal for tanks and other armoured combat vehicles [336]. Additionally, advancements in armour design, such as the V-shaped underbody of Mine Resistant Ambush Protected Vehicles (MRAPs), significantly improve blast protection by deflecting explosive forces away from the vehicle [337, 338]. The effectiveness of these protective measures is critical, as they directly influence the survivability of the crew in combat situations [339].

Mobility is another vital aspect of armoured vehicles. Their design allows for rapid manoeuvring across various terrains, which is essential for breaking through enemy lines and repositioning during combat operations. Research indicates that the optimization of suspension systems in wheeled armoured vehicles enhances their mobility and stability, enabling them to perform effectively in diverse operational environments [340, 341]. The ability to traverse rough terrains while maintaining operational speed is crucial for the success of military missions [342].

Firepower is a defining characteristic of armoured vehicles, which are often equipped with advanced weaponry to support infantry operations. The integration of gun turrets on these vehicles allows for effective engagement of enemy forces while providing cover for dismounted troops [342]. The combination of mobility and firepower enables armoured units to execute coordinated attacks, thereby increasing their tactical advantage on the battlefield [343].

The concept of combined arms operations is significantly enhanced by the use of armoured vehicles. These vehicles can work in conjunction with infantry, artillery, and air support to create a formidable fighting force. The synergy between different military branches allows for more effective combat strategies, as armoured vehicles can provide direct fire support while infantry units secure and hold territory [343, 344]. This integration is essential for modern military operations, where adaptability and coordination are key to success.

Below are notable examples from around the world, highlighting their roles and how they are used.

1. Tanks

Examples:

- **M1 Abrams (USA):** Equipped with a 120mm smoothbore gun, advanced targeting systems, and robust armour. It is used to dominate streets, neutralize fortified positions, and provide overwatch for infantry.

Figure 46: An M1A2 Abrams tank assigned to the 98th Cavalry Regiment, fires a tank round at a target during a live fire qualifications at the Udairi Multi Purpose Range Complex, Kuwait, August, 2023.The M1A2 System Enhancement Package Version 2 (SEPv2) is an upgraded version of the M1 Abrams main battle tank.It features improved electronics, including updated command, control, and communication systems.(U.S. Army Photo by Sgt. Joaquin Vasquez-Duran). Sgt. Joaquin Vasquez-Duran, Public domain, via Wikimedia Commons.

- **T-90 (Russia):** Features reactive armour and a 125mm gun, effective in suppressing enemy strongholds and clearing pathways for infantry.

- **Merkava IV (Israel):** Designed for urban warfare with heavy frontal armour, advanced electronics, and an internal troop compartment to carry infantry if needed.

Usage:

Tanks play a crucial role in urban warfare by providing direct fire support to infantry forces. Their firepower is used to neutralize enemy strongpoints, such as fortified positions, bunkers, and armoured threats, ensuring that advancing infantry can move with reduced resistance. The ability of tanks to engage these targets with precision and overwhelming force makes them an invaluable asset in the close-quarters environment of urban combat.

The heavy armour of tanks is another significant advantage, allowing them to withstand attacks from small arms, improvised explosive devices (IEDs), and other threats commonly encountered in urban settings. This resilience enables tanks to lead the advance through enemy-held streets, acting as a shield for infantry units and providing a secure base from which operations can progress.

In addition to their mobility and firepower, tanks are strategically positioned at key intersections within urban areas. By occupying these locations, they dominate the surrounding space and deny the enemy the ability to manoeuvre freely. This capability to control critical points in the urban landscape enhances the overall effectiveness of combined arms operations, ensuring coordinated advances and securing the ground for subsequent actions.

2. Infantry Fighting Vehicles (IFVs)

Examples:

- **Bradley Fighting Vehicle (USA):** Armed with a 25mm chain gun, TOW missile launchers, and capable of transporting infantry squads.

- **BMP-2 (Russia):** Features a 30mm autocannon and anti-tank guided missiles (ATGMs), with room for infantry.

Figure 47: BMP-2 of Russian Ground Forces. Mil.ru, CC BY 4.0, via Wikimedia Commons.

- **Puma (Germany):** One of the most advanced IFVs, featuring modular armour, a 30mm cannon, and advanced optics.

Usage:

Infantry Fighting Vehicles (IFVs) play an essential role in urban combat by accompanying infantry units and serving as both a fire support platform and a means of troop transport. These versatile vehicles are designed to operate in close coordination with dismounted infantry, providing them with the necessary firepower and protection to navigate the complexities of urban terrain.

Equipped with advanced weapons systems, IFVs are highly effective at engaging lightly fortified positions, neutralizing snipers, and combating enemy vehicles. Their firepower complements infantry operations by suppressing or eliminating threats that could hinder the progress of ground troops. This capability ensures that infantry can manoeuvre with greater confidence and efficiency, even in heavily contested environments.

One of the key advantages of IFVs is their ability to facilitate the rapid deployment of infantry squads directly into high-risk areas. By delivering troops swiftly and safely into contested zones, IFVs minimize exposure to enemy fire and enable infantry units to establish a foothold in urban operations. This combination of mobility, firepower, and protection makes IFVs indispensable in modern urban warfare scenarios.

3. Armoured Personnel Carriers (APCs)

Examples:

- **M113 (USA):** A widely used APC, providing basic protection and mobility for infantry.

- **BTR-80 (Russia):** Amphibious and lightly armed, used for transporting infantry and resupply in urban areas.

- **Boxer (Germany/Netherlands):** Modular design with advanced protection for transporting infantry under fire.

Figure 48: Boxer Commando Post of Royal Netherlands Army, German Forces day, Field Marshal Rommel barracks, Augustdorf 2017. Boevaya mashina, CC BY-SA 4.0, via Wikimedia Commons.

Usage:

Armored Personnel Carriers (APCs) are vital assets in urban combat, primarily designed to transport troops safely through contested areas. While not as heavily armed as Infantry Fighting Vehicles or tanks, APCs provide essential mobility and protection for infantry units operating in environments where heavy resistance may not be anticipated. Their design emphasizes safeguarding personnel against small arms fire, shrapnel, and improvised explosive devices (IEDs), ensuring that troops can reach their objectives with minimal risk.

In addition to troop transport, APCs are frequently utilized for casualty evacuation and logistical support. Their protective capabilities allow them to traverse hazardous zones to retrieve wounded personnel or deliver critical supplies to forward units. This role is especially important in urban settings, where the dense environment can make conventional supply lines vulnerable to disruption.

APCs offer quick mobility, allowing forces to move rapidly between key points in the urban battlefield. However, their role is generally limited to avoiding direct engagements with enemy forces, as their defensive firepower is designed more for suppression and self-defence than for offensive operations. This careful balance of mobility, protection, and support functions makes APCs indispensable for sustaining infantry operations in urban warfare.

4. Mine-Resistant Ambush Protected Vehicles (MRAPs)

Examples:

- **Cougar (USA):** Designed to withstand IEDs and ambushes, heavily used in Iraq and Afghanistan.

- **Oshkosh M-ATV (USA):** Optimized for mobility in rugged and urban terrains.

- **RG-35 (South Africa):** Provides superior mine resistance and protection against small arms fire.

Figure 49: BAe Systems RG35 MPV. Bob Adams from George, South Africa, CC BY-SA 2.0, via Wikimedia Commons.

Usage:

Mine-Resistant Ambush Protected (MRAP) vehicles play a crucial role in urban warfare, particularly in environments with a high risk of improvised explosive devices (IEDs). These vehicles are specifically designed to transport infantry and supplies safely through areas where such threats are prevalent. Their robust construction and specialized features make them an essential asset for minimizing casualties and maintaining operational momentum.

One of the key design features of MRAPs is their V-shaped hull, which is engineered to deflect blasts from underneath the vehicle. This design significantly enhances crew survivability by mitigating the impact of explosions. This capability is particularly critical in urban combat scenarios, where narrow streets and dense infrastructure often conceal explosive threats.

In urban warfare, MRAPs are employed not only for troop and supply transport but also for securing routes and providing logistical support under fire. Their ability to navigate contested areas while offering protection against IEDs makes them indispensable for maintaining the flow of resources and personnel to forward positions. Additionally, their presence serves as a psychological deterrent to adversaries, reinforcing the operational effectiveness of infantry and support units in urban environments.

5. Specialized Urban Combat Vehicles

Examples:

- **Achzarit (Israel):** An APC based on a converted T-55 chassis, heavily armoured for urban operations.

Figure 50: Achzarit. gkirok, CC BY-SA 3.0, via Wikimedia Commons.

- **Namer (Israel):** Built on the Merkava tank chassis, combining heavy protection and infantry transport capabilities.

- **Centauro B1 (Italy):** A wheeled tank destroyer with an autoloaded 105mm gun, ideal for urban reconnaissance and fire support.

Usage:

Specialized vehicles such as the Achzarit and Namer are purpose-built for the unique challenges of urban warfare, where narrow streets and close-quarters engagements dominate the battlefield. These vehicles are specifically designed to navigate tight urban environments while maintaining high levels of protection for the troops they carry. Their robust armour enables them to engage enemy positions effectively, providing crucial support for dismounted infantry as they advance or secure contested areas.

Heavy armoured vehicles like the Achzarit and Namer not only shield infantry from small arms fire and shrapnel but also offer a mobile platform for firepower that can suppress or neutralize enemy positions. This capability is particularly valuable in urban settings, where adversaries often fortify positions within buildings or other dense structures.

In addition to troop carriers, vehicles such as the Centaur play a pivotal role in urban reconnaissance and mobile firepower support. These vehicles are equipped to provide intelligence on enemy movements and positions while engaging threats quickly and effectively. Their mobility and firepower allow them to operate efficiently in the complex terrain of urban warfare, bridging the gap between infantry and heavier support elements like tanks. Together, these specialized vehicles ensure that ground forces can operate with increased effectiveness and survivability in the demanding conditions of urban combat.

6. Wheeled Armoured Vehicles

Examples:

- **Stryker (USA):** Modular design with multiple variants, including a mobile gun system (MGS) for urban fire support.

- **Pandur II (Austria):** A versatile wheeled vehicle used for troop transport and reconnaissance.

- **Patria AMV (Finland):** Highly manoeuvrable and heavily armed for urban engagements.

Figure 51: Patria AMV-Svarun, SKOV (Middle Wheeled Armoured Vehicle) XC 400, 8 x 8 Svarun. MORS, CC BY 3.0, via Wikimedia Commons.

Usage:

Wheeled vehicles are highly valued in urban warfare due to their superior manoeuvrability and speed on paved streets. Unlike tracked vehicles, wheeled platforms can navigate tight urban environments more efficiently, making them an ideal choice for operations where rapid response and flexibility are critical. Their ability to move quickly through city streets allows them to provide timely support to infantry and adapt to the dynamic nature of urban combat.

These vehicles often serve as versatile platforms, carrying infantry squads or specialized equipment tailored to the mission. Some are outfitted with mortar systems to provide indirect fire support, while others house command modules that enable effective coordination and communication among units in the field. Their adaptability ensures that they can fulfill a variety of roles, from troop transport to logistical support, while maintaining operational effectiveness in dense urban areas.

Wheeled vehicles like the Stryker Mobile Gun System (MGS) variants bring additional capabilities to urban warfare by providing mobile firepower. Equipped with powerful weapon systems, these vehicles can engage enemy positions in tight, confined spaces where larger, less agile platforms might struggle. Their combination of speed, adaptability, and firepower makes wheeled vehicles indispensable assets in achieving tactical objectives in urban combat scenarios.

Special Considerations for Close Air Support in Urban Terrain

Close Air Support (CAS) in urban environments poses distinct challenges that necessitate meticulous planning and execution. The densely packed urban terrain, characterized by narrow streets, tall buildings, and complex layouts, significantly complicates visibility and manoeuvrability for aircraft. These factors require specialized tactics and coordination to ensure effective CAS while minimizing collateral damage and fratricide. The urban landscape can obscure pilots' views of targets, making precise target identification crucial. Ground forces must provide accurate target information, including grid coordinates and laser designations, to assist pilots in overcoming these challenges. The use of Precision-Guided Munitions (PGMs) is essential in urban CAS, as they enable pinpoint strikes that reduce the risk of unintended damage to nearby structures and civilian casualties [345-347].

Coordination between ground forces and CAS assets is critical in urban operations. The close proximity of friendly units to enemy positions increases the risk of catastrophic outcomes from miscommunication. Forward Air Controllers (FACs) or Joint Terminal Attack Controllers (JTACs) play a vital role in ensuring that pilots receive real-time updates on the ground situation, marking targets using laser or smoke, and confirming clearance to engage. Effective communication systems, such as encrypted radios, are essential for maintaining clear lines of coordination [348-350]. The integration of advanced technologies, including real-time surveillance and data sharing, enhances situational awareness and operational effectiveness in urban CAS missions [351, 352].

The unique characteristics of urban terrain necessitate adaptations in flight profiles and attack strategies. Aircraft often operate at lower altitudes to maintain line-of-sight with targets, which increases their vulnerability to small arms fire and man-portable air defence systems (MANPADS). Pilots must employ tactics such as high-speed passes, steep dive angles, and rapid egress manoeuvres to minimize exposure while delivering effective support. Rotary-wing aircraft, particularly attack helicopters, are advantageous in urban CAS due to their ability to hover and manoeuvre in tight spaces, providing sustained firepower [353-355].

Mitigating collateral damage and ensuring compliance with international humanitarian law are paramount in urban CAS missions. The densely populated nature of urban areas necessitates thorough risk assessments before engaging targets. Commanders and pilots must carefully consider the potential impact of each strike against operational objectives. Techniques such as pre-mission reconnaissance, real-time surveillance using drones, and employing low-yield or non-explosive ordnance can help minimize unintended consequences [356-358].

Weather and environmental conditions further complicate CAS in urban terrain. Factors such as wind patterns, urban heat islands, and restricted airspace due to tall buildings can affect the trajectory and effectiveness of munitions. Pilots must account for these variables when planning their attacks and coordinate with ground forces to ensure that the timing and delivery of support align with the operational tempo [359-361].

Close Air Support (CAS) equipment varies globally, with each nation employing platforms and systems tailored to their operational needs. CAS equipment includes aircraft, munitions, targeting systems, and communication tools designed to deliver precise firepower in support of ground forces. Below are specific examples of CAS equipment and their applications worldwide:

Aircraft for Close Air Support

A-10 Thunderbolt II (USA): The A-10, or "Warthog," is an American CAS aircraft renowned for its survivability, loitering capability, and heavy armament, including the 30mm GAU-8/A Avenger rotary cannon. It is used to destroy armoured vehicles, fortified positions, and troop concentrations while operating at low altitudes. Its robust design allows it to absorb significant damage and remain operational.

Figure 52: An A-10 Thunderbolt II, assigned to the 74th Fighter Squadron, Moody Air Force Base, GA, returns to mission after receiving fuel from a KC-135 Stratotanker, 340th Expeditionary Air Refueling Squadron, over the skies of Afghanistan in support of Operation Enduring Freedom. Master Sgt. William Greer, U.S. Air Force, Public domain, via Wikimedia Commons.

Sukhoi Su-25 (Russia): The Su-25 "Frogfoot" is Russia's dedicated CAS platform, equipped with a wide range of unguided and guided munitions. It is primarily used to engage enemy positions, tanks, and infrastructure. The Su-25 has been extensively used in conflicts such as the Syrian Civil War, providing support to ground forces in urban and rural settings.

Panavia Tornado GR4 (UK): Used by the Royal Air Force until its retirement in 2019, the Tornado GR4 was a multirole aircraft that excelled in CAS missions. It employed precision-guided munitions like the Brimstone missile and Paveway laser-guided bombs to strike enemy positions with minimal collateral damage.

Dassault Rafale (France): The Rafale is a versatile multirole aircraft capable of CAS. Equipped with advanced targeting systems and precision-guided munitions, it has been deployed in operations in Mali and the Middle East to support ground troops against insurgent forces.

Figure 53: Dassault Rafale - Indian Air Force (IAF). Dylan Agbagni (CC0), CC0, via Wikimedia Commons.

JAS 39 Gripen (Sweden): The Gripen, used by Sweden and other nations, is a lightweight fighter capable of providing CAS. It employs a mix of precision-guided bombs and air-to-ground missiles to neutralize enemy threats, even in contested environments.

Rotary-Wing Platforms

AH-64 Apache (USA): The Apache attack helicopter is widely used for CAS, providing direct fire support with its 30mm M230 chain gun, Hydra 70 rockets, and

Hellfire missiles. It is highly effective in urban warfare due to its ability to hover and engage targets at close range.

Mil Mi-24 Hind (Russia): The Mi-24 serves as both a troop transport and attack helicopter, armed with cannons, rockets, and anti-tank guided missiles. It has been extensively used in conflicts across Africa and the Middle East, providing CAS for ground forces.

Figure 54: Czech Republic Air Force Mil Mi-24V Hind E ILA Berlin 2016. Julian Herzog, CC BY 4.0, via Wikimedia Commons.

Eurocopter Tiger (Germany/France): The Tiger is a modern attack helicopter equipped with advanced targeting systems and weaponry like Hellfire missiles and rocket pods. It is used for CAS in challenging environments, including mountainous terrain and urban settings.

Munitions

AGM-114 Hellfire Missile (USA): The Hellfire is a precision-guided air-to-ground missile used by platforms like the AH-64 Apache and MQ-9 Reaper drones. It is effective against armoured vehicles, bunkers, and personnel, providing highly targeted CAS.

Brimstone Missile (UK): The Brimstone missile, used by the Tornado and Typhoon aircraft, is a low-collateral, precision-guided weapon ideal for urban CAS. It is designed to engage moving and stationary targets in complex environments.

KAB-500L (Russia): A laser-guided bomb used by aircraft like the Su-25 and Su-34, the KAB-500L is employed for precision strikes on fortified positions, bridges, and enemy vehicles.

GBU-12 Paveway II (USA): This laser-guided bomb is used worldwide by NATO forces for CAS missions. It delivers precision strikes on targets in urban and open environments.

Figure 55: Marines and Sailors transport GBU-12 Paveway II laser-guided bombs across the flight deck of the forward-deployed amphibious assault ship USS Essex (LHD 2). Essex is part of the forward-deployed Essex Amphibious Ready

Group and is participating in Valiant Shield 2010, a joint-service exercise designed to enhance interoperability between U.S. forces. U.S. Navy photo by Mass Communication Specialist 3rd Class Casey H. Kyhl, Public domain, via Wikimedia Commons.

Targeting and Communication Systems

LITENING Targeting Pod (USA): Used on platforms like the F-16, A-10, and Tornado, the LITENING pod provides high-resolution imagery and laser guidance for precision munitions. It is crucial for target acquisition in CAS missions.

Figure 56: Northrop+Grumman AN/AAQ-28 LITENING AT, S/N 7514, mounted on Portuguese Air Force F-16AM Reg 15101 on display at NATO Days 2023. Boevaya mashina, CC BY-SA 3.0, via Wikimedia Commons.

Sniper Advanced Targeting Pod (USA): The Sniper pod is employed by aircraft like the F-15 and F-16. It offers long-range identification and tracking of targets, enabling effective CAS in urban and rural environments.

Advanced Field Artillery Tactical Data System (AFATDS) (USA): This digital command-and-control system integrates CAS coordination with ground forces, ensuring precise targeting and minimizing risks to friendly units.

Raven B RQ-11 (USA): This small UAV provides real-time reconnaissance and targeting data to support CAS missions. It is particularly valuable in urban warfare, where visibility is often limited.

Specific Use Cases

Battle of Fallujah (2004): In Iraq, the A-10 Thunderbolt II and AH-64 Apache provided critical CAS during intense urban combat. They targeted insurgent strongholds, vehicles, and snipers to support advancing U.S. Marines.

Syrian Civil War (2011–Present): Russian Su-25s and Mi-24 Hinds have been instrumental in urban operations, providing CAS to Syrian government forces in cities like Aleppo. Precision-guided munitions and rotary-wing platforms played key roles in these engagements.

Operation Serval (Mali, 2013): French Rafales provided CAS to ground forces combating insurgents in urban areas. Using precision munitions, they neutralized enemy positions while minimizing civilian casualties.

The diverse range of CAS equipment employed globally highlights the importance of tailored solutions for different operational contexts. From heavy bombers to UAVs, the integration of advanced technology and weaponry ensures effective support for ground forces, particularly in the complex and dynamic environment of urban warfare.

Integration of Special Operations Forces in Urban Campaigns

Special Operations Forces (SOF) play a significant role in urban campaigns, leveraging their specialized training, unique skill sets, and advanced technology to perform missions that conventional forces may not be able to accomplish effectively. Urban warfare presents complex challenges such as dense infrastructure, high civilian populations, and limited manoeuvrability, all of which demand precision and adaptability. SOF units are uniquely equipped to address these challenges through a combination of direct action, intelligence gathering, and support to conventional forces.

One of the primary roles of SOF in urban campaigns is conducting direct action missions, including raids, hostage rescues, and the neutralization of high-value targets. These missions often occur in environments where the enemy has fortified positions or blended into civilian populations, making precision and stealth essential. SOF teams utilize advanced reconnaissance tools, such as unmanned aerial vehicles (UAVs) and human intelligence networks, to plan and execute their operations with minimal collateral damage. Their ability to operate covertly allows them to strike critical targets without escalating broader conflicts or compromising operational security.

Figure 57: U.S. and Australian special operations forces (SOF) conduct a high-altitude low-opening (HALO) parachute jump from a Royal Australian Air Force (RAAF) C-27J Spartan of the 35 Squadron during Talisman Sabre in Queensland,

Intelligence gathering and reconnaissance are other key areas where SOF units excel in urban warfare. These forces often deploy ahead of conventional units to map the battlefield, identify enemy positions, and assess threats. Their expertise in close-quarters reconnaissance is invaluable in urban environments, where the complex layout of buildings and underground systems can obscure traditional intelligence sources. SOF teams often use high-tech surveillance equipment, including thermal imaging and ground-penetrating radar, to detect hidden threats such as booby traps or enemy ambushes.

Another crucial aspect of SOF integration in urban campaigns is their role in training and supporting local forces. Many urban conflicts involve insurgencies or civil wars, where the cooperation of local security forces is essential for long-term success. SOF units often train, advise, and assist these forces, providing them with the skills and tactics necessary to operate effectively in urban settings. This approach not only strengthens local capabilities but also fosters goodwill and collaboration between allied forces, which is critical for maintaining stability in post-conflict environments.

SOF units also provide logistical and operational support to conventional forces during urban campaigns. Their ability to operate independently and behind enemy lines allows them to secure supply routes, coordinate airstrikes, and disrupt enemy logistics. In situations where conventional forces face stiff resistance, SOF teams can infiltrate enemy-controlled areas to carry out sabotage operations, weaken enemy morale, or create diversions that facilitate broader military objectives.

Coordination with conventional forces is another essential element of SOF operations in urban campaigns. These forces often act as a precision tool, complementing the brute strength of conventional units. For example, SOF might infiltrate a high-risk building to neutralize snipers or secure a foothold for larger forces to advance. Effective integration requires seamless communication and clear command structures to ensure that SOF operations align with the overall campaign strategy.

The use of advanced technology further enhances the effectiveness of SOF in urban warfare. From UAVs for real-time surveillance to precision-guided munitions for surgical strikes, SOF units rely on cutting-edge equipment to navigate and dominate the urban battlefield. This technology, combined with their rigorous training and adaptability, makes SOF indispensable in overcoming the unique challenges of urban warfare.

Chapter 6

COMMUNICATION AND COMMAND IN URBAN WARFARE

Importance of Communication within and between Units in Urban Settings

Effective communication within and between military units is critical in urban warfare due to the inherent complexities and dynamic nature of urban environments. Urban settings are characterized by confined, multi-dimensional spaces and significant civilian populations, which necessitate precise coordination, situational awareness, and rapid decision-making. Failures in communication can lead to disjointed operations, increased risks of fratricide, and higher civilian casualties, making robust communication systems and protocols indispensable [25, 180].

Within-unit communication is essential for ensuring that individual squads, fire teams, and platoons can coordinate their movements, share situational updates, and execute tasks efficiently. Urban environments often present fragmented battlefields where line-of-sight is obstructed by buildings, rubble, or underground infrastructure. In such scenarios, maintaining clear and constant communication among team members enables synchronized actions, such as breaching buildings, clearing rooms, or securing intersections. Team leaders rely on real-time updates from their subordinates to assess threats and adjust tactics, ensuring cohesive unit operations even in high-stress environments [236, 362].

Between-unit communication is equally critical, as urban operations typically involve multiple units working simultaneously on interconnected objectives. Infantry units may need to coordinate with armoured vehicles, engineers, artillery support, or air assets to execute complex operations like isolating enemy strongholds or securing key infrastructure. For instance, a platoon clearing a building may need to request suppressive fire from a nearby machine gun team or call for an airstrike to neutralize entrenched enemy positions. Effective communication ensures that these requests are executed promptly and accurately, reducing delays and enhancing overall mission effectiveness [35, 51].

The multidimensional nature of urban combat—spanning ground, underground, and aerial domains—further underscores the importance of communication. Units operating in different vertical layers, such as rooftop positions or subterranean tunnels, must remain in constant contact to avoid friendly fire and ensure synchronized actions. Communication also facilitates the integration of specialized forces, such as snipers, UAV operators, or reconnaissance teams, into broader operations. By sharing intelligence and operational updates, these units contribute to a comprehensive understanding of the battlefield, enabling more informed decision-making [21, 236].

The presence of civilians in urban areas adds another layer of complexity, requiring units to communicate not only with each other but also with humanitarian organizations, local authorities, and civilian populations. Clear communication protocols help minimize collateral damage by ensuring that civilians are evacuated safely and that operations are conducted in compliance with rules of engagement. For example, a unit may need to coordinate with local police to secure evacuation routes or with humanitarian agencies to provide aid to displaced populations. Miscommunication in such scenarios can lead to confusion, delays, and unintended harm to civilians [23, 115].

Advanced communication technologies play a vital role in overcoming the challenges of urban warfare. Secure radios, satellite communications, and networked systems allow units to maintain connectivity even in areas with poor signal strength or high electronic interference. Digital platforms enable the sharing of maps, live video feeds, and other critical data in real-time, enhancing situational awareness and facilitating rapid decision-making. However, these technologies must be supported by robust training and contingency plans to address potential issues such as equipment failure, signal jamming, or cyberattacks [363, 364].

In addition to technological solutions, standardized communication protocols and clear command structures are essential for ensuring seamless coordination. Units must use consistent terminology, call signs, and reporting formats to avoid misunderstandings and delays. Commanders at all levels should prioritize communication in their planning, conducting regular briefings and debriefings to ensure that all personnel understand the mission objectives and their roles within the broader operation [47, 365].

The evolution of urban warfare communication systems is deeply rooted in the complexity and dynamic nature of modern combat environments. Over time, these systems have transitioned from rudimentary methods of communication, such as basic radio systems, to highly sophisticated, interconnected networks capable of handling the intricacies of urban operations. Urban warfare demands seamless, reliable communication to coordinate diverse military units, maintain situational awareness, and respond rapidly to threats. This evolution has been driven by technological advancements, lessons learned from past conflicts, and the unique challenges of operating in densely populated, infrastructure-heavy environments [366].

One of the most critical aspects of modern urban warfare communication systems is the integration of surveillance technologies. Drones, cameras, and sensors have revolutionized how information is gathered on the battlefield. These tools provide real time intelligence, allowing commanders to monitor enemy movements, assess the terrain, and anticipate potential threats. For example, during the Battle of Fallujah, drones provided live video feeds that helped troops identify and neutralize enemy strongholds in real-time. This capability enhances situational awareness and enables precise decision-making in environments where line-of-sight is often obstructed [366].

Encryption technologies have also become a cornerstone of urban warfare communication. With the increased risk of cyber threats and signal interception in modern conflicts, secure communication is paramount. Encryption ensures that sensitive data, including troop movements and strategic plans, remains confidential and protected from adversaries. End-to-end encryption across communication channels, whether voice, video, or data, prevents unauthorized access and ensures the integrity of shared information. This technology is vital in safeguarding operational secrecy and maintaining a tactical advantage [366].

Integrated command centres represent another major advancement in urban warfare communication systems. These hubs consolidate data from various sources, including surveillance tools, ground forces, and aerial units, into a centralized platform. Command centres enable real-time coordination, allowing commanders to direct operations across multiple dimensions, including ground, air, and subterranean levels. The centralized nature of these centres ensures that information flows efficiently between units, reducing delays and enhancing mission execution. During Operation Defensive Shield, Israeli Defense Forces demonstrated the effectiveness of this integration by coordinating ground assaults with air support and intelligence gathering, achieving operational objectives with precision [366].

Despite these advancements, deploying communication systems in urban warfare remains fraught with challenges. Dense urban infrastructure, including tall buildings, underground tunnels, and narrow streets, can obstruct signals and create dead zones. Adversaries may employ electronic jamming or cyberattacks to disrupt communication networks, further complicating operations. To mitigate these issues, modern systems incorporate redundant pathways, such as satellite communications and signal repeaters, to maintain connectivity even in challenging conditions. Additionally, robust training programs prepare personnel to operate and troubleshoot communication systems under high-stress conditions, ensuring readiness for unforeseen disruptions [366].

Interoperability among various communication platforms is another critical element. Urban warfare often involves coordination between multiple units, such as infantry, armour, engineers, and air support, each relying on distinct communication tools. Ensuring these systems work seamlessly together allows for unified action and prevents miscommunication. For instance, during the Siege of Sarajevo, decentralized communication networks enabled local forces and allied units to collaborate effectively despite the siege's prolonged and chaotic nature [366].

Future trends in urban warfare communication systems are expected to focus on further integration of artificial intelligence (AI), enhanced 5G connectivity, and quantum cryptography. AI will facilitate real-time data analysis, enabling faster decision-making and improved response times. 5G networks will provide robust, high-speed connections, even in dense urban environments, while quantum cryptography will ensure unbreakable encryption, bolstering the security of communication systems against sophisticated cyber threats [366].

Urban warfare communication systems also face significant ethical and legal considerations, especially regarding civilian populations. The use of surveillance tools raises privacy concerns, while the deployment of advanced targeting technologies requires strict adherence to international laws, such as the Geneva Conventions, to minimize collateral damage. Ensuring that communication systems comply with these frameworks is essential for maintaining legitimacy and upholding humanitarian standards in urban conflicts [366].

Tools for Maintaining Situational Awareness: Maps, Drones, Sensors

Maintaining situational awareness in urban warfare is critical for ensuring the safety of personnel, achieving mission objectives, and minimizing collateral damage. Urban environments, characterized by dense populations, complex infrastructure, and restricted visibility, demand advanced tools and technologies to provide real-time information. Among the most essential tools for maintaining situational awareness are maps, drones, and sensors, each contributing unique capabilities to the operational picture.

Maps have long been a foundational tool for situational awareness, offering a visual representation of terrain, infrastructure, and strategic locations. In urban warfare, maps are not limited to traditional topographical layouts but are increasingly augmented with detailed satellite imagery, 3D models, and overlays of critical information. Digital maps are particularly useful, as they can be updated in real-time to reflect changes on the battlefield, such as newly identified enemy positions, areas of destruction, or civilian evacuation routes. Commanders rely on these maps for planning movements, identifying key objectives, and coordinating between units. Integrated mapping systems allow for data sharing across command centres and field units, ensuring a unified understanding of the operational environment.

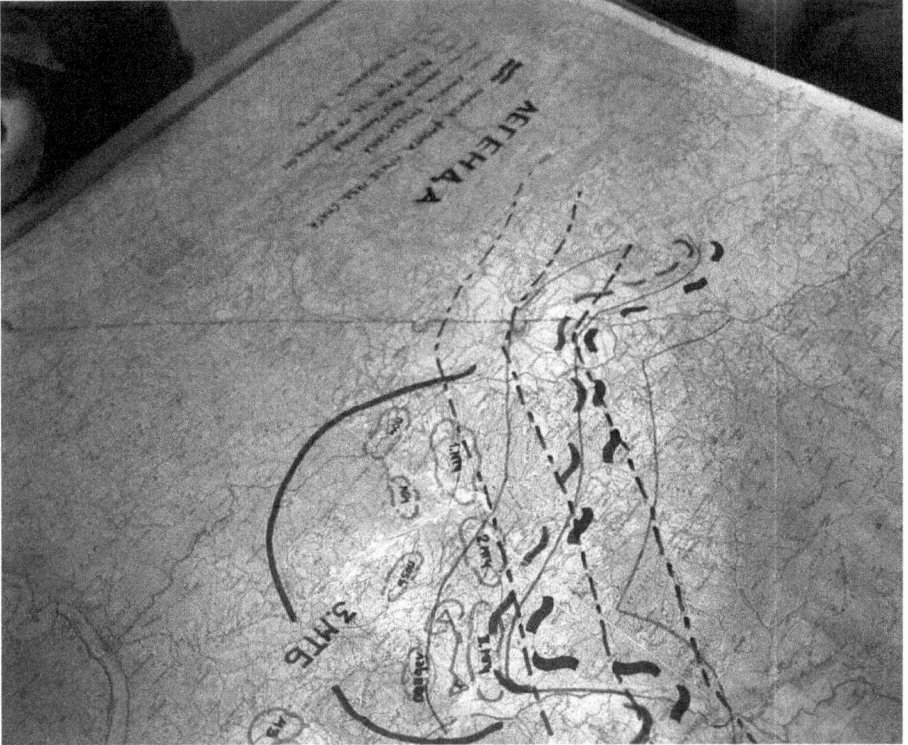

Figure 58: A Serbian Army topographical map that shows the confrontation lines separating Serbian and Muslim forces and current land mine fields. The U.S. National Archives, Public Domain, via Picryl.

Types of Military Maps Used in Urban Warfare -

Topographical Maps: These maps offer detailed depictions of terrain features, including elevation, roads, and rivers. In urban settings, they are often enhanced with overlays that illustrate the layout of buildings and key landmarks, which are vital for understanding the operational environment. The integration of such features allows military planners to assess potential advantages or disadvantages posed by the terrain [367].

City Plans and Street Maps: More granular than topographical maps, city plans focus on urban infrastructure, detailing roads, intersections, and critical installations such as hospitals and power stations. This information is indispensable for route planning and operational logistics, allowing military units to navigate effectively through urban landscapes [368].

3D Urban Models: Digital 3D models provide a comprehensive view of the urban environment, showcasing building heights and spatial relationships among urban features. These models are particularly beneficial for planning aerial operations and rooftop insertions, as they allow for a more nuanced understanding of the operational space [369].

Operational Maps: Tailored for specific missions, operational maps include tactical overlays indicating unit positions, engagement areas, and logistics points. These maps are updated in real-time, reflecting changes on the ground and enhancing situational awareness during operations [23].

Thematic Maps: These maps focus on specific data sets, such as population density and potential hazard zones. They are crucial for planning operations that consider civilian safety and compliance with rules of engagement, thereby minimizing civilian casualties [370].

Geospatial Intelligence Maps: Generated from satellite imagery and aerial reconnaissance, these maps provide real-time updates about the operational area, helping military planners identify changes in the urban landscape, such as new obstacles or enemy fortifications [371].

Underground Maps: Urban areas often feature extensive underground networks. Specialized maps that detail these networks are essential for planning subterranean operations, which are increasingly relevant in modern urban warfare [372].

Drones are indispensable tools in urban warfare for enhancing situational awareness. Equipped with high-resolution cameras, thermal imaging, and advanced sensors, drones provide real-time aerial views of urban landscapes. These unmanned systems can navigate confined spaces, rooftops, and streets that are otherwise inaccessible or dangerous for ground troops. Drones are particularly effective for reconnaissance missions, enabling forces to identify enemy movements, fortified positions, and potential threats without risking human lives. Their ability to transmit live video feeds to command centres and field operators allows for dynamic decision-making and immediate responses to emerging situations. Furthermore, drones can carry additional payloads such as communication relays, extending the range of other tools like radios and sensors in urban environments where signal interference is common.

Sensors complement maps and drones by providing ground-level intelligence and enhancing the ability to detect hidden threats. Sensors come in various forms, including motion detectors, acoustic sensors, and chemical detectors. Motion sensors are deployed to monitor enemy movements or secure perimeters, providing early warnings of an approaching threat. Acoustic sensors can detect gunfire, explosions, or other combat-related sounds, pinpointing their origin even in complex urban environments. Chemical sensors play a critical role in identifying the presence of hazardous substances, such as chemical weapons or gas leaks, which are significant risks in urban warfare. Combined with other tools, sensors ensure comprehensive situational awareness, especially in areas where visual observation is limited.

The integration of these tools—maps, drones, and sensors—creates a synergistic approach to maintaining situational awareness. For example, drones equipped with sensors can provide detailed environmental data while transmitting it to digital maps used by commanders. This integration allows forces to analyse multiple data points simultaneously, creating a cohesive and actionable picture of the battlefield. Such capabilities are vital in urban warfare, where the complexity of the terrain and the presence of civilians require precise and informed operations.

In addition to their individual and combined benefits, these tools face challenges, such as the risk of jamming, data overload, or reliance on infrastructure that may not be readily available in a war zone. To mitigate these issues, forces must train extensively in the use of these technologies and establish redundancies, such as manual navigation techniques and alternative communication methods. Proper training ensures that operators can maximize the potential of these tools, even under adverse conditions.

Chain of Command and Control in Complex Environments

The chain of command and control is the structured system of authority and decision-making that ensures effective leadership, coordination, and accountability within military operations, particularly in complex environments like urban warfare or multinational missions. This system allows commanders at all levels to delegate responsibilities, communicate objectives, and adapt to dynamic operational conditions while maintaining unity of effort.

In complex environments, where missions often involve multiple units, branches, and sometimes allied forces, the chain of command becomes the backbone of operational success. It provides a clear framework for decision-making, ensuring that orders flow downward efficiently while information and feedback move upward seamlessly. This hierarchical structure ensures that every unit, from the smallest fire team to the overarching command, operates cohesively toward shared objectives. Leaders at each level are empowered to make decisions within their scope of authority, enabling flexibility and responsiveness to evolving situations.

The importance of a well-defined chain of command is heightened in complex environments due to the challenges posed by fragmented battlefields, multi-level engagement zones, and the integration of specialized units like engineers, air support, and medical teams. For instance, in urban warfare, small-unit leaders might independently execute tactical manoeuvres within their assigned sectors while maintaining alignment with the larger strategic plan dictated by higher command. This decentralized execution relies on the overarching chain of command to synchronize actions and avoid duplication of effort or conflicting priorities.

Control mechanisms within the chain of command are equally vital. Control involves monitoring operations, ensuring compliance with orders, and making adjustments based on situational changes. This is achieved through robust communication systems, regular reporting, and real-time data analysis. In modern military operations, command posts or headquarters act as nerve centres where commanders process intelligence, track unit movements, and issue directives. These control hubs leverage advanced technologies like satellite communications, digital mapping, and surveillance feeds to maintain situational awareness across dispersed units.

The principle of mission command, a philosophy often integrated into modern command and control systems, emphasizes trust and initiative. It allows subordinates to exercise judgment and adapt to unforeseen circumstances while adhering to the intent of higher command. In complex environments, this approach fosters agility and empowers leaders at every level to respond effectively to emerging threats or opportunities. For example, a platoon commander may adjust a mission's approach based on real-time intelligence, knowing it aligns with the overarching objective.

Joint and coalition operations introduce additional layers of complexity to the chain of command. Coordination between different national forces, each with its own doctrines, communication systems, and cultural considerations, requires clear lines of authority and robust liaison mechanisms. In such scenarios, unified command structures or frameworks like NATO's command hierarchy ensure interoperability and prevent miscommunication. Shared protocols, common operating procedures, and integrated command centres play a critical role in maintaining cohesion among multinational forces.

NATO's command hierarchy is a robust and multilayered structure designed to effectively coordinate and manage multinational military operations. This hierarchical framework is essential for ensuring that member nations can collaborate seamlessly across various missions, reflecting a commitment to operational efficiency and adaptability in dynamic environments. The structure is characterized by clear lines of authority and decision-making processes that facilitate unity of effort among member states, which is crucial for the success of NATO operations [373, 374].

At the apex of this command hierarchy is the North Atlantic Council (NAC), which serves as NATO's primary decision-making body. Comprising representatives from all member states, the NAC is responsible for setting overall strategic objectives and providing political direction for NATO operations. While it does not engage in tactical decision-making, the NAC's role in defining the scope, goals, and rules of engagement for military missions is critical. Decisions within the NAC are made by consensus, reflecting the collective agreement of member states, which underscores the alliance's commitment to cooperative governance [373].

Operational control of NATO missions is vested in the Military Committee (MC), which includes the chiefs of defence from each member nation or their representatives. The MC acts as the primary link between the NAC and NATO's military commands, providing strategic guidance and translating political decisions into military directives. This ensures that operations align with the overarching goals set by the NAC, thereby maintaining coherence in NATO's military endeavours [373, 374].

Beneath the Military Committee are two strategic-level commands: Allied Command Operations (ACO) and Allied Command Transformation (ACT). ACO, based at the Supreme Headquarters Allied Powers Europe (SHAPE) in Mons, Belgium, is responsible for planning and executing NATO operations. The Supreme Allied Commander Europe (SACEUR), typically a U.S. general or admiral, leads ACO and possesses operational authority over NATO forces deployed in missions. Meanwhile, ACT, located in Norfolk, Virginia, focuses on training, modernization, and developing future capabilities to maintain NATO's strategic edge. This dual command structure allows NATO to address both current operational needs and future challenges effectively [373, 374].

At the operational level, ACO oversees a network of joint force commands (JFCs) located in Brunssum (Netherlands), Naples (Italy), and Norfolk (USA). These commands are tasked with managing specific missions or regions, ensuring that regional operations align with NATO's broader strategic objectives. For instance, the JFC in Naples often focuses on operations in the Mediterranean and North Africa, while Brunssum supports missions in Northern Europe and beyond. This regional command structure enhances NATO's ability to respond to diverse operational demands [373, 374].

The tactical-level command structures further translate operational plans into actionable directives. This level includes specialized commands, such as the NATO Response Force (NRF), which is designed for rapid deployment in crises. Subordinate commands encompass air, land, maritime, and special operations components, which execute missions on the ground, at sea, or in the air. These units operate under the authority of their respective JFCs, ensuring that they meet mission-specific requirements while adhering to NATO's strategic framework [373, 374].

A defining feature of NATO's command hierarchy is its emphasis on interoperability among member states. Forces contributed by different nations are integrated into a unified command structure, necessitating shared communication protocols, standardized procedures, and joint training exercises. This integration is vital for ensuring that diverse contributions from member states can operate seamlessly together, thereby enhancing the overall effectiveness of NATO missions [373, 374].

Flexibility is another critical characteristic of NATO's command hierarchy. Although the structure is hierarchical, it is designed to adapt to the specific demands of various missions. For example, during large-scale operations like the International Security Assistance Force (ISAF) in Afghanistan, the hierarchy expanded to include regional commands overseeing operations in different parts of the country. Conversely, in smaller missions, such as disaster response or training initiatives, the command structure is streamlined to match the scale and complexity of the operation [373, 374].

Furthermore, NATO's command hierarchy incorporates liaison officers and embedded personnel to facilitate communication and coordination between member states and partner nations. These roles are essential in multinational operations, where differences in language, culture, and military doctrine could pose significant challenges. Liaison officers ensure that all contributing forces are aligned with NATO's objectives and operational plans, thereby enhancing the effectiveness of joint missions [373, 374].

The chain of command also addresses the critical challenge of accountability in complex environments. Clear delineation of roles and responsibilities ensures that every action is traceable to a specific individual or unit. This accountability fosters discipline, reinforces operational integrity, and ensures compliance with rules of engagement and international laws. For instance, if civilian casualties occur during an operation, the chain of command helps identify decision points and accountability at each level.

Operation Iraqi Freedom (OIF), conducted from 2003 to 2011, serves as a significant case study in understanding the effectiveness of the chain of command and control in a complex military environment. The operation involved coalition forces from multiple nations, necessitating a well-structured command hierarchy and the integration of various military capabilities to achieve strategic objectives in an urbanized battlefield.

The command structure of OIF was initiated by the U.S. Central Command (CENTCOM), which oversaw the entire operation at a strategic level. CENTCOM delegated operational control to the Combined Joint Task Force-7 (CJTF-7), responsible for coordinating ground operations across Iraq. This hierarchical organization extended down to division-level commands, brigade combat teams

(BCTs), battalions, companies, platoons, and individual squads, allowing for clear delineation of responsibilities and objectives at each level of command [375]. For instance, in Baghdad, specific sectors were assigned to individual BCTs, which were further subdivided among battalions and companies, ensuring that each unit could focus on its mission while aligning with the overall strategic plan [375].

Urban warfare in Baghdad required the integration of diverse military capabilities, including infantry, armour, engineers, artillery, and air support. This integration was crucial for effective operations, as demonstrated when an infantry battalion collaborated with armoured units to clear insurgents from neighbourhoods. Engineers played a vital role in handling explosive ordnance disposal (EOD), while artillery provided suppressive fire on enemy positions. Such operations were meticulously synchronized through the chain of command, ensuring cohesive action among all units involved [375, 376]. For example, a company commander could coordinate with an armoured platoon to breach fortified insurgent positions, request artillery support, and rely on engineers to clear mines, all while utilizing UAVs for real-time surveillance [375, 376].

The principles of mission command were essential in enabling flexibility during OIF. In urban areas like Sadr City, platoon leaders often had to make immediate tactical decisions based on rapidly changing situations. Empowered by a clear understanding of their battalion's objectives, these leaders could adapt their tactics to respond to emerging threats, such as sniper fire or unexpected civilian movements [375, 377]. This decentralized execution allowed ground units to react dynamically without waiting for direct orders, which was critical in the fast-paced environment of urban warfare [375, 377].

Effective communication was facilitated through integrated command centres, such as those located in Baghdad's Green Zone. These centres acted as hubs for real-time coordination, processing intelligence from various sources, including UAVs and ground reports. Commanders utilized secure communication systems to issue updated directives and adjust operations as new threats emerged [375, 378]. For instance, if an infantry platoon encountered a heavily armed insurgent group, the company commander could quickly call for precision airstrikes or tank support, while simultaneously relaying updates on the situation back to the command centre for broader operational adjustments [375, 378].

The multinational nature of the coalition forces presented challenges in ensuring interoperability among different military doctrines and communication systems.

The NATO command framework, along with liaison officers, facilitated communication and coordination, ensuring a unified effort among allied forces [375, 379]. Accountability was rigorously maintained throughout the chain of command, particularly in instances of civilian casualties during airstrikes, where investigations traced the decision-making process to ensure compliance with rules of engagement [375, 379].

The chain of command and control during Operation Iraqi Freedom illustrated how a structured system could effectively manage the complexities of urban warfare. It enabled the integration of diverse military capabilities, decentralized execution by small units, and rapid adaptation to dynamic conditions. Despite facing significant challenges, including insurgent tactics and civilian considerations, the command structure ensured a coordinated and disciplined approach to achieving mission objectives [375, 380].

Dealing with Communication Disruptions Due to Urban Infrastructure

Dealing with communication disruptions caused by urban infrastructure presents a significant challenge in urban warfare. The dense urban environment, characterized by tall buildings, underground networks, and crowded streets, creates substantial obstacles for reliable communication. These physical barriers can block, reflect, or absorb signals, leading to degraded communication systems that may endanger military operations. The propagation characteristics of communication signals in urban settings, particularly in the UHF band, are heavily influenced by these environmental factors, which can result in weak or lost signals during military operations [329]. This interference complicates coordination among units, intelligence relay, and support requests, necessitating advanced communication strategies.

To mitigate these challenges, military units are increasingly employing advanced technologies and strategic planning. Satellite communication (SATCOM) systems serve as a reliable alternative, allowing direct communication with command centres or other units outside the urban area, thereby bypassing local obstructions [381]. Additionally, mesh networks are utilized to create a decentralized communication web, where nodes such as soldiers, vehicles, or drones relay messages across the network. This approach ensures that information reaches its destination even when direct communication lines are compromised, as highlighted by Etefia et al. [382] in their analysis of military communication networks. The integration of these technologies is crucial for maintaining operational effectiveness in complex urban environments.

The deployment of unmanned systems, particularly drones, has emerged as an effective solution to urban communication challenges. Drones equipped with communication relays can operate above the urban landscape, acting as mobile signal boosters or relay stations. This capability enhances connectivity between units dispersed throughout the city and provides an aerial perspective for improved situational awareness [383]. Ground-based signal repeaters strategically placed can further extend the range of communication systems, ensuring that units remain connected despite the barriers posed by urban infrastructure [180].

Redundancy in communication systems is essential for ensuring operational continuity. Military units are advised to carry multiple types of communication devices, including radios operating on different frequencies, satellite phones, and encrypted digital communication platforms. This redundancy allows for a seamless transition between systems in the event of a failure [381]. Commanders often establish pre-defined communication protocols, such as fallback frequencies and alternative signalling methods, to maintain coordination even when primary systems are disrupted [180].

Leveraging existing civilian infrastructure is another adaptive strategy that military forces can employ. Utilizing the city's communication networks, such as cell towers and Wi-Fi systems, can augment military communication capabilities. However, this approach necessitates stringent security measures to prevent enemy interception or cyberattacks (Sampaio, 2016). In scenarios where urban infrastructure is compromised, deploying mobile communication hubs, such as vehicle mounted systems or portable field antennas, can ensure effective communication in degraded environments [381].

Training and preparation are critical in addressing the unpredictable nature of urban combat. Soldiers must be adept at operating and troubleshooting various communication devices under stress, as well as executing mission objectives with minimal or intermittent communication (López-Rodríguez, 2024). Pre-mission planning typically includes establishing clear communication plans, designated check-in times, and predetermined rally points to mitigate the impact of potential disruptions [381].

Finally, maintaining communication security in urban environments is paramount. The complexity of urban infrastructure increases the risk of electronic eavesdropping or jamming by adversaries. Military units rely on encrypted communication systems and signal-jamming countermeasures to protect the integrity and confidentiality of their communications [381]. Signals intelligence (SIGINT) teams are also crucial in detecting and neutralizing enemy efforts to disrupt communication networks [381].

Case Studies: Successes and Failures in Urban Communication Strategies

These case studies illustrate how communication strategies can impact operations, emphasizing the importance of preparation, adaptability, and the integration of advanced technologies.

Battle of Fallujah (2004): The Second Battle of Fallujah during the Iraq War stands as a landmark example of effective urban communication strategies. U.S. forces utilized a combination of cutting-edge technologies, robust planning, and redundancy to maintain cohesion during one of the most intense urban battles of the 21st century.

Integrated command centres played a critical role in the success of the operation. These centres consolidated intelligence from drones, satellite imagery, and ground reports to provide real-time situational awareness to commanders. Communication networks were designed with redundancy, including the use of satellite communications (SATCOM) to overcome the dense urban environment that often disrupted radio signals.

Additionally, small-unit leaders were empowered through mission command principles, allowing decentralized decision-making that aligned with overall strategic objectives. Teams relied on encrypted radios and secured data networks to share updates, request airstrikes, and coordinate movements. Despite the complexity of the battlefield, effective communication ensured synchronized operations between infantry, armour, engineers, and air support, leading to the successful clearing of insurgent strongholds.

Operation Protective Edge (2014): The Israeli Defense Forces (IDF) demonstrated a high degree of communication efficiency during Operation Protective Edge in Gaza. Urban combat in Gaza's densely populated areas required precise coordination to minimize collateral damage while achieving military objectives.

The IDF leveraged advanced communication technologies, including encrypted mobile devices and drone-based surveillance systems, to maintain real-time communication with ground units. Integrated command centres allowed for seamless coordination between infantry, air support, and artillery. Civil-military communication also played a pivotal role, with the IDF issuing warnings to civilians through SMS, phone calls, and leaflets before conducting strikes.

The use of drones as communication relays enabled uninterrupted communication across the urban battlefield, overcoming physical barriers posed by dense infrastructure. These measures contributed to operational success while mitigating the risk of civilian casualties in a highly scrutinized conflict.

Failure: Battle of Grozny (1994-1995): The Russian forces' assault on Grozny during the First Chechen War highlights the catastrophic consequences of poor communication strategies in urban warfare. Russian units faced significant communication breakdowns due to inadequate planning and technological shortcomings.

Troops lacked secure and reliable communication systems, relying on outdated radios that were easily intercepted by Chechen forces. This enabled the enemy to anticipate Russian movements and mount effective ambushes. Additionally, poor coordination between units led to disjointed attacks, with armour units advancing without sufficient infantry support, making them vulnerable to guerrilla tactics and anti-tank ambushes.

The lack of centralized command and control further exacerbated the situation. Units operated with little situational awareness, leading to friendly fire incidents and significant casualties. This failure underscores the importance of integrating modern communication technologies and robust planning to maintain cohesion in urban combat.

Mogadishu (1993): The Battle of Mogadishu, commonly known as "Black Hawk Down," revealed critical failures in communication that contributed to mission difficulties. During the U.S. operation to capture Somali warlord Mohamed Farrah

Aidid's lieutenants, urban terrain and rapid escalation exposed gaps in communication strategies.

Disruptions in communication networks led to a lack of coordination between ground forces, helicopter crews, and command elements. Critical information regarding enemy reinforcements and changing battlefield dynamics was delayed or failed to reach the troops in time. Additionally, reliance on radio systems that were prone to jamming in urban settings left units isolated and vulnerable.

The lack of interoperability between U.S. forces and UN contingents further compounded the issue, as language barriers and differing communication protocols hampered joint operations. This failure demonstrated the need for interoperable communication systems and contingency plans to handle disruptions.

Operation Mosul (2016-2017): The battle to retake Mosul from ISIS showcased how modern communication strategies could address the challenges of urban warfare. Iraqi forces, supported by U.S. and coalition partners, employed advanced communication systems to synchronize their efforts in a densely populated and contested urban environment.

A multi-layered communication strategy was key to success. Iraqi ground forces used encrypted radios to coordinate movements and share intelligence, while drones provided real-time reconnaissance and acted as communication relays. Coalition forces integrated their command centres with Iraqi operations, ensuring seamless coordination of airstrikes, artillery support, and logistics.

Moreover, civilian communication channels were utilized to disseminate evacuation instructions and counter ISIS propaganda. The integration of modern technology and close collaboration between international and local forces demonstrated the effectiveness of comprehensive communication strategies in urban combat.

The analysis of urban warfare communication strategies reveals critical lessons drawn from both successful and failed military operations. These lessons emphasize the importance of technological integration, redundancy and adaptability, coordination and interoperability, and civil-military communication.

Technological Integration is paramount in modern urban warfare, as evidenced by operations in Fallujah and Mosul. These operations showcased the effective use of advanced technologies such as encrypted networks, drones, and centralized command centres, which significantly enhanced situational awareness and operational efficiency. The integration of such technologies allows for real-time data sharing and improved decision-making capabilities, which are essential in the chaotic environment of urban combat [384, 385]. The ability to leverage these technologies not only facilitates better communication among units but also enhances the overall effectiveness of military operations in complex urban terrains [386].

Redundancy and Adaptability are also crucial in urban environments where communication systems can be easily disrupted. Successful operations have demonstrated the necessity of employing multiple communication methods and fallback protocols to ensure continuous operational capability. For instance, the integration of various communication technologies can mitigate the impact of potential failures in primary systems, thereby maintaining command and control during critical phases of operations [387]. This adaptability is essential for military planners to anticipate and respond to the unique challenges posed by densely populated areas, where traditional communication lines may be compromised.

Coordination and Interoperability are highlighted by the failures observed in Grozny and Mogadishu, where lack of clear communication and coordination among units led to operational setbacks. These instances underscore the critical need for interoperable systems that facilitate seamless communication and collaboration between different military branches and allied forces [388, 389]. The establishment of standardized protocols and interoperable technologies is vital for ensuring that all units can effectively share information and coordinate their actions, thereby enhancing overall mission success [390, 391].

Civil-Military Communication plays a significant role in urban operations, as demonstrated in Gaza, where effective communication with civilian populations helped mitigate collateral damage and maintain public support. Establishing clear lines of communication with civilians can enhance trust and cooperation, which are essential for successful military operations in urban settings [392, 393]. By integrating civil-military communication strategies, military planners can better navigate the complexities of urban warfare, ensuring that operations are conducted with minimal impact on civilian life [394, 395].

Chapter 7

PSYCHOLOGICAL WARFARE AND CIVILIAN INTERACTION

Understanding Civilian Dynamics and Influence on Urban Operations

U nderstanding civilian dynamics and their influence on urban operations is a critical aspect of modern military strategy. Urban environments are often densely populated, making civilians an integral factor in planning and executing operations. Their presence introduces both challenges and opportunities, requiring careful navigation to balance military objectives with ethical and humanitarian considerations.

In urban warfare, civilians significantly influence the operational environment. Their movements, needs, and interactions with military forces and opposing factions shape the flow of operations. Civilians may inadvertently hinder operations by occupying key infrastructure or moving through conflict zones, creating logistical and tactical challenges. Conversely, civilians can serve as valuable sources of intelligence, providing insight into enemy positions, local geography, and potential threats. Understanding these dynamics helps military forces plan operations that minimize disruption to civilian life while maximizing mission success.

The future urban environment presents significant operational challenges and strategic imperatives, as urban areas increasingly become focal points for population growth, economic activity, and potential conflict. Recognized by NATO and reflected in operational frameworks like the UK's Future Operating Environment 2035, urban terrain demands a sophisticated approach due to its complexity, density, and dynamic interactions. Cities, as strategic assets, present unique challenges stemming from their intricate terrain, significant civilian populations, and vital infrastructure, all of which influence the conduct and outcome of military operations [396].

Urban areas, with their interconnected systems, impose substantial physical and cognitive demands on military forces. The dense and multi-dimensional nature of urban terrain includes subterranean, surface, and aerial components, all complicated by natural geographic elements like rivers or coastlines. This physical complexity is mirrored by the human dimension, where civilian populations of varied affiliations interact with military and insurgent actors. Civilians may be neutral, supportive, or opposed to military operations, often acting out of survival or coercion. Their movements, needs, and interactions amplify the difficulty of maintaining situational awareness and managing the impact of operations [396].

Historically, military strategies have sought to avoid urban combat due to its inherent difficulties. However, as urbanization accelerates and cities increasingly serve as centres of political, economic, and social activity, avoiding urban engagements has become impractical. Adversaries often exploit urban environments, blending with civilian populations, using infrastructure for tactical advantages, and employing asymmetric and hybrid warfare tactics. This operational complexity necessitates a multidisciplinary approach, integrating direct combat capabilities with humanitarian and civil-military planning [396].

Urban operations require a deep understanding of interconnected urban systems, encapsulated in NATO's urban quad model. This framework addresses the physical environment, human population, infrastructure, and the information systems that underpin urban functionality. The concept underscores the need for military planners to consider the intricate dynamics of these systems to protect civilians and maintain operational effectiveness. In scenarios such as the urban littoral, where cities interface with coastal zones and waterways, operations must also address unique challenges like flooding, disease risk, and critical chokepoints for global trade [396].

NATO's Urban Quad Model is a conceptual framework designed to understand and navigate the complexities of urban environments during military operations. It emphasizes the interaction between four critical elements: physical systems, human populations, infrastructure, and information systems. This interconnected model helps military planners analyse and address the unique challenges of urban areas, enabling them to develop strategies that account for the multi-dimensional nature of cities. Each component plays a distinct role in shaping the urban environment and its operational dynamics.

The physical system represents the tangible aspects of an urban area, including its built environment and natural surroundings. This encompasses buildings, streets, subterranean networks like tunnels and sewers, and super-surface elements such as rooftops and aerial zones. The physical system is characterized by dense, layered terrain that complicates visibility, movement, and combat operations. Urban sprawl and informal settlements further increase the complexity by creating "edgeless" zones where rural and urban areas merge. Coastal cities introduce additional challenges through their littoral zones, including waterways, ports, and flood-prone areas. For military forces, navigating and controlling this multi-dimensional terrain requires careful planning, specialized equipment, and adaptability.

The human system focuses on the population living within the urban area. Urban populations are diverse, with varying allegiances, needs, and behaviours. Civilians may be neutral, supportive, or hostile toward military forces, and their actions can significantly influence operations. Factors like population density, movement patterns, and sociocultural dynamics shape the human system. Civilians can be directly impacted by conflict, whether as bystanders, forced participants (e.g., human shields), or through displacement. Understanding the human system is essential for minimizing civilian casualties, maintaining legitimacy, and achieving operational objectives.

The infrastructure system includes the essential services and facilities that sustain urban life, such as transportation networks, communication systems, utilities, and public services. This system is highly interconnected, meaning that damage to one component can cascade through others, affecting the city's overall functionality. For example, the destruction of a power grid can disrupt water supplies, healthcare services, and transportation. Military operations in urban areas must account for these interdependencies to avoid unintended humanitarian crises and ensure the protection of critical infrastructure.

The information system encompasses the networks and technologies that enable communication, data exchange, and decision-making within the urban environment. This includes both physical elements like telecommunications infrastructure and digital systems such as cyber networks and social media platforms. In modern cities, information systems are integral to daily life and military operations alike. Disruptions to these systems can cripple civilian and military functions, while adversaries may exploit them for propaganda, misinformation, or sabotage. Understanding and securing the information system is crucial for maintaining situational awareness and operational control.

The Urban Quad Model emphasizes that these four systems are not isolated but are interconnected and dynamic. Changes in one system can ripple through the others, creating complex, multi-order effects. For instance, an airstrike that destroys a bridge (physical system) may disrupt civilian evacuation routes (human system), cut off utility lines (infrastructure system), and hinder communication networks (information system). Military planners using the Urban Quad Model aim to anticipate and mitigate such effects, ensuring that operations are effective while minimizing harm to civilians and essential urban functions.

The technological evolution of urban environments further complicates military operations. Smart cities and networked infrastructures introduce vulnerabilities to cyberattacks and information manipulation, increasing the risks to both civilian populations and military objectives. Adversaries can weaponize these networks, targeting essential services or spreading disinformation. Consequently, protecting and managing urban information systems becomes as critical as safeguarding physical infrastructure [396].

Historical examples like the battles of Mosul and Marawi illustrate the profound challenges of urban warfare. The liberation of Mosul demonstrated the tactical complexities of high-intensity urban combat, including efforts to protect civilians amidst the enemy's use of human shields. The battle highlighted the importance of controlling critical infrastructure and leveraging information warfare. Similarly, the battle of Marawi underscored the necessity of pre-emptive civilian evacuation, integration of local and international efforts, and the importance of mitigating long-term urban devastation [396].

As the global population continues to urbanize, military and civilian institutions must prioritize resilience and preparedness for urban conflict. This includes ensuring continuity of essential services, robust civil-military cooperation, and adherence to international humanitarian laws and urban-specific rules of engagement. Emerging technologies and threats necessitate continuous adaptation of doctrines and policies to protect civilians and maintain strategic and operational effectiveness in the increasingly complex future urban landscape [396].

The behaviour of civilian populations is often influenced by the actions of military forces. Demonstrating respect for local customs, providing humanitarian aid, and protecting civilian lives can foster goodwill and cooperation, reducing the likelihood of hostility or insurgent recruitment. Conversely, excessive collateral damage, disregard for cultural sensitivities, or failure to address civilian needs can erode trust and lead to resistance, complicating operations and prolonging conflict. This underscores the importance of adopting a people-centric approach in urban environments.

Civilians are often deeply affected by the presence of military forces and the ongoing conflict. They may be displaced from their homes, face shortages of basic necessities, or experience psychological trauma. Military forces must anticipate these outcomes and establish systems to mitigate them, such as creating safe corridors for evacuation, providing medical aid, and coordinating with humanitarian organizations. These efforts not only fulfill ethical obligations but also stabilize the operational environment, enabling more effective military actions.

Urban operations are further complicated by the presence of adversaries embedded within civilian populations. Enemy forces may use civilians as human shields, conduct operations from civilian infrastructure, or exploit the population to gather intelligence. This creates a challenging scenario where military forces must differentiate between combatants and non-combatants, often in rapidly changing and high-pressure situations. Adhering to rules of engagement and leveraging advanced technologies, such as facial recognition or UAV surveillance, can aid in making these critical distinctions.

Coordination with local authorities and community leaders is essential for understanding and influencing civilian dynamics. Local leaders often hold the trust of the population and can mediate between civilians and military forces, facilitating communication and cooperation. Collaborating with these leaders helps military forces align their operations with civilian needs and expectations, minimizing friction and misunderstandings.

Effective communication with civilians is another critical element. Military forces must convey their intentions clearly, explain safety protocols, and provide updates on ongoing operations to build trust and ensure civilian compliance with evacuation orders or curfews. Communication channels such as radio broadcasts, leaflets, or social media platforms can be used to reach a wide audience and maintain transparency.

Legal, Policy and Strategic Frameworks for Minimizing Harm to Civilians

The Law of Armed Conflict (LOAC) represents a framework designed to balance military necessity with the principle of humanity, ensuring that armed engagements are conducted ethically and within established legal parameters. It is codified in treaties such as the 1949 Geneva Conventions and their 1977 Additional Protocols, as well as in customary international law. LOAC governs not only the direct conduct of hostilities but also the planning and execution of military operations, with a focus on minimizing harm to civilians and civilian infrastructure [397].

At its core, LOAC is built upon three fundamental principles: distinction, proportionality, and precautions. The principle of distinction requires parties to a conflict to clearly differentiate between civilians, civilian objects, and legitimate military targets. Attacks must be directed solely at military objectives, and indiscriminate attacks—those that fail to distinguish between military and civilian entities—are strictly prohibited. Proportionality further limits military actions by prohibiting attacks that would cause civilian harm disproportionate to the anticipated military advantage. Precaution mandates that all feasible measures must be taken to avoid or minimize incidental civilian harm, including altering or cancelling attacks if the harm would outweigh the military gain [397].

LOAC extends its obligations to both offensive and defensive operations. For attackers, precautionary measures include advance risk assessments, warnings to civilians, careful selection of attack methods, and timing adjustments. Defenders are similarly required to protect civilians under their control by issuing warnings of incoming attacks, relocating civilians away from military objectives, and avoiding the placement of such objectives in densely populated areas. These measures underscore the comprehensive and reciprocal responsibilities imposed by LOAC on all parties involved in a conflict [397].

Urban warfare presents unique challenges to LOAC compliance due to the dense intermingling of military and civilian elements. The complex urban terrain often complicates the identification of legitimate targets, while adversaries may exploit civilians as human shields or embed military objectives within civilian areas. LOAC explicitly prohibits the use of human shields and tactics such as starvation to achieve military objectives. Encirclement and siege tactics, while lawful when targeting enemy forces, must not deliberately deprive civilians of essential supplies [397].

The evolving nature of warfare has also introduced additional layers of complexity. Urban operations increasingly blur the lines between military and police functions, requiring coordination between armed forces, law enforcement, and special units to adhere to international human rights law. The rise of cyber and space operations further challenges traditional applications of LOAC, necessitating adaptations to ensure emerging technologies comply with legal and ethical standards. For instance, automated decision-making tools and accelerated operational tempos demand robust mechanisms to assess potential humanitarian impacts [397].

LOAC compliance is not merely a battlefield responsibility; it is a foundational element of operational planning and execution. Personnel at all levels must be trained in LOAC principles, ethical judgment, and mission-specific rules of engagement to minimize civilian harm. This includes preparations for non-combat activities such as establishing bases, conducting searches, or coordinating with humanitarian organizations. Legal obligations extend beyond combat scenarios to encompass the overall protection of civilians and critical infrastructure in conflict zones [397].

In recent years, militaries and organizations like NATO have implemented policies to mitigate civilian harm. For example, NATO's International Security Assistance Force (ISAF) in Afghanistan introduced tactical directives to reduce civilian casualties. Similarly, the African Union Mission in Somalia restricted the use of indirect fire in populated areas. NATO's Protection of Civilians (PoC) policy, adopted in 2016, represents a significant commitment to minimizing civilian harm and fostering secure environments. This policy has been complemented by practical tools such as NATO's PoC Handbook and further developments in response to conflicts in Ukraine and other regions [397].

The United States has also taken significant steps with the 2022 Civilian Harm Mitigation Action and Response Plan (CHMR-AP), which aims to integrate civilian harm mitigation across all aspects of military operations. This plan includes the creation of a civilian protection center of excellence, overhauls in doctrine and training, and strengthened collaboration with international allies. Additionally, the 2022 Political Declaration to Limit Humanitarian Suffering from the Use of Explosive Weapons in Populated Areas, signed by 83 nations, reflects a global consensus on the need to protect civilians in contemporary conflict [397].

Dealing With Hostages, Human Shields, and Insurgent Tactics

Dealing with hostages, human shields, and insurgent tactics in urban warfare presents a multifaceted challenge that necessitates a careful balance between military objectives and the protection of civilian lives. Hostage situations frequently arise as insurgents leverage civilians to gain strategic advantages over military forces. The complexity of these scenarios is heightened by the potential for hostages to be held in fortified positions or dispersed across various locations, complicating rescue operations. Military responses must prioritize civilian safety while neutralizing threats, which requires extensive intelligence gathering to ascertain the number of hostages, their locations, and the insurgents' capabilities and intentions [398, 399]. Special operations forces are often deployed for these missions due to their specialized training in precision operations and their proficiency in navigating high-risk environments [400]. Additionally, negotiation teams play a crucial role in these scenarios, aiming to de-escalate tensions and secure the release of hostages without resorting to violence [398].

The use of human shields by insurgents poses another significant challenge in urban warfare. This tactic involves placing civilians near military targets to deter attacks or to exploit civilian casualties for propaganda purposes. Human shields may be coerced, voluntary, or unaware of their role, complicating military responses [399]. Addressing this tactic requires strict adherence to the principles of the Law of Armed Conflict (LOAC), particularly the principles of distinction, proportionality, and precaution [399]. Military forces must employ advanced surveillance and intelligence techniques to identify and isolate human shields while targeting insurgent positions with precision-guided munitions (Wang et al., 2015). This often necessitates delaying operations or accepting higher risks for military personnel to avoid civilian casualties, underscoring the moral complexities inherent in such situations [399].

The principle of distinction requires parties to a conflict to differentiate between combatants and civilians, as well as between military objectives and civilian objects. Combatants and military objectives may lawfully be targeted, while civilians and civilian objects (such as homes, schools, and hospitals) must be spared from direct attack.

In practice, this principle imposes an obligation on military forces to:

1. Identify and verify targets as legitimate military objectives before launching an attack.

2. Avoid deliberate targeting of civilians or objects that are not contributing to the military effort.

3. Take into account the dual-use nature of certain infrastructure (e.g., bridges or communication networks) and assess whether their targeting is justified under military necessity.

For example, in urban warfare, distinguishing between armed insurgents and civilians who might be coerced into cooperating with them becomes a critical and challenging task. Failure to apply the principle of distinction can result in civilian casualties, undermine mission objectives, and lead to violations of international law.

The principle of proportionality prohibits attacks that may cause incidental harm to civilians or civilian objects that would be excessive in relation to the direct and concrete military advantage anticipated. This principle recognizes that some civilian harm may be unavoidable in armed conflict but seeks to limit it as much as possible.

To comply with proportionality:

1. Commanders must weigh the expected military gain against potential harm to civilians and civilian infrastructure.

2. Operations must be planned and executed to minimize incidental damage while achieving legitimate military objectives.

3. Excessive harm to civilians relative to the anticipated military benefit renders the attack unlawful.

For instance, targeting an enemy command centre in an urban area might be lawful if the military advantage outweighs potential collateral damage. However, using a large explosive weapon that risks destroying an entire neighbourhood for a minor tactical gain would violate the principle of proportionality.

The principle of precaution requires all feasible measures to be taken to avoid or minimize harm to civilians and civilian objects before, during, and after military operations. This obligation applies to both attacking and defending parties in a conflict.

For the attacking party, precaution includes:

1. Verifying targets as military objectives before engaging.

2. Using precision-guided munitions or other appropriate means to limit collateral damage.

3. Providing advance warnings to civilians, where feasible, to allow them to evacuate potential areas of harm.

4. Adjusting or cancelling attacks if civilian harm would be disproportionate.

For the defending party, precaution involves:

1. Avoiding placing military objectives near densely populated civilian areas.

2. Evacuating civilians from areas likely to be targeted.

3. Clearly marking protected areas, such as hospitals or cultural sites, to prevent accidental targeting.

For example, in an urban environment, an attacking force might issue warnings through leaflets, radio broadcasts, or text messages to civilians before conducting airstrikes on a building used by enemy forces. The defending force, on the other hand, has a responsibility to avoid using hospitals or schools as military command centres.

Together, these principles guide military operations to ensure compliance with international humanitarian law:

- Distinction ensures that only legitimate targets are attacked.

- Proportionality limits the scale and intensity of attacks to prevent excessive harm.

- Precaution mandates active measures to protect civilians and civilian objects.

By adhering to these principles, military forces can conduct operations in a manner that reduces unnecessary suffering, maintains the legitimacy of their actions, and upholds international legal and moral standards.

Effective hostage rescue operations hinge on the collection of accurate and actionable intelligence. This includes identifying the location of hostages and captors, understanding the physical layout of the area, and assessing the captors' capabilities and intentions. Advanced surveillance technologies, such as drones and ground sensors, play a crucial role in gathering this intelligence [400]. Collaboration with local informants can also yield critical insights that inform operational planning. Detailed plans must be developed to guide the operation, incorporating contingencies to minimize risks to hostages and civilians [400, 401].

Hostage rescue operations in urban settings are typically executed by specialized forces trained in close-quarters combat and negotiation. These units are equipped with advanced weaponry and breaching tools suited for confined environments. Coordination among various military and law enforcement units is essential to secure the perimeter and neutralize threats effectively [400]. In multinational operations, establishing unified command structures is vital to ensure seamless communication and operational effectiveness [401].

Negotiation is often the preferred initial strategy in hostage situations, aiming to resolve conflicts without resorting to violence. Trained negotiators work to establish communication with captors, gather information about their demands, and buy time for potential rescue operations [398]. Psychological operations may also be employed to influence captors, such as disseminating disinformation to create confusion or exploiting their vulnerabilities [398].

When a rescue operation is deemed necessary, tactical precision is critical to prevent harm to hostages. Urban warfare presents unique challenges, including narrow streets and potential civilian presence. Key tactics include dynamic entry, where teams breach locations quickly to neutralize threats, and simultaneous actions to prevent captors from regrouping [398, 400]. The use of non-lethal methods, such as tear gas or stun grenades, can incapacitate captors while minimizing risks to hostages [398, 400].

Hostage operations must comply with international humanitarian law, emphasizing the principles of distinction, proportionality, and precaution. This includes avoiding unnecessary harm to hostages and civilians and ensuring that the use of force is justified and proportional to the threat [402, 403]. Providing immediate medical care and psychological support to rescued hostages is also essential to address the trauma experienced during captivity [401, 402].

Rescued hostages often require immediate medical and psychological care, as the trauma from captivity and the rescue operation can have lasting effects. Long-term support systems should be established to aid recovery, and post-operation debriefings are crucial for analysing the mission and identifying lessons learned for future protocols [402].

Urban warfare exacerbates the difficulties of hostage situations due to the proximity of civilians, increasing the risk of collateral damage, and the potential for captors to utilize underground or concealed locations [402, 404]. The use of hostages as human shields or propaganda tools heightens the stakes of any rescue attempt, necessitating meticulous planning and execution [402, 404].

The liberation of Mosul from ISIS (2016–2017) serves as a pertinent case study, illustrating the complexities of hostage rescue in urban combat. ISIS frequently employed civilians as human shields, necessitating careful planning and coordination among coalition forces. The use of drones for surveillance and precision airstrikes to neutralize threats was critical, although challenges such as booby traps and underground tunnels complicated rescue missions [400, 401].

Insurgents in urban environments frequently utilize asymmetric tactics that exploit the complexities of the terrain and the presence of civilians. These tactics may include ambushes, sniper attacks, and the deployment of improvised explosive devices (IEDs) [399]. The blending of insurgents with the civilian population complicates the ability of military forces to distinguish between combatants and non-combatants, necessitating a combination of intelligence-led operations, advanced surveillance technologies, and robust rules of engagement to ensure compliance with LOAC while maintaining operational effectiveness [405]. For instance, drones equipped with thermal imaging can be instrumental in detecting hidden enemy movements, while specialized units trained in counter-insurgency can conduct targeted operations to neutralize threats without causing undue harm to civilians [406].

In addition to tactical measures, effective psychological operations (PSYOPS) and information campaigns are essential to counter insurgent narratives and build trust with the civilian population. Insurgents often manipulate public perception by emphasizing civilian casualties or portraying themselves as protectors of the local populace [399]. Military forces must counter these narratives by demonstrating their commitment to civilian protection, providing humanitarian aid, and respecting cultural norms [399]. Transparent communication with the public, both locally and internationally, is critical for maintaining legitimacy and support for military operations [399].

Coordination with non-governmental organizations (NGOs), humanitarian agencies, and local authorities is also vital in these scenarios. Such partnerships can facilitate the evacuation of civilians, provide medical care and shelter for hostages post-release, and help rebuild trust within affected communities [407]. Negotiating humanitarian corridors and ceasefires can allow civilians, including hostages and human shields, to escape conflict zones safely, further emphasizing the need for a comprehensive approach to urban warfare [407].

Ultimately, the success of military operations involving hostages, human shields, and insurgent tactics hinges on a balanced approach that integrates tactical precision, legal and ethical considerations, and strategic communication. Military forces must adapt to the dynamic and morally complex challenges of urban warfare, leveraging advanced technologies, specialized training, and comprehensive planning to achieve their objectives while upholding humanitarian principles [399].

Psychological Operations (PSYOPS) to Demoralize or Influence Enemy Forces

Psychological Operations (PSYOPS) play a pivotal role in contemporary military strategy, particularly in efforts to demoralize and influence enemy forces. These operations are designed to manipulate perceptions, emotions, and behaviours through a variety of techniques, including propaganda dissemination, misinformation, and the exploitation of cultural vulnerabilities. The overarching goal of PSYOPS is to undermine the enemy's morale, disrupt their decision-making processes, and ultimately compel surrender or reduce their capacity to resist.

The effectiveness of PSYOPS is rooted in their ability to exploit psychological weaknesses within enemy ranks. Targeted messaging can convince enemy soldiers that their situation is hopeless, thereby fostering despair and encouraging desertion. Research indicates that psychological tactics, such as the planned use of propaganda, are integral to influencing the opinions and behaviours of opposition groups, which is essential for achieving military objectives [408]. Moreover, historical examples demonstrate the utility of PSYOPS in warfare; during Operation Desert Storm, coalition forces effectively used leaflets and broadcasts to demoralize Iraqi troops, leading to significant surrenders [409].

Modern PSYOPS have evolved significantly with advancements in technology, particularly in the digital realm. The rise of social media and the internet has transformed how psychological warfare is conducted, allowing for rapid dissemination of disinformation and propaganda on a global scale. For example, during the ongoing Russia-Ukraine conflict, both sides have utilized social media to shape narratives and influence public opinion, showcasing the contemporary relevance of PSYOPS in modern warfare [410]. The ability to manipulate information digitally enhances the reach and impact of psychological operations, making them a formidable tool in contemporary conflicts.

Operation Desert Storm (1991): During the Gulf War, coalition forces employed extensive PSYOPS to demoralize Iraqi troops. Leaflets dropped over enemy positions warned soldiers of impending airstrikes and promised humane treatment if they surrendered. Radio broadcasts and loudspeaker messages reinforced these themes, leading to mass defections and surrenders among Iraqi forces.

NATO Operations in Kosovo (1999): NATO used PSYOPS to influence Serbian forces and civilians during the Kosovo conflict. Radio broadcasts and leaflets were employed to undermine support for the Serbian government and military, emphasizing NATO's commitment to avoiding civilian harm while highlighting the inevitability of military defeat.

Russia-Ukraine Conflict (Ongoing): Both sides in the ongoing conflict have leveraged PSYOPS extensively. Social media platforms have become battlegrounds for narratives aimed at demoralizing the enemy and rallying domestic and international support. Russia has used misinformation campaigns to spread confusion and fear, while Ukraine has employed targeted messaging to highlight the incompetence and losses of Russian forces.

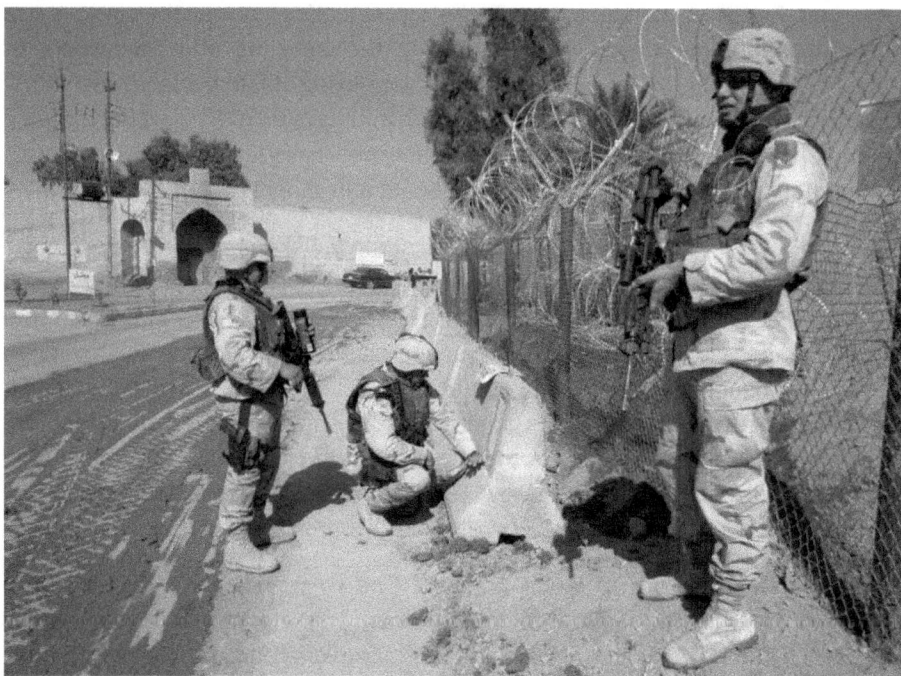

Figure 59: US Army (USA) Soldiers assigned to the 350th Tactical Psychological Operations Company, Tactical PsyOps Team 15-24, attached to the 1-4 Cavalry, 1ST Infantry Division put up posters of wanted anti Iraqi criminals in Ad Duluyiah, Iraq, during Operation IRAQI FREEDOM. The U.S. National Archives, Public Domain, National Archives and Defense Visual Information Distribution Service.

Key techniques employed in PSYOPS include the dissemination of propaganda through various mediums, such as leaflets, radio broadcasts, and social media. These messages are often tailored to exploit specific cultural or ideological vulnerabilities within the enemy's ranks. For instance, PSYOPS can highlight internal divisions or grievances to foster dissent and weaken cohesion among enemy forces [408]. Additionally, misinformation campaigns can create confusion and compel adversaries to make poor tactical decisions, further undermining their operational effectiveness [409].

Figure 60: Italian Army - 28th Regiment "Pavia" PSYOPS exercise with the US Army's AFRICOM affiliated 7th Psychological Operations Battalion. esercito.difesa.it, CC BY 2.5, via Wikimedia Commons.

Despite their effectiveness, PSYOPS also present ethical challenges. The use of misleading information can backfire, damaging the credibility of the originator and complicating post-conflict reconciliation efforts. Therefore, it is crucial for military planners to balance the effectiveness of psychological operations with adherence to ethical standards and international law [411]. The potential for unintended consequences necessitates a careful approach to the planning and execution of PSYOPS, ensuring that operations do not inadvertently escalate conflicts or harm civilian populations.

Countermeasures against psychological operations (PSYOPS) targeting military and civilian populations are critical to maintaining morale, operational security, and decision-making integrity. Effective countermeasures require a combination of technological, strategic, and psychological resilience.

Countering enemy propaganda requires swift detection, analysis, and a tailored response to neutralize its effects. Intelligence teams must monitor the dissemination channels used by adversaries, such as leaflets, broadcasts, and social media, to identify key themes and vulnerabilities exploited by the messages. Rapid dissemination of counter-narratives is essential to contest the influence of propaganda.

For example, if leaflets suggest hopelessness or offer promises of safe surrender, military commanders can reinforce troop morale by communicating accurate updates about strategic progress, reinforcing unity, and highlighting enemy misinformation. Public affairs teams and military leadership can directly address rumours spread via radio or social media, using verified information to debunk claims of internal strife or exaggerated losses. Creating channels for soldiers to receive trusted information helps prevent enemy narratives from taking hold.

Counteracting misinformation and deception demands robust intelligence and counterintelligence measures. Military units must verify reports of enemy movements, supply shortages, or other critical information through multiple independent sources before acting. This reduces the likelihood of falling victim to false information designed to mislead.

Technological tools such as satellite imagery, UAV surveillance, and electronic intelligence (ELINT) are essential for verifying enemy claims or movements. Real-time analysis and dissemination of accurate intelligence ensure that tactical decisions are grounded in fact rather than manipulated data. Defensive deception measures, such as feeding false information back to adversaries or staging counter-deceptions, can also confuse their efforts, mitigating the effectiveness of their PSYOPS campaigns.

To counter attempts to exploit cultural or psychological divisions, military leadership must focus on fostering cohesion and addressing potential internal grievances proactively. Initiatives to build trust within diverse ranks and reinforce shared values and mission objectives are vital.

Psychological training programs can inoculate personnel against adversary tactics that seek to manipulate cultural or ideological fault lines. By emphasizing inclusivity and the importance of loyalty, unity, and mission focus, commanders can reduce the effectiveness of enemy attempts to exploit divisions. Additionally, clear communication from leadership addressing specific issues raised in enemy PSYOPS campaigns prevents these messages from gaining traction among troops.

Countering digital PSYOPS requires advanced cybersecurity and strategic communication capabilities. Military and civilian cybersecurity teams must monitor and protect communication channels from infiltration, manipulation, or hacking by adversaries. This includes encrypting sensitive communications, using secure messaging platforms, and employing real-time monitoring systems to detect and neutralize cyber threats.

Disinformation campaigns on social media can be mitigated through coordinated efforts to flag and debunk fake news, doctored videos, and misleading content. Military public relations teams must maintain an active online presence to provide accurate information and counteract adversary narratives. Platforms with rapid-response teams can prevent the spread of demoralizing or divisive content by identifying and addressing hostile actors early.

Enemy efforts to demoralize through displays of strength or success can be countered by maintaining transparency and showcasing the resilience and achievements of one's own forces. Regularly updating personnel with accurate reports of operational progress, success stories, and leadership messages builds confidence and unity.

Tactical psychological efforts can reinforce this by highlighting the failures and vulnerabilities of the enemy. For example, distributing visual evidence of adversary setbacks, such as failed operations or captured equipment, counters their attempts to intimidate. Positive messaging aimed at allied and neutral audiences can also prevent enemy PSYOPS from gaining influence among broader populations.

The most effective countermeasures integrate proactive planning and education. Personnel must be trained to recognize PSYOPS tactics and understand their objectives to resist manipulation. Creating a culture of resilience, trust in leadership, and the capacity for critical analysis among soldiers and civilians undermines the efficacy of psychological operations. Regular communication and leadership presence reinforce morale and ensure that enemy PSYOPS campaigns are met with informed and decisive responses.

Humanitarian Considerations and Minimizing Collateral Damage

Urban warfare poses significant challenges due to the complexities of densely populated environments, where the intermingling of combatants and civilians complicates military operations. The humanitarian considerations in such contexts are paramount, as they aim to protect civilian populations from the direct and indirect consequences of armed conflict. Civilians often find themselves trapped in conflict zones, unable to evacuate due to logistical, cultural, or safety barriers. This situation necessitates a robust commitment to humanitarian principles, which seek to mitigate suffering and ensure access to essential services, such as water, electricity, and healthcare, even amidst active hostilities [412].

Figure 61: PORT-AU-PRINCE, Haiti (Jan. 17, 2010) An air crewman assigned to the Chargers of Helicopter Sea Combat Squadron (HCS) 26 drops humanitarian aid to earthquake victims in Port-au-Prince. Carrier Air Wing (CVW) 17 embarked aboard USS Carl Vinson (CVN 70) is conducting humanitarian and disaster relief operations as part of Operation Unified Response. U.S. NAVY, Public Domain, via Picryl.

The presence of civilians in urban warfare complicates military strategies, particularly when combatants exploit this situation by using civilians as human shields. This tactic not only endangers civilian lives but also challenges military forces to balance operational objectives with the imperative to minimize civilian harm. The principles of distinction, proportionality, and precautions, as outlined in International Humanitarian Law (IHL), serve as critical guidelines for military planners. As discussed earlier, these principles mandate that combatants differentiate between military targets and civilians, ensure that any incidental civilian harm is not excessive in relation to the military advantage gained, and take all feasible precautions to minimize civilian casualties [413, 414].

To effectively mitigate collateral damage in urban warfare, military forces can employ a range of technological, tactical, and operational strategies. Precision-guided munitions, for instance, significantly reduce the risk of civilian casualties and damage to critical infrastructure. The integration of unmanned aerial vehicles (UAVs) and advanced surveillance technologies enhances situational awareness and targeting accuracy, allowing for more informed decision-making in complex urban environments [415, 416]. Furthermore, restricting the use of high-yield explosive weapons in populated areas and employing non-lethal technologies can help achieve military objectives while minimizing physical destruction and civilian harm [417].

Operational planning also plays a crucial role in minimizing collateral damage. Timing military operations to coincide with periods of reduced civilian presence, providing advance warnings to civilians, and establishing "no-strike" lists for critical infrastructure are essential strategies. Humanitarian pauses and corridors can facilitate the evacuation of civilians and the delivery of aid, thereby reducing the likelihood of civilian casualties during active combat [412, 415]. Close coordination with humanitarian organizations is vital to ensure that civilian needs are addressed during military operations, fostering a collaborative approach that integrates humanitarian considerations into operational planning [413].

Training military personnel on the principles of IHL and ethical decision-making is critical to ensuring compliance with humanitarian standards. Educating soldiers and commanders on identifying civilian risks in urban environments and implementing rules of engagement that prioritize civilian protection can significantly enhance the ethical conduct of military operations [414]. Despite these efforts, the reality of urban warfare often results in unintended civilian harm, underscoring the importance of post-conflict accountability mechanisms and continuous learning from past operations to refine strategies aimed at reducing collateral damage [412, 417].

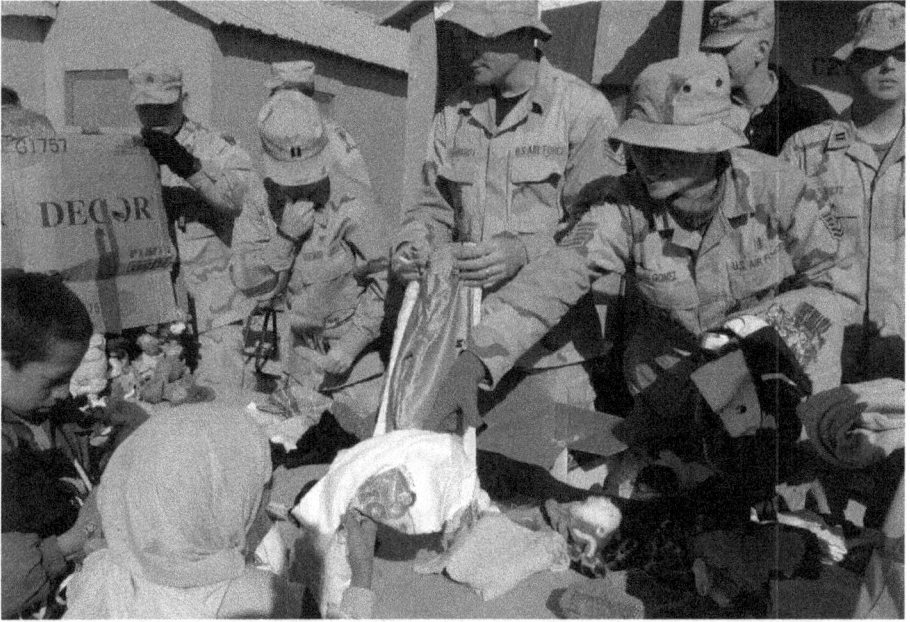

Figure 62: Soldiers from Task Force Strength, Bagram Air Base gives clothes to Afghan children during a joint humanitarian aid mission with the Egyptian Army at Bagram, Afghanistan on Nov. 17, 2005. Task Force Strength does humanitarian missions along with coalition forces as part of Operation Enduring Freedom. The U.S. National Archives, Public Domain, via Picryl.

Building Relationships with Local Populations for Intelligence

In urban warfare and counterinsurgency operations, building relationships with local populations is a cornerstone of successful intelligence gathering. The dynamic and complex nature of urban environments, where combatants are often interspersed with civilians, makes traditional intelligence methods insufficient. Establishing trust and fostering cooperation with local communities allows military and security forces to gain critical insights into enemy activities, local dynamics, and potential threats.

Local populations are often the most reliable source of ground-level information in urban settings. Their intimate knowledge of the terrain, social networks, and daily patterns provides invaluable context that cannot be replicated by surveillance technology or remote intelligence operations. Civilians often observe enemy movements, recognize outsiders, or notice changes in their environment that signal potential threats. Harnessing this knowledge requires a rapport based on trust, respect, and mutual benefit.

Failing to build these relationships can isolate security forces, leaving them vulnerable to misinformation or ambushes. Moreover, alienating local populations risks driving them toward supporting adversaries, either out of necessity, coercion, or resentment toward military operations perceived as invasive or harmful.

Establishing trust and credibility between security forces and local populations is crucial for effective governance and community safety. Trust serves as the foundation for cooperation and is built through consistent, respectful interactions that prioritize the community's well-being. Actions that foster trust include protecting civilians during military operations, providing humanitarian assistance, and treating individuals with respect, regardless of their background. For instance, the military's ability to protect civilians during operations is paramount; failure to do so can lead to a significant erosion of trust within the community [418]. Furthermore, humanitarian efforts, such as medical aid and infrastructure repair, are essential in demonstrating a genuine commitment to the community's welfare, thereby enhancing trust [419].

Credibility is equally important in this context. Security forces must fulfill their promises to the local population; for example, if they assure protection in exchange for cooperation, any failure to deliver on this promise can lead to disillusionment and a loss of support from the community [420]. Studies have shown that public trust in military forces can fluctuate based on their perceived effectiveness and the fulfillment of their commitments [421]. In post-authoritarian societies, trust in the military is often influenced by the public's overall satisfaction with political processes and their knowledge of security matters [418]. Thus, a lack of transparency or perceived ineffectiveness can significantly diminish trust, highlighting the importance of accountability and consistent communication [422].

Moreover, the dynamics of trust-building are complex and influenced by various factors, including the community's historical experiences with security forces and the broader socio-political context [423]. Trust can be reinforced through repeated positive interactions and the establishment of a reliable presence within the community [424]. For instance, military operations that prioritize civilian safety and demonstrate respect for local customs can enhance the credibility of security forces, fostering a more cooperative relationship with the population [421].

Cultural understanding and sensitivity are vital components in effective relationship-building, particularly in contexts involving military personnel and local populations. A deep comprehension of local cultures, traditions, and norms is essential to avoid missteps that could alienate communities and hinder intelligence efforts. Misunderstanding or disrespecting cultural practices can lead to significant barriers in communication and cooperation, which are crucial for successful operations in diverse environments [425]. Training military personnel to engage with various communities requires an emphasis on cultural intelligence (CQ), which is defined as the capability to function effectively in culturally diverse situations [425]. This training should include awareness of one's own cultural biases and how they may affect interactions with others [426, 427].

In societies characterized by strong tribal or clan structures, engaging with community leaders or elders is often the most effective strategy for gaining acceptance and access to critical information [428]. Recognizing the influence of religious or cultural figures can also facilitate dialogue and cooperation, enhancing the effectiveness of military operations [429]. The importance of cultural intelligence is underscored by research indicating that individuals with higher CQ are better equipped to navigate complex cultural landscapes, leading to improved communication and relationship-building [425].

Figure 63: U.S. Army Engineers build upon the San Rafael School as part of the Beyond The Horizon Operation in San Rafael, Guatemala, April 25, 2016. Museo Nacional del Prado, Public Domain, via National Archives and Defense Visual Information Distribution Service.

Establishing structured community engagement programs is another effective strategy for bridging the gap between security forces and local populations. Such programs can include regular meetings, where civilians can voice their concerns and discuss their needs, thereby fostering a sense of inclusion and trust [428]. Civil-military projects, such as building schools or improving infrastructure, demonstrate goodwill and a commitment to the community's welfare, which can significantly enhance the relationship between military forces and local populations [430]. Furthermore, employing locals as translators or guides not only provides economic benefits but also deepens the community's investment in the mission's success, reinforcing the importance of cultural sensitivity in operational contexts [428].

Figure 64: U.S. Soldiers work together on a construction project at Thung Song Honsa School in Chachoengsao province, Kingdom of Thailand, Feb. 7, 2018. Chachoengsao locals joined Royal Thai, U.S. and Indonesian Armed Forces working the construction project for Exercise Cobra Gold 18. Humanitarian civic assistance projects conducted during the exercise support the needs and humanitarian interests of the Thai people. Royal collection of the United Kingdom, Public Domain, via National Archives and Defense Visual Information Distribution Service.

Feedback and grievance redressal mechanisms are essential components of these community engagement programs, ensuring that civilian concerns are addressed promptly and fairly [428]. This two-way approach to communication promotes trust and commitment, which are critical for effective collaboration between military forces and civilian communities [428]. Ultimately, the integration of cultural understanding and sensitivity into military training and operations can lead to more effective engagement with local populations, fostering cooperation and enhancing overall mission success.

The establishment and management of Human Intelligence (HUMINT) networks within local populations are critical for military operations, particularly in conflict zones. These networks rely on informants who provide valuable information about enemy activities, necessitating a careful approach to ensure their safety, anonymity, and motivation.

To protect informants from potential retaliation, military forces must implement robust measures that prioritize their safety. This includes creating a secure environment where informants can communicate without fear of exposure. Moffett et al. [431] emphasize the importance of rapport and trust in handling covert human intelligence sources, noting that the handler's personality traits and the informant's motivations significantly influence the effectiveness of these relationships. Additionally, providing appropriate incentives, such as financial compensation or community benefits, can enhance informants' willingness to cooperate. This aligns with findings from Hart et al. [432], which highlight the necessity of valuing contributions to foster a sense of security and commitment among informants.

Moreover, the verification of information is paramount to prevent the spread of disinformation. Cross-referencing intelligence from multiple sources helps to ensure accuracy and build credibility within the HUMINT network. The development of systems like Proppy, which monitors and analyses news sources for propagandistic content, illustrates the importance of employing technology to combat misinformation [433]. This technological approach can be complemented by traditional methods of intelligence gathering, which require a nuanced understanding of local dynamics and cultural contexts.

Proppy is an advanced platform designed to monitor and analyse news sources for propagandistic content, utilizing artificial intelligence (AI) and natural language processing (NLP) to identify and categorize potential propaganda within various information sources. In the contemporary media landscape, where disinformation campaigns and biased reporting can significantly influence public opinion and political decisions, such tools are critical for promoting media literacy and safeguarding democratic processes [433, 434].

Proppy employs sophisticated algorithms to detect various propaganda techniques, including emotional manipulation, one-sided arguments, and loaded language. This capability is supported by research indicating that effective propaganda often involves exaggeration, misinformation, or selective presentation of facts [433, 435]. The system's ability to analyse text for these signs is crucial for identifying content that may have propagandistic intent.

The platform continuously monitors a wide array of news outlets, blogs, and social media platforms, providing comparative analyses of how different sources report on the same events. This feature is essential for discerning bias and understanding the propagandistic approaches employed by various media [433]. By clustering articles based on their likelihood of containing propagandistic content, Proppy helps users identify discrepancies in reporting across sources [436].

Proppy utilizes NLP techniques to detect persuasive language patterns and rhetorical strategies commonly found in propaganda. This includes assessing the tone and sentiment of news articles to flag potential biases. Research has shown that understanding the linguistic features of propaganda can enhance the detection of biased content [437].

The tool categorizes news items based on the likelihood of containing propagandistic content, employing a scoring system to quantify the level of bias. This systematic approach allows for a nuanced understanding of how propaganda manifests in different news articles [433, 438].

Users can receive real-time alerts about emerging propagandistic trends, along with detailed reports that provide insights into propaganda strategies and thematic patterns. This capability is particularly valuable for educators and researchers aiming to foster media literacy [433, 439].

Proppy integrates with fact-checking databases to verify the accuracy of claims and cross-references information from multiple sources. This feature is vital for highlighting discrepancies and contradictions in reporting, thereby enhancing the reliability of information consumed by users [433, 434].

In urban warfare, Proppy and similar tools for monitoring and analyzing propagandistic content play a critical role in shaping military operations, safeguarding civilian populations, and countering adversarial narratives. The complexity of urban environments, where civilians and combatants are often intermingled, makes the informational battlefield as pivotal as the physical one. Proppy excels in detecting and analyzing enemy propaganda, identifying false narratives designed to instill fear, sow distrust, or provoke resistance. This intelligence enables timely countermeasures, such as disseminating accurate information through trusted local channels like community leaders or social media platforms, thereby reducing the impact of misinformation.

Proppy also helps protect civilian perceptions by monitoring the media and messaging to which local populations are exposed. By identifying harmful or manipulative narratives, it allows military forces to tailor their communication strategies to address civilian concerns, reinforcing trust and credibility. Analyzing propaganda trends, Proppy aids in crafting campaigns that emphasize the legitimacy and humanitarian goals of military operations. This bolsters the population's confidence in the mission and counters the adversary's attempts to destabilize the civilian-military relationship.

The tool enhances Psychological Operations (PSYOPS) by providing deep insights into the cultural and emotional triggers exploited by enemy propagandists. These insights enable the development of counter-narratives that demoralize enemy forces or win over the local populace. Proppy's ability to analyze themes within propaganda ensures that messaging resonates with specific demographic or cultural groups, enhancing its effectiveness. Additionally, Proppy strengthens operational security by identifying propaganda patterns that might reveal leaks or adversarial knowledge of allied strategies, enabling forces to address vulnerabilities proactively and maintain control of the informational landscape.

In multinational campaigns or coalitions, Proppy harmonizes messaging between different allied nations, ensuring consistent narratives and avoiding contradictions that adversaries could exploit. It helps assess the impact of allied communications to ensure alignment with strategic goals. By preparing troops with insights into expected propaganda trends and techniques, Proppy enhances readiness and situational awareness, making it a vital tool in pre-mission planning. In a hypothetical scenario, Proppy might identify insurgent propaganda falsely claiming that allied forces intend to destroy civilian infrastructure. In response, allied forces could deploy a multi-pronged communication strategy, leveraging Proppy's analysis to debunk these claims, reassure the population, and maintain their operational advantage.

Despite these strategies, several challenges can impede the establishment of effective HUMINT networks. Distrust and fear among civilians, stemming from past experiences with military forces or insurgents, can hinder cooperation. Chatfield et al. [440] discuss how terrorist organizations exploit social media to spread propaganda, which can exacerbate fears and reinforce existing biases against military forces. Competing loyalties in divided societies further complicate relationship-building, as individuals may feel torn between allegiance to their community and the military [441].

Communication barriers, including language differences and cultural misunderstandings, can also obstruct effective dialogue. Cultural awareness is important for military effectiveness, as it facilitates better interactions with local populations [442]. Furthermore, hostile propaganda from adversaries can undermine trust, portraying military forces as oppressors rather than protectors [443]. Addressing these challenges requires a patient and adaptable approach, emphasizing transparency and consistent engagement with local communities.

The strategic benefits of building strong relationships with local populations extend beyond intelligence gathering. Such relationships foster a supportive environment where civilians feel invested in the success of military operations, ultimately enhancing operational effectiveness. Cross-cultural competence is vital for military personnel to navigate complex social landscapes and build trust [444]. Additionally, cooperation with local populations can counter insurgent narratives, as civilians who feel valued and protected are less likely to support adversaries, thereby weakening their operational base [445].

Chapter 8

LOGISTICS AND SUSTAINMENT IN URBAN WARFARE

Supply Challenges in Urban Environments: Food, Water, Ammunition, and Medical Supplies

Logistics and sustainment in urban warfare present complex challenges, as dense, contested, and often damaged urban environments make the delivery and distribution of essential supplies a formidable task. Effective supply chain management is critical to ensuring that military forces maintain operational effectiveness and that civilian populations are not left without basic necessities. Food, water, ammunition, and medical supplies represent the core elements of logistical planning in these environments, each with unique challenges requiring careful consideration.

Food supply in urban warfare involves balancing the nutritional needs of troops with the logistical constraints of transporting and storing provisions in congested areas. Urban combat often involves prolonged engagements, requiring a steady supply of rations to sustain soldiers operating under high-stress conditions. The risk of supply convoys being ambushed or delayed is heightened in urban environments, where adversaries can exploit narrow streets and obstructed visibility. To mitigate these risks, forces often rely on prepositioned stockpiles, compact and high-calorie ration packs, and decentralized distribution networks that minimize the movement of large, vulnerable convoys.

Figure 65: Marines with Combat Logistics Battalion 2, Combat Logistics Regiment 15, unload supplies during their 32-hour combat logistics patrol, Oct. 24, 2012. CLB-2 Marines have operated in Helmand province for the last three months with few incidents. Defense Visual Information Distribution Service, Public Domain, via GetArchive.

Food security in urban warfare contexts is particularly precarious. Urban areas often rely on complex supply chains that can be disrupted by conflict, leading to shortages and increased prices for essential goods. Research indicates that urban agriculture can play a vital role in enhancing food security by shortening supply chains and providing fresh produce directly to urban populations [446]. However, the effectiveness of urban agriculture is often limited by the availability of land and resources, which can be further strained during conflicts [447]. Additionally, the planning and regulation of food distribution become paramount, as vulnerable populations may face barriers to accessing food due to logistical challenges or infrastructural damage [447].

From a technical perspective, food logistics rely on durable, lightweight, and nutrient-dense ration packs designed to meet soldiers' caloric and dietary needs in austere conditions. Meals Ready-to-Eat (MREs) are commonly used, as they are vacuum-sealed to ensure long shelf life and resistance to harsh environmental conditions. They often include chemical heaters that activate with water, allowing troops to have hot meals without needing cooking equipment.

Strategically, food supplies are pre-positioned at forward operating bases (FOBs) or secured urban areas to minimize the need for transportation during combat operations. Supply routes are planned to avoid contested areas, leveraging intelligence on enemy activity to protect convoys. In urban conflicts, where roads are often obstructed or heavily monitored by adversaries, decentralized distribution networks and drones are increasingly used to deliver rations directly to troops in isolated positions. Collaboration with local suppliers can also supplement military rations while supporting community relations.

Water logistics pose an equally critical challenge, as access to potable water is vital for troop health and hygiene. Urban combat often disrupts municipal water systems, rendering traditional supply lines unreliable. Forces must address these disruptions by transporting water into the battlefield or employing purification systems to treat available sources. Mobile water purification units, which can filter and disinfect water from contaminated or damaged urban reservoirs, are a valuable asset in these scenarios. The need for large quantities of water further complicates logistics, as water is heavy and voluminous, making its transportation particularly burdensome in densely built-up areas.

Figure 66: A tactical water purification system or (TWPS) set. Defense Visual Information Distribution Service, Public Domain, via Picryl.

Technically, water logistics incorporate mobile purification systems capable of treating contaminated sources. These include reverse osmosis units, ultraviolet sterilization systems, and chemical treatments that ensure potable water. Lightweight hydration systems, such as camelbacks, enable soldiers to carry water efficiently, while larger storage systems, like collapsible water bladders, are used at bases.

Strategically, water sources are secured early in operations to reduce dependency on transported supplies. Engineers often repair or safeguard municipal water systems in urban settings, or establish new wells and purification stations. Planners prioritize water points based on proximity to combat zones and the expected consumption rates of personnel. Airlifting water via helicopters or using convoys with armoured tankers ensures delivery in areas with compromised road access. Real-time monitoring of water usage and supply levels is critical to prevent shortages, especially during prolonged engagements.

Ammunition supply chains in urban warfare are also fraught with challenges. The dense urban environment complicates the movement of military supplies, requiring careful planning and execution to ensure that troops have the necessary resources to sustain operations. The reliance on local logistics networks can lead to vulnerabilities, especially if these networks are disrupted by enemy action or civilian interference [17]. Moreover, the need for rapid resupply in urban combat scenarios necessitates the development of decentralized logistics strategies that can respond quickly to changing battlefield conditions [448].

Ammunition supply in urban warfare demands a highly responsive logistics network. The close-quarters nature of urban combat often leads to higher-than-average ammunition consumption, as engagements are frequent and intense. Resupplying frontline units can be perilous due to the proximity of enemy forces and the difficulty of manoeuvring supply vehicles through narrow, rubble-filled streets. To overcome these challenges, modern forces employ innovative methods such as drone-based delivery systems, which can rapidly transport small quantities of ammunition to isolated or heavily engaged units. Additionally, prepositioned ammunition caches within secure urban strongholds can reduce reliance on resupply convoys.

Ammunition logistics are technically demanding due to the diversity of munitions required, including small arms rounds, grenades, mortar shells, and specialized weapons. Smart inventory systems using RFID tagging and barcoding track ammunition stocks and facilitate rapid resupply. Vehicles and containers designed for safe storage and transport prevent accidental detonations and simplify distribution.

Strategically, ammunition caches are established in secure, forward positions to reduce the need for frequent resupply missions. Combat units are equipped with "basic loads" calculated based on expected engagement intensity, and reserves are allocated at higher levels of command. In high-risk areas, drones and unmanned ground vehicles (UGVs) are increasingly employed to deliver small loads of ammunition directly to units under fire. Real-time communications between logistics hubs and combat units ensure that resupply missions are prioritized based on operational urgency.

Figure 67: A weapons cache of Rocket Propelled Grenades (RPG), mortar rounds and various munitions, found near A'Namaneah, Iraq, by US Marine Corps (USMC) personnel with Regimental Combat Team 1 (RCT 1), in support of Operation IRAQI FREEDOM. . The U.S. National Archives, Public Domain, via National Archives and Defense Visual Information Distribution Service.

Medical supplies are essential for treating combat injuries and preventing disease outbreaks among troops and civilians. Urban warfare often results in high casualty rates due to the density of combat and the increased likelihood of collateral damage. The destruction of urban infrastructure can exacerbate medical challenges, as hospitals and clinics may be damaged or inaccessible. Military medical units must operate close to the frontlines, setting up field hospitals in secure locations while ensuring a steady flow of medical supplies, such as bandages, surgical kits, and medications. Evacuating wounded personnel to higher-level care facilities is another critical concern, often requiring secure routes and air or ground transport capabilities. The integration of military and civilian medical resources can enhance the effectiveness of medical logistics, but this requires careful coordination and planning to avoid exacerbating existing vulnerabilities in urban healthcare systems [449]. Collaboration with humanitarian organizations and NGOs facilitates access to additional medical resources and expertise when civilian casualties overwhelm military medical capabilities.

Portable field hospitals and modular medical facilities are designed for rapid deployment and scalability, providing essential care close to the frontline. Cold chain technologies ensure the preservation of temperature-sensitive items like vaccines and blood. Strategically, medical supply chains are integrated into overall operational planning to ensure casualty care aligns with the tempo of combat operations. Pre-positioning medical supplies at casualty collection points and stabilization units reduces the time required to treat and evacuate the wounded. Evacuation plans include securing routes for ground ambulances or using helicopters for aerial medevac missions.

Figure 68: An aerial view of a field hospital erected during Exercise WOUNDED EAGLE '83. The exercise is designed to test the Civilian-Military Contingency Hospital System, a program for transferring wartime casualties to civilian hospitals in the event they overflow military facilities. The U.S. National Archives, Public Domain, via Picryl.

The interconnected nature of urban infrastructure adds further complexity to logistics and sustainment. Damaged roads, bridges, and supply depots can disrupt entire logistics chains, necessitating alternative delivery methods and routes. Forces must also contend with the presence of civilian populations, whose needs for food, water, and medical care must be balanced against military priorities to maintain ethical standards and comply with international humanitarian law. This dual burden requires careful coordination with humanitarian organizations and local authorities to avoid exacerbating civilian suffering.

Ultimately, successful logistics and sustainment in urban warfare depend on adaptability, innovation, and meticulous planning. Forces must employ a combination of traditional supply chain methods and advanced technologies, such as unmanned delivery systems, portable water purification units, and mobile medical facilities. By addressing these logistical challenges proactively, military operations in urban environments can maintain operational momentum while minimizing harm to civilian populations.

Resupply Strategies, Field Hospitals, and Casualty Evacuation

Resupply Strategies

Resupply strategies are critical for maintaining operational readiness in military engagements, particularly in remote and hostile environments. These strategies encompass various methods to ensure personnel have access to essential supplies, including food, water, ammunition, and equipment, while minimizing risks to supply chains.

Aerial resupply is a vital method that utilizes helicopters, cargo planes, or drones to deliver supplies directly to forward operating bases (FOBs) or isolated units. This approach is especially effective in urban or mountainous terrains where ground routes may be compromised. The use of precision delivery systems, such as air-dropping supplies or landing them directly, helps to minimize collateral damage and loss during operations [450].

Figure 69: Bundles of bottled water attached to parachutes fall from an aircraft during an aerial resupply on Combat Outpost Herrera, Paktiya province, Afghanistan, Oct. 15, 2009. The U.S. Army, CC BY 2.0, via Wikimedia Commons.

Recent studies have highlighted the increasing reliance on unmanned aerial vehicles (UAVs) for medical logistics and blood product transport in combat zones, showcasing their potential for rapid and efficient resupply in austere environments [450]. Furthermore, the development of autonomous aerial supply delivery systems, such as the Squad Operations Advanced Resupply (SOAR) UAS, demonstrates a novel solution for small unit resupply in contested areas [451].

Figure 70: A joint tactical autonomous air resupply systems (JTAARS) carries a small package during the Manoeuvre Fires Integrated Experiment (MFIX) at Fort Sill. Defense Visual Information Distribution Service, Public Domain, via National Archives and Defense Visual Information Distribution Service.

Ground-based resupply remains the backbone of military logistics, relying on armoured vehicles and convoy strategies to transport supplies. These convoys are heavily guarded, with escorts providing overwatch to mitigate potential threats. The planning of routes incorporates real-time intelligence to avoid ambushes, ensuring the safety of personnel and supplies [452]. In urban settings, smaller and more agile vehicles are often employed to navigate narrow streets and circumvent blockades, enhancing the effectiveness of ground resupply operations [452]. The integration of automatic ammunition resupply systems for artillery has also been noted as a significant advancement in logistics, allowing for rapid replenishment and improved combat effectiveness [452].

Strategically placed stockpiles in secured locations near expected operational areas are essential for reducing dependency on continuous supply lines. These caches allow for quick access to critical resources, thereby enhancing operational efficiency. The establishment of redundancy within this system is crucial, as it accounts for the potential loss of one or more supply points, ensuring that operations can continue uninterrupted [452].

In complex operational environments, decentralized resupply empowers small units to procure resources autonomously. This can involve local sourcing, collaboration with allied forces, or utilizing drones and autonomous ground vehicles for last-mile delivery [452]. The decentralized approach not only enhances operational flexibility but also allows for a more responsive logistics framework that can adapt to changing battlefield conditions [452].

Effective resupply strategies are underpinned by robust policies and processes. Demand forecasting plays a critical role in predicting the needs of troops based on mission profiles, environmental conditions, and combat intensity, ensuring that resupply efforts are neither excessive nor inadequate [452]. Additionally, supply chain security is paramount; protecting supply routes with layered defences—including patrols, checkpoints, and counter-IED measures—mitigates risks associated with resupply operations [452]. Continuous monitoring of logistics personnel using real-time data to track inventory and supply movements allows for dynamic adjustments to plans in response to delays or losses, further enhancing the resilience of supply chains [452].

Field Hospitals

Field hospitals serve as critical mobile medical facilities designed to deliver rapid, high-quality healthcare in proximity to conflict zones. Their primary function is to bridge the gap between immediate point-of-injury care and comprehensive hospital treatment, ensuring that casualties receive timely stabilization and life-saving interventions. The design and deployment of these facilities are tailored to meet the urgent needs of military and civilian populations in conflict areas, reflecting a strategic approach to emergency medical care.

Field hospitals are characterized by their modular and scalable design, which allows for swift deployment in various operational contexts. They are typically established near transportation hubs or casualty collection points to ensure quick access for injured individuals. This strategic positioning facilitates the efficient transfer of patients from the battlefield to medical care. The layout of field hospitals includes dedicated sections for triage, surgery, intensive care, and recovery, which enhances their operational flexibility and responsiveness to varying medical needs in crisis situations [453, 454].

A key feature of field hospitals is their triage systems, which prioritize patients based on the severity of their injuries and the likelihood of survival. This systematic approach ensures optimal resource allocation and effective management of medical supplies, which is crucial in high-stress environments [455]. Surgical units within these hospitals are equipped with advanced medical technologies to address trauma cases, including gunshot wounds, burns, and blast injuries. Furthermore, field hospitals are designed to provide specialized care, such as psychological support, infectious disease control, and maternity care, thereby addressing the diverse needs of affected populations [456, 457].

Effective resource allocation is vital in field hospitals, where supplies such as medical equipment, blood, and pharmaceuticals are meticulously managed to prevent shortages and wastage. Integration with local healthcare systems is also essential, particularly in urban warfare scenarios, where collaboration with civilian hospitals and medical professionals can enhance the overall care provided to affected populations [458, 459]. Additionally, field hospitals play a crucial role in evacuation coordination, acting as stabilization points that prepare patients for transfer to larger medical facilities for further treatment [454, 460].

Casualty Evacuation

Casualty evacuation (CASEVAC) is a critical process in military and emergency medical services, focusing on the timely transportation of injured personnel from the battlefield to medical facilities. This process is essential for improving survival rates among wounded soldiers and civilians, as it ensures that they receive immediate medical attention. The methods of evacuation can be categorized into aerial, ground, and marine CASEVAC, each serving specific operational needs.

Aerial CASEVAC is predominantly executed using helicopters, such as the Black Hawk and Chinook, which are vital for rapid evacuation in high-risk or inaccessible areas. The use of aerial support not only enhances the speed of evacuation but also minimizes exposure to hostile forces, thereby increasing the chances of survival for critically injured personnel [461]. Historical data from the Korean War illustrates the effectiveness of helicopter evacuations, which significantly improved survival rates compared to previous conflicts where such methods were not utilized [461]. Ground CASEVAC, on the other hand, employs armoured ambulances or multipurpose vehicles in scenarios where air access is limited. These ground routes are often secured by escort units to mitigate the risk of further casualties during transport [462]. In littoral zones, marine CASEVAC utilizes boats or amphibious vehicles, particularly in urban conflicts near waterways, highlighting the adaptability of evacuation methods to various operational environments [462].

Figure 71: Marines of Regimental Combat Team 5, transport a non-ambulatory patient via litter, outside of Fallujah, Iraq in 2006. Patrick Smith. Flickr Gallery, CC BY 3.0, via Wikimedia Commons.

The policies and processes governing CASEVAC are designed to optimize the efficiency and safety of evacuations. Clear evacuation routes are meticulously planned to minimize exposure to enemy fire, leveraging intelligence to identify the safest paths [462]. Triage at the point of injury is critical, as it involves immediate assessment and stabilization of casualties to ensure that only those who can survive transport are evacuated [463]. Effective communication systems are also essential, enabling real-time coordination between field units, evacuation teams, and medical facilities to prioritize resources and streamline the evacuation process [464].

Strategically, the integration of technology plays a significant role in enhancing CASEVAC operations. Advanced tracking systems allow for real-time monitoring of casualties and the availability of evacuation assets, facilitating swift decision-making [465]. Furthermore, CASEVAC teams are embedded within combat operations to ensure rapid responses during engagements, underscoring the importance of preparedness and training for all personnel involved in the evacuation process [462]. Rigorous training protocols for medics, pilots, and other involved personnel are crucial to ensure readiness under combat conditions, which is vital for maintaining operational effectiveness [462].

Establishing Secure Supply Lines in Densely Populated Areas

Establishing secure supply lines in densely populated areas is a critical logistical challenge in modern warfare. Urban environments present unique obstacles, such as restricted movement, civilian congestion, potential adversary ambushes, and infrastructural limitations. To ensure the uninterrupted flow of essential supplies like food, water, medical resources, and ammunition, comprehensive planning and adaptive strategies must be employed.

The first step in establishing secure supply lines is the meticulous planning of routes. Urban areas often have a labyrinth of narrow streets, overpasses, and subterranean pathways, making navigation complex. Planners must use advanced mapping technologies, such as satellite imagery, drone surveillance, and ground reconnaissance, to chart routes that avoid chokepoints, heavily populated zones, and areas controlled by adversaries. Redundant routes are essential to mitigate the risks of blockades or ambushes. These alternative pathways ensure that supplies can still reach their destination if the primary route is compromised.

Supply lines in urban warfare require robust security to safeguard them against enemy attacks and looting. Armoured convoys are often used to transport critical supplies, with security personnel providing perimeter defence. Convoys are typically escorted by quick-response units, drones, or reconnaissance vehicles that scout ahead to identify threats such as ambushes, improvised explosive devices (IEDs), or sniper positions. In especially high-risk areas, supply missions might rely on night operations to minimize visibility and reduce the risk of enemy engagement.

Checkpoints are established along the route to monitor and secure critical junctures. These checkpoints serve as hubs for quick repairs, medical assistance, or temporary stockpiles. Soldiers manning the checkpoints are equipped with surveillance technology and communication systems to maintain situational awareness and relay information back to command centres.

In densely populated areas, managing civilian interactions is a crucial aspect of securing supply lines. The presence of civilians can both complicate logistics and provide opportunities for intelligence gathering. Military personnel must establish trust with the local population by communicating their intentions clearly, minimizing collateral damage, and providing humanitarian assistance where feasible. Creating designated civilian-free zones along supply routes can reduce congestion and enhance security.

Civil-military cooperation is vital to address community concerns and mitigate disruptions to daily life. Local authorities and leaders are often engaged to facilitate coordination and ensure that supply activities are not perceived as hostile. This collaboration helps prevent sabotage or interference from the civilian population, some of whom may sympathize with adversaries.

Technology plays a pivotal role in establishing secure supply lines in urban settings. Autonomous vehicles and drones are increasingly used for last-mile delivery, bypassing physical obstacles and reducing risks to personnel. These technologies can navigate through narrow streets or even deliver supplies directly to buildings under siege. Real-time communication systems allow logistics teams to track supply movements, predict potential threats, and adapt to changes in the environment dynamically.

Cybersecurity measures are also critical, as adversaries may attempt to disrupt supply operations through hacking or electronic warfare. Protecting communication networks and logistical databases ensures that supply lines remain operational and resilient against sabotage.

Supply line security is intrinsically linked to the success of broader military operations. Commanders must integrate logistics planning with tactical and strategic goals to ensure seamless coordination. For example, military units advancing into an urban area may secure key infrastructure, such as roads and bridges, as part of their mission to facilitate the movement of supplies. Similarly, aerial surveillance and intelligence gathering are coordinated with supply missions to pre-emptively neutralize threats.

Urban warfare often involves rapidly changing conditions, including shifting frontlines, damaged infrastructure, and fluctuating civilian dynamics. Supply lines must be adaptable, with contingency plans to address disruptions. Pre-positioning stockpiles in secure locations reduces the dependency on continuous resupply missions, while air or sea routes can be used as alternatives when ground routes are compromised.

Establishing secure supply lines, minimizing civilian harm, and ensuring effective military operations in urban warfare require an intricate understanding of the risks, strategies, and considerations inherent to densely populated and interconnected environments. The complexity of urban areas poses unique challenges that demand adaptive policies, meticulous planning, and consistent monitoring. Among the primary concerns is the increased risk of civilian harm from explosive weapons, due to factors such as weapon inaccuracies, targeting errors, and inadequate operational assessments [466].

Urban areas inherently amplify risks due to the density of population, the proximity of civilians to military targets, and the interconnected nature of urban infrastructure. Random and systemic errors in weapon systems can lead to unintended consequences, including collateral damage to critical infrastructure like hospitals, water supplies, and communication networks. Wide-area effects, especially from weapons with large payloads, exacerbate this risk when used in environments where civilians and military targets are in close proximity. Furthermore, urban environments often feature complex underground and multi-dimensional structures that challenge accurate targeting and heighten the risk of harm to non-combatants [466].

Strategically addressing these risks requires integrating the protection of civilians into the core operational framework. Military planners must ensure that protection mandates are explicitly outlined, encompassing both immediate response strategies and longer-term recovery objectives. Effective coordination between military forces, humanitarian organizations, and NGOs is vital to creating a unified approach that prioritizes civilian safety. Arms control measures, such as safe disposal of munitions, education on explosive risks, and the regulation of weapons use, further enhance efforts to mitigate civilian harm. These policies must be reinforced by rigorous dialogue between political and military leaders to align resourcing with operational objectives [466].

On the ground, securing urban supply lines and reducing civilian harm involves addressing several logistical and operational challenges. Military forces must assess the terrain's three-dimensional nature, account for the interplay of physical and social systems, and develop pre-emptive strategies to address adversary tactics such as using human shields or embedding within civilian populations. This necessitates comprehensive training for troops, focusing on urban warfare dynamics, the effects of explosive weapons, and compliance with International Humanitarian Law (IHL). Training facilities must simulate urban conditions to prepare forces for real-world challenges, emphasizing the need for pre-deployment assessments and in-theatre evaluations [466].

Partnering with non-state armed groups (NSAGs) and host governments adds layers of complexity. Effective engagement requires clear policies to ensure compliance with IHL, risk assessments of partners, and strategies for demobilization and reintegration. Outreach programs promoting IHL adherence can foster cooperation, reduce harm, and enhance operational effectiveness. Establishing secure supply lines in urban environments also involves creating mechanisms for real-time civilian feedback, facilitating the evacuation of non-combatants, and developing adaptable strategies to maintain critical infrastructure and essential services during conflict [466].

Ultimately, addressing the challenges of urban warfare requires an iterative approach that combines proactive planning, robust training, adaptive policy-making, and constant evaluation. By embedding the protection of civilians into operational frameworks and leveraging collaborative efforts across military and civilian domains, the complexities of urban conflict can be navigated more effectively, ensuring both operational success and the minimization of harm to non-combatants [466].

Vehicle and Equipment Maintenance Challenges in Urban Combat

Vehicle and equipment maintenance in urban combat environments presents a myriad of challenges that are distinct from those encountered in more traditional warfare settings. The urban landscape, characterized by dense infrastructure, narrow streets, and multi-story buildings, creates a complex and dynamic battlefield that significantly stresses military vehicles and equipment. The need for effective maintenance in these conditions is paramount for sustaining operational readiness, ensuring troop safety, and maintaining the momentum of military operations [5, 36].

One of the primary challenges faced in urban combat is the physical wear and tear on vehicles and equipment. The confined nature of urban terrain often necessitates frequent sharp turns, rapid accelerations, and decelerations, as well as navigation over rubble and uneven surfaces. Such conditions increase the risk of damage to critical components like tires, suspensions, and undercarriages. For instance, armoured vehicles may experience excessive strain on their tracks or wheels, and their engines are often pushed to their limits due to stop-and-go movements typical in urban warfare [467, 468]. The presence of sharp objects and collapsed infrastructure can lead to punctured tires and damaged hydraulic systems, necessitating immediate repairs to maintain operational capability [468].

Moreover, the threat posed by improvised explosive devices (IEDs) and other urban-specific hazards complicates maintenance efforts further. IEDs are frequently concealed within urban debris, posing a constant risk to vehicles and equipment. Even armoured vehicles, designed to withstand explosions, can suffer damage from shockwaves and shrapnel, which can adversely affect sensitive electronics and communication systems [5, 368]. The volatile nature of urban battlefields often limits access to secure maintenance facilities, making frequent inspections and repairs challenging [5, 36].

Logistical constraints also represent a significant hurdle in urban combat. Military units often find themselves isolated from their main logistical hubs, complicating the transportation of spare parts and specialized repair equipment. Narrow streets and blocked pathways can impede supply convoys, while ongoing enemy fire may force units to abandon damaged vehicles [467, 468]. Furthermore, the need for rapid mobility in urban operations reduces the time available for thorough inspections and repairs, increasing the risk of minor issues escalating into critical failures [468].

Figure 72: A U.S. Marine Corps Motor Transport Mechanic with 3rd Battalion, 1st Marine Regiment, assigned to Special Purpose Marine-Air Ground Task Force – Crisis Response - Central Command, removes an old 12-volt battery for a Mine-Resistant Ambush-Protected vehicle in Kuwait, Nov 07, 2020. Defense Visual Information Distribution Service, Public Domain, via Picryl.

The maintenance of advanced technologies in military vehicles and equipment adds another layer of complexity. Modern military systems often incorporate sophisticated electronics, sensors, and communication systems that are vulnerable to electromagnetic interference, dust, and moisture. For example, GPS systems may malfunction due to signal disruptions caused by dense urban structures, while sensors can become obstructed or damaged by debris [5, 368]. Maintaining these systems requires specialized knowledge and equipment, which may not always be readily available in forward-operating areas [469].

Figure 73: US Marine Corps (USMC) Maintenance Mechanics assigned to Combat Service Support Battalion (CSSB) 18, 1ST Force Services Support Group (FSSG), work to repair a MK-23 Medium Tactical Vehicle Replacement (MTVR), 7-ton cargo truck, at the Logistics Support Area (LSA), located at Camper Viper, Iraq, during Operation IRAQI FREEDOM. The U.S. National Archives, Public Domain, via National Archives and Defense Visual Information Distribution Service.

To address these challenges, militaries must adopt adaptive maintenance strategies. Deploying mobile repair teams equipped with modular repair kits allows for on-the-spot fixes, thereby reducing downtime and minimizing the need to transport vehicles to distant maintenance depots [467, 468]. Pre-positioning spare parts and repair resources in secure urban locations can enhance logistical efficiency, while utilizing drones or other reconnaissance tools can assist in identifying and assessing damage in hazardous areas [5]. Training maintenance crews to operate under combat conditions and equipping them with the skills to perform rapid repairs on advanced systems is equally essential [468].

Additionally, robust preventive maintenance programs must be integrated into urban combat strategies. Regular inspections, scheduled servicing, and proactive replacement of high-wear components can significantly reduce the likelihood of critical failures during operations [467, 468]. Effective communication between frontline units and logistical support teams is crucial for prioritizing maintenance needs and ensuring that resources are allocated effectively [467, 468].

Examples of Logistical Strategies in Prolonged Urban Sieges

Prolonged urban sieges present extreme logistical challenges, requiring innovative strategies to sustain military operations while minimizing harm to civilians. Historical and modern examples illustrate how logistics can be managed under these conditions to ensure the availability of food, water, ammunition, medical supplies, and fuel, while also addressing the complexities of urban environments.

1. The Siege of Sarajevo (1992–1996)

During the Bosnian War, Sarajevo was under siege for nearly four years. The logistics involved innovative and improvised methods to sustain the civilian population and defending forces:

- **Supply Lines via Tunnels:** The Sarajevo Tunnel, also known as the Tunnel of Hope, was constructed beneath the airport to bypass Serbian forces' blockade [470]. This tunnel enabled the transport of food, medical supplies, ammunition, and fuel into the city while evacuating civilians and the injured [471].

- **Airdrops:** International humanitarian efforts included NATO-organized airlift operations, delivering food and medical supplies. These operations faced challenges like precision in drop zones due to the urban density and hostile anti-aircraft fire [472].

- **Local Resource Mobilization:** Defenders and civilians resorted to using local resources, such as dismantling wooden structures for fuel and repurposing materials for defensive purposes, to compensate for the scarcity of external supplies [471].

2. The Siege of Leningrad (1941–1944)

During World War II, the German blockade of Leningrad (modern-day St. Petersburg) exemplified the critical role of logistics in prolonged sieges:

- **Lake Ladoga Lifeline:** The "Road of Life," a supply route over the frozen Lake Ladoga, was established to transport food, fuel, and medicine into the city during winter [473]. Despite constant aerial and artillery attacks, the route provided a vital lifeline.

- **Resource Rationing:** Strict rationing systems were implemented, prioritizing food and medical supplies for soldiers and critical civilian services [474]. This helped sustain the population through severe shortages.

- **Urban Agriculture:** Citizens were encouraged to grow vegetables and raise small livestock within the city limits, supplementing scarce food supplies and reducing dependence on external resources [475].

3. The Battle of Mosul (2016–2017)

The Iraqi Army and coalition forces faced significant logistical challenges during the prolonged urban battle to retake Mosul from ISIS:

- **Forward Operating Bases (FOBs):** Small, secure FOBs were established close to combat zones to reduce the time and risk associated with transporting supplies. These bases acted as hubs for distributing food, water, and ammunition to frontline units.

- **Air-Dropped Supplies:** Coalition forces employed helicopters and drones to deliver critical supplies to isolated units in the densely populated and heavily contested areas of western Mosul [476].

- **Coordination with Civilian Agencies:** Humanitarian corridors were established to evacuate civilians and provide food, water, and medical aid to displaced populations, ensuring that military supply lines remained focused on operational needs [477].

4. The Siege of Aleppo (2012–2016)

In the Syrian Civil War, the siege of Aleppo highlighted the use of both traditional and unconventional logistical strategies by various factions:

- **Tunnelling for Supplies:** Rebel forces used tunnels to smuggle food, weapons, and reinforcements into besieged areas. These tunnels also served as evacuation routes for wounded fighters and civilians [478].

- **Air Supply Drops:** The Syrian government utilized helicopters and aircraft to deliver supplies to isolated government-held zones. However, these efforts were often limited by contested airspace and anti-aircraft fire.

- **Negotiated Access:** Humanitarian organizations negotiated temporary ceasefires to deliver essential supplies to civilians trapped in the siege. These efforts required significant coordination with all parties involved in the conflict [479].

5. The Battle of Stalingrad (1942–1943)

The German and Soviet forces demonstrated contrasting logistical strategies during the battle, a prolonged siege within a heavily urbanized environment:

- **Soviet Supply Chains:** The Soviet defenders relied on supply lines across the Volga River to transport reinforcements, food, and ammunition into the city [480]. Despite heavy German bombardment, these routes were critical to maintaining the defence.

- **Encirclement Tactics:** The Soviets implemented a logistical strategy of encircling German forces during the counter-offensive, cutting off German supply lines and leading to severe shortages of food, ammunition, and fuel within the besieged German-held areas of the city.

- **Resource Scavenging:** Both sides scavenged materials from the urban environment, including repurposing rubble for fortifications and salvaging abandoned equipment for repairs and ammunition [481].

Key Takeaways from Logistical Strategies in Urban Sieges

1. **Adaptability:** Tunnels, air-dropped supplies, and makeshift routes exemplify the need for innovative and flexible logistical approaches to counter urban terrain and blockades.

2. **Civil-Military Coordination:** Successful operations often involved close coordination with civilian agencies and international organizations to address humanitarian needs and preserve limited military resources.

3. **Local Resource Utilization:** Urban sieges demand creative use of available resources, from urban agriculture to scavenging materials for survival and defence.

4. **Protection of Supply Lines:** Ensuring the security of supply routes, whether by air, ground, or underground, is critical to sustaining forces and civilian populations.

These examples demonstrate how logistical ingenuity and adaptability are essential for sustaining operations in prolonged urban sieges while minimizing civilian suffering.

Chapter 9

TECHNOLOGICAL INNOVATIONS IN URBAN WARFARE

Role of Modern Technology: Drones, AI, Robotics, and Cyber Capabilities

Modern technology has transformed the nature of warfare, particularly in urban combat and high-stakes operations. The integration of drones, artificial intelligence (AI), robotics, and cyber capabilities has introduced unprecedented precision, efficiency, and adaptability to military strategies. These technologies address the unique challenges posed by urban environments, enhance situational awareness, and provide new tools to mitigate risks while maintaining operational effectiveness.

Drones

Drones, also known as unmanned aerial vehicles (UAVs), have revolutionized modern warfare by providing aerial surveillance, intelligence gathering, and precision strike capabilities without endangering personnel. In urban combat, drones are used to monitor enemy movements, map complex terrains, and identify potential threats. Equipped with high-resolution cameras, thermal imaging, and sensors, drones can navigate through narrow alleys, monitor rooftops, and explore subterranean spaces, offering a comprehensive view of the battlefield.

Additionally, armed drones deliver precision strikes on high-value targets while minimizing collateral damage. They allow for engagement from a safe distance, which is critical in densely populated urban settings. Swarming drone technologies further enhance tactical options, enabling coordinated operations that overwhelm enemy defences. These capabilities make drones indispensable for both offensive and defensive strategies in modern warfare.

Unmanned Aerial Vehicles (UAVs), or drones, have evolved significantly over the past century, becoming pivotal in modern warfare. Initially utilized for reconnaissance and target acquisition, their roles have expanded to include surveillance, precision strikes, electronic warfare, and naval and ground-based operations. This transformation has been fuelled by technological advancements and operational requirements, particularly in high-intensity conflicts like the war in Ukraine. Here, drones have demonstrated their value in asymmetric warfare, disrupting traditional tactics with cost-effective and versatile capabilities. The proliferation of UAVs highlights the intersection of precision weaponry and robotics, emphasizing the need for nations like Australia to integrate these technologies into their defence strategies and procedures [482].

The war in Ukraine has particularly underscored the utility and limitations of drones in modern combat. Small drones have proven indispensable for reconnaissance and rapid artillery targeting, significantly reducing response times and increasing precision. Meanwhile, larger drones like the Bayraktar TB2, initially effective in providing intelligence, surveillance, and reconnaissance (ISR) and tactical strikes, faced challenges against sophisticated air defence systems. This has shifted focus toward smaller, cost-effective drones capable of operating below detection thresholds. Ukrainian innovations, including multi-purpose drones equipped with payloads like electronic warfare (EW) and geospatial intelligence systems, have demonstrated the need for adaptability in drone technologies to counter evolving battlefield dynamics [482].

Figure 74: Bayraktar TB2 of Ukrainian Air Force. Ministry of Defence of Ukraine, CC BY 4.0, via Wikimedia Commons.

Drones are also reshaping the strategic calculus by displacing human roles in surveillance and direct combat, reducing risks to personnel. Their ability to operate persistently and autonomously over extended periods enhances battlefield awareness and operational efficiency. For example, drones have been instrumental in detecting enemy movements, supporting precision artillery strikes, and even disrupting adversary morale through harassment. Innovations in naval drones, such as Ukraine's "Sea Baby" capable of delivering explosive payloads, have extended drone utility to maritime domains, challenging traditional naval operations and achieving significant strategic effects, such as the disruption of Russia's Black Sea fleet [482].

However, the rapid proliferation of drones has also spurred advancements in counter-drone technologies. Electronic warfare (EW), jamming systems, and counter-uncrewed aerial systems (C-UAS) have become critical in neutralizing the threat posed by UAVs. Techniques like jamming enemy communication links, deploying kinetic interceptors, and utilizing directed energy weapons are evolving to counteract the low-cost, high-impact nature of drones. Additionally, the economic implications of drone warfare are significant; the cost-efficiency of drones imposes financial burdens on adversaries reliant on expensive countermeasures. This dynamic underscores the importance of innovation, adaptation, and the development of doctrines to effectively integrate UAVs and countermeasures into modern military operations. As technology advances, future drones integrated with artificial intelligence (AI), swarming capabilities, and autonomous decision-making will further transform warfare, demanding continuous evolution in strategies and policies [482].

Artificial Intelligence (AI)

AI has become a cornerstone of modern military operations, offering tools for rapid data analysis, decision-making, and predictive modelling. In urban warfare, where the environment is dynamic and information-dense, AI enables commanders to process vast amounts of data in real time. AI-powered systems analyse satellite imagery, surveillance feeds, and open-source intelligence to identify patterns, detect anomalies, and predict enemy behaviour.

AI also enhances autonomous systems, such as drones and ground robots, enabling them to operate with minimal human intervention. These systems can perform tasks like threat detection, reconnaissance, and even logistics support, allowing human operators to focus on strategic decisions. Machine learning algorithms continuously adapt to new information, improving the accuracy of targeting and minimizing civilian harm. Furthermore, AI aids in electronic warfare by identifying and countering cyber threats, jamming enemy communications, and protecting allied networks.

Artificial Intelligence (AI) in urban warfare marks a significant transformation in military strategies and operations. The integration of advanced AI technologies enables armed forces to operate more effectively in the intricate and dynamic environments of densely populated urban areas, where conventional warfare tactics often fall short. By leveraging AI, military personnel gain enhanced situational awareness, allowing for better navigation and response to the unique challenges posed by urban combat scenarios [483].

A core strength of AI in urban warfare lies in its ability to perform real-time data analysis and predictive modelling. These capabilities empower commanders to process vast amounts of information from multiple sources, enabling precise anticipation of enemy movements and the potential impacts on civilian populations. Such insights allow military strategies to be adapted dynamically, ensuring that operations achieve their objectives while minimizing collateral damage and safeguarding civilian lives [483].

AI also plays a pivotal role in the development and deployment of autonomous systems, including drones and ground robots. These systems, powered by AI, are capable of performing diverse tasks such as surveillance, reconnaissance, and logistics with minimal human intervention. Their ability to operate independently ensures that critical missions are executed efficiently and accurately, even in the complex and often unpredictable terrain of urban settings [483].

As AI continues to evolve, its applications in urban warfare are poised to expand, reshaping the nature of military engagements. By enhancing operational precision, enabling smarter decision-making, and improving resource allocation, AI is not only revolutionizing current combat strategies but also setting the foundation for future advancements in urban military operations [483].

Military applications of artificial intelligence (AI) have become an area of intense focus, promising transformative impacts on operational efficiency, strategic decision-making, and combat readiness. In the last year, AI capabilities have advanced significantly, especially in generative AI technologies like large language models (LLMs). These advancements have democratized AI access, enabling not just the U.S. military but also potential adversaries to leverage cutting-edge technology. As AI redefines the nature of military dominance, shifting the emphasis from sheer manpower to algorithmic prowess, it becomes imperative for the U.S. military to adapt, innovate, and integrate AI into its operations while addressing ethical, security, and regulatory challenges [484].

Artificial intelligence, defined as the development of systems capable of performing tasks traditionally requiring human intelligence, has permeated various military domains. AI's evolution, particularly in natural language processing (NLP) and computer vision, has expanded its application from routine tasks to highly complex operations. NLP advancements enable machines to comprehend and respond to human language, while computer vision breakthroughs enhance image and video analysis. These developments, combined with AI's capabilities in autonomous systems, strategic analysis, and combat simulation, position it as a cornerstone of modern military operations [484].

The U.S. military has long been a pioneer in adopting AI, initially for data processing and battlefield simulations. Over time, AI has expanded to encompass diverse roles such as drone operations, threat monitoring, and cybersecurity. AI's ability to rapidly analyse vast datasets, predict outcomes, and optimize strategies has made it indispensable for tasks like target recognition and decision-making under pressure. However, these advancements also bring challenges, particularly as adversaries gain access to similar technologies, emphasizing the need for robust countermeasures and ethical oversight [484].

AI's military applications span a wide spectrum. In warfare systems, AI integrates into weapons, sensors, and navigation systems, enhancing operational precision and reducing human error. For example, the Pentagon's updated autonomous weapons policy underscores the safe and ethical use of AI-driven weaponry. Drone swarms, inspired by natural systems like bee colonies, exemplify AI's potential in autonomous coordination, enabling dynamic, multi-agent operations with minimal human intervention. These swarms can relay critical information, adapt to evolving threats, and optimize mission outcomes [484].

Strategic decision-making is another area where AI excels. By aggregating and analysing data from diverse sources, AI supports military commanders in making informed decisions under high-stress conditions. Generative AI enhances this process by creating simulations, identifying patterns, and presenting actionable insights in a conversational format. Similarly, in data processing and research, AI streamlines the analysis of large datasets, extracting relevant information to inform strategy and intelligence. Its ability to uncover hidden patterns and simulate scenarios equips military leaders with a comprehensive understanding of complex situations [484].

AI also revolutionizes combat simulation and training, providing realistic, risk-free environments for soldiers to hone their skills. Advanced simulations replicate battlefield conditions, allowing for strategic experimentation and skill enhancement without the associated dangers. Furthermore, AI's role in target recognition improves precision in identifying threats, while threat monitoring bolsters situational awareness through real-time analysis of operational environments [484].

In cybersecurity, AI offers robust defence mechanisms against sophisticated cyberattacks. It identifies vulnerabilities, detects anomalies, and predicts potential threats, ensuring the integrity of critical systems. AI's capabilities in transportation and logistics further enhance military efficiency by optimizing routes and pre-emptively addressing operational challenges. Lastly, in casualty care and evacuation, AI assists medics by providing real-time analysis and treatment recommendations based on extensive medical datasets, helping save lives in high-stress environments [484].

While AI holds transformative potential, it also introduces significant challenges. The accessibility of generative AI to adversaries increases the risk of malicious uses, such as the creation of deceptive content or cyber warfare tactics. Ethical concerns also arise, particularly regarding autonomous weapons and AI's role in decision-making. Ensuring transparency, accountability, and alignment with international humanitarian laws is critical as AI systems take on more responsibilities in combat and strategic planning.

Another challenge lies in the speed of AI advancements, which often outpace regulatory frameworks. The military must proactively address these gaps by developing comprehensive policies that govern AI's use, emphasizing safety, security, and ethical compliance. Investments in counter-AI measures, including robust electronic warfare capabilities and AI-driven defence systems, are essential to maintain a competitive edge[484].

AI's trajectory in military applications points toward increasing integration and sophistication. The development of artificial intelligence-enabled drones, swarming technologies, and fully autonomous systems highlights the potential for AI to reshape military doctrines and tactics. Advances in multimodal AI, which combines text, image, audio, and video inputs, will further enhance operational capabilities, enabling seamless human-AI collaboration [484].

Incorporating AI into military operations requires a balanced approach, leveraging its strengths while addressing its limitations. As nations like the United States invest in AI research and development, maintaining a technological edge will depend on continual innovation, ethical stewardship, and strategic foresight. The lessons learned from AI's current applications will inform its future integration, ensuring that it serves as a force multiplier for U.S. military objectives while safeguarding against emerging threats [484].

Artificial intelligence (AI) and machine learning are poised to become increasingly integrated with other cutting-edge technologies, such as quantum computing, advanced sensors, and the Internet of Things (IoT). Quantum computing holds the potential to exponentially enhance processing power, allowing AI algorithms to analyse vast datasets and tackle highly complex scenarios with unprecedented speed and precision. Meanwhile, advanced sensors and IoT devices will provide a constant influx of real-time data, significantly improving the accuracy and situational relevance of AI-driven intelligence and decision-making systems [483].

As these technologies become more ingrained in military operations, they will bring forth critical ethical and operational challenges. Among these are concerns about bias in AI algorithms, which could lead to flawed decisions, as well as the necessity for robust cybersecurity measures to protect against adversarial threats. Additionally, the deployment of autonomous weapons systems raises profound ethical questions, particularly regarding their alignment with international laws and moral principles [483].

To address these challenges, it will be essential to ensure that AI systems are designed and implemented with transparency and accountability. Building safeguards to align AI applications with ethical standards and legal frameworks will be a foundational requirement. By navigating these issues carefully, the integration of AI with emerging technologies can drive transformative advances while maintaining operational integrity and adherence to global norms [483].

Robotics

Robotics has introduced a new dimension to urban warfare, providing tools to navigate hazardous terrains, disarm explosives, and support logistics. Ground robots equipped with cameras, sensors, and manipulator arms are used for reconnaissance in areas too dangerous for human soldiers, such as tunnels, collapsed buildings, or heavily mined zones. These robots can detect and neutralize improvised explosive devices (IEDs), reducing risks to personnel.

In combat scenarios, armed robotic systems offer firepower support and can operate in swarms to suppress enemy positions. Logistics robots, including autonomous vehicles, streamline supply chains by delivering ammunition, food, and medical supplies to frontline units in urban environments where traditional transportation methods may be hindered. The use of robotics not only enhances operational efficiency but also reduces the risk to human life in high-threat areas.

Ground robots represent a transformative leap in military technology, significantly enhancing operational efficiency in modern warfare, particularly in urban environments. These autonomous or remotely operated machines are designed to navigate complex and hazardous terrains while performing a wide range of tasks, including reconnaissance, surveillance, and bomb disposal. Their ability to operate in dense and unpredictable urban settings makes them invaluable for enhancing situational awareness and mission effectiveness.

Robots like the iRobot PackBot have proven indispensable in scenarios where human soldiers face significant risks. Equipped with advanced sensors and cameras, these machines can gather and relay real-time intelligence to command centers, enabling informed decision-making. By performing reconnaissance and handling hazardous tasks, these robots reduce the need for human presence in dangerous areas, thereby minimizing casualties and protecting soldiers from harm.

Figure 75: U.S. Marine Corps Sgt. Brett Hughes, an explosive ordnance disposal (EOD) technician with 2nd EOD Company, 8th Engineer Support Battalion, prepares an irobot Packbot for reconnaissance during an exercise on Camp Lejeune. Defense Visual Information Distribution Service.

Another notable example is the Boston Dynamics Spot robot, which demonstrates exceptional versatility and adaptability. Capable of traversing rough and uneven terrain, Spot can undertake missions such as perimeter security and environmental monitoring. Its deployment in various operational scenarios highlights the integration of AI and robotics in urban warfare, allowing ground forces to extend their reach and effectiveness without exposing personnel to direct threats.

As conflicts increasingly shift to urbanized landscapes, the use of ground robots is set to grow. Their ability to augment human capabilities while taking on high-risk tasks underscores their importance in evolving military strategies. Ground robots are not only reshaping battlefield dynamics but also redefining the role of technology in ensuring operational success and safeguarding human lives in the complex arenas of modern warfare.

Cyber Capabilities

Cyber capabilities have become integral to modern military operations, particularly in urban warfare, where the digital landscape often intersects with the physical domain. Cyber tools are used to protect critical infrastructure, secure communications, and disrupt enemy networks. Offensive cyber operations can target enemy command and control systems, disable surveillance networks, or manipulate adversarial communications to create confusion and misdirection.

Cyber capabilities also play a significant role in information warfare. By infiltrating enemy systems, military forces can spread disinformation, erode morale, and influence public opinion. Conversely, defensive cyber strategies safeguard against hacking attempts, protect sensitive data, and ensure the integrity of military operations. The integration of cyber capabilities with AI further enhances their effectiveness, enabling real-time threat detection and automated responses to cyberattacks.

Integrating Modern Technologies for Strategic Advantage

The synergy between drones, AI, robotics, and cyber capabilities creates a powerful toolkit for modern military forces. These technologies complement each other, enhancing situational awareness, operational precision, and adaptability in urban environments. For example, drones equipped with AI-powered sensors can identify enemy positions, while robotic systems neutralize threats on the ground. Meanwhile, cyber tools secure communication channels and disrupt enemy operations, ensuring mission success.

The adoption of these technologies also emphasizes the importance of ethical considerations and compliance with international humanitarian law. As modern technology continues to evolve, military forces must balance operational objectives with the need to protect civilian populations and infrastructure. By leveraging drones, AI, robotics, and cyber capabilities, militaries can navigate the complexities of urban warfare while minimizing risks and maximizing strategic outcomes.

The integration of decentralized artificial intelligence (AI) with emerging technologies like 5G, the Internet of Things (IoT), and edge computing represents a groundbreaking transformation in military capabilities. This convergence establishes a robust infrastructure that significantly enhances the efficiency, adaptability, and operational effectiveness of military forces. By combining these technologies, military operations can achieve new levels of precision, coordination, and responsiveness [485].

5G technology is a pivotal advancement for military operations, providing unparalleled speed and bandwidth for data transmission. It enables vast amounts of information to be transferred almost instantaneously, facilitating real-time communication even in remote or challenging environments. This capability is especially critical in scenarios where timely intelligence can determine mission success or failure. For instance, real-time data from reconnaissance drones or battlefield sensors can be immediately relayed to ground forces, allowing them to adjust strategies and respond to threats with precision and speed [485].

The IoT amplifies these advantages by creating an interconnected network of devices and sensors across the battlefield. IoT systems collect and transmit data from diverse sources, such as vehicles, equipment, and personnel, providing military commanders with a comprehensive and up-to-date view of the operational environment. This connectivity enables advanced monitoring, such as tracking troop movements, assessing environmental conditions, and predicting equipment maintenance needs. The result is more informed decision-making and enhanced situational awareness, giving military leaders a strategic edge [485].

Edge computing complements these technologies by processing data closer to its source rather than relying solely on centralized data centres. This localized processing reduces latency, ensuring critical information is acted upon in real time. In military contexts, this speed is invaluable. For example, an AI system at the edge can analyse sensor data from the battlefield and deliver actionable insights to commanders instantly, enabling rapid responses to emerging threats and evolving conditions without delays associated with transmitting data to and from central servers [485].

The integration of decentralized AI with 5G, IoT, and edge computing fosters unparalleled real-time decision-making and coordination. By providing military leaders with the most current and accurate data, these technologies empower swift, informed decisions that are essential in dynamic and high-pressure environments. Enhanced connectivity also facilitates seamless collaboration among different units, allowing ground forces, aerial units, and other operational branches to share intelligence and coordinate manoeuvres effectively. This capability strengthens overall mission execution, leveraging the unique strengths of each operational component [485].

Moreover, this synergy ensures greater interoperability between decentralized AI systems and existing military infrastructure. Seamless communication between advanced and legacy systems is essential to maximizing the potential of these technologies and avoiding inefficiencies or delays. As military operations evolve to meet the complexities of modern warfare, the ability to integrate these cutting-edge technologies into cohesive systems will be vital [485].

Ultimately, the convergence of decentralized AI, 5G, IoT, and edge computing is reshaping how military operations are conducted. This transformative integration not only enhances operational efficiency and situational awareness but also ensures a more agile, informed, and coordinated approach to achieving mission objectives. By staying at the forefront of technological innovation, military forces can adapt to emerging challenges and maintain a decisive edge in the battlespace of the future [485].

Surveillance Systems and Intelligence-Gathering Tools

Urban warfare poses unique challenges that necessitate advanced surveillance systems and intelligence-gathering tools to ensure operational success. In densely populated environments, the lines between civilians and combatants blur, creating complexities for distinguishing threats while minimizing collateral damage. Surveillance and intelligence tools provide critical support to military forces, enhancing situational awareness, enabling precision targeting, and facilitating rapid decision-making.

Urban surveillance systems are designed to operate effectively in the confined, multilayered, and often chaotic terrain of cities. These systems include aerial drones, ground-based sensors, and integrated camera networks. Unmanned Aerial Vehicles (UAVs), such as quadcopters or fixed-wing drones, are pivotal in providing real-time reconnaissance over dense cityscapes. Equipped with high-resolution cameras, thermal imaging, and advanced sensors, drones can detect enemy movements, monitor key infrastructure, and identify potential ambush points. Their ability to hover or manoeuvre through narrow urban spaces makes them indispensable for surveillance missions.

Ground-based sensors further augment urban surveillance by monitoring and collecting data on enemy movements and activities. These sensors can include acoustic detectors, seismic sensors, and radar systems capable of penetrating walls and underground structures. They provide valuable insights into subterranean activities, such as tunnel construction or concealed enemy operations, which are common in urban warfare. Coupled with integrated camera networks positioned throughout urban areas, ground-based surveillance systems offer continuous monitoring of critical zones, ensuring that military forces remain informed about evolving threats.

In urban warfare, intelligence gathering is essential for understanding the operational environment and anticipating enemy tactics. Military forces employ a combination of human intelligence (HUMINT), signals intelligence (SIGINT), and geospatial intelligence (GEOINT) to gather actionable information.

HUMINT involves leveraging local populations, informants, and undercover operatives to collect on-the-ground intelligence. Establishing relationships with local communities is critical to this effort, as civilians often have firsthand knowledge of enemy activities and can provide valuable insights. However, reliance on HUMINT requires careful vetting processes to avoid misinformation or compromised sources.

SIGINT, on the other hand, captures electronic communications and intercepts signals to uncover enemy plans and movements. In urban settings, this involves monitoring mobile phone networks, radio transmissions, and encrypted digital communications. Modern signal interception tools equipped with artificial intelligence (AI) can decode patterns, detect anomalies, and filter relevant data from large volumes of intercepted information.

GEOINT uses satellite imagery, UAV feeds, and other geospatial data to create detailed maps of urban environments. This intelligence aids in identifying critical infrastructure, escape routes, and areas of strategic importance. It also supports operations by providing accurate 3D models of cityscapes, enabling planners to simulate missions and anticipate challenges.

The effectiveness of surveillance and intelligence-gathering tools in urban warfare lies in their integration. Modern systems aggregate data from various sources into centralized platforms that utilize AI and machine learning for analysis. These platforms process vast amounts of information to generate actionable insights, such as identifying enemy strongholds or predicting civilian movement patterns. Real-time data analysis ensures that commanders have up-to-date intelligence, enabling them to make informed decisions rapidly.

For example, a combined system might use drones to surveil a target area, sensors to detect movement, SIGINT to monitor communications, and GEOINT to analyse terrain. AI systems would then compile and cross-reference this information to provide a comprehensive understanding of the operational landscape. This integration reduces the risk of overlooking critical details and ensures that military actions are precise and well-informed.

Despite their advantages, the use of surveillance and intelligence-gathering tools in urban warfare raises ethical and operational challenges. Protecting civilian privacy and ensuring compliance with international humanitarian laws are paramount concerns. Systems must be designed to differentiate between combatants and non-combatants to minimize harm. Additionally, the dense electromagnetic spectrum in urban environments can interfere with surveillance equipment, while adversaries may employ countermeasures such as jamming or decoys to evade detection.

Surveillance in urban warfare represents a cornerstone of modern military intelligence operations, addressing the unique challenges posed by densely populated and infrastructure-rich environments. The complexity of urban settings requires robust surveillance strategies to maintain situational awareness, support operational decisions, and reduce collateral damage. These environments are characterized by multi-layered terrains, blending surface streets and parks, subterranean tunnels and subways, and supersurface elements like rooftops and airspace. This multi-dimensional nature, combined with dense infrastructure, often obstructs signals, disrupts line-of-sight systems, and creates blind spots for traditional surveillance tools [486].

The density of urban populations further complicates surveillance efforts, as civilians and combatants can intermingle, making threat differentiation exceedingly difficult. High civilian density amplifies the risk of collateral damage, necessitating precise and reliable surveillance mechanisms. Additionally, urban infrastructure itself poses challenges to communication systems. Elements such as tall buildings, power lines, and industrial machinery can interfere with communication channels like FM/UHF radios, complicating coordination and intelligence gathering [486].

To counter these challenges, a variety of surveillance tools and techniques are employed. Human Intelligence (HUMINT) plays a pivotal role by utilizing informants, local authorities, and linguists to gather actionable intelligence. HUMINT provides critical insights into local dynamics, cultural sensitivities, and concealed threats. Imagery Intelligence (IMINT) leverages drones, satellites, and reconnaissance aircraft to produce aerial imagery, enabling real-time monitoring of urban zones and threat identification. Signals Intelligence (SIGINT) focuses on electronic communication analysis to expose insurgent networks and pinpoint threats, while Geospatial Intelligence (GEOINT) integrates geospatial data with satellite imagery to produce detailed urban maps, aiding in the identification of key infrastructure and mobility corridors [486].

Surveillance in urban operations serves multiple applications. Real-time threat detection capabilities allow forces to identify and address immediate dangers such as enemy positions or improvised explosive devices (IEDs). Surveillance tools also help monitor civil considerations, using frameworks like ASCOPE (Areas, Structures, Capabilities, Organizations, People, Events) to understand and adapt to civilian environments. The intelligence gathered supports tactical operations by enabling precise strikes and safer navigation of complex terrains, ultimately minimizing risks to civilian populations. Additionally, surveillance evaluates demographic and infrastructural data such as traffic patterns and energy networks, providing critical insights for operational planning [486].

Advanced technologies further enhance urban warfare surveillance. Unmanned Aerial Systems (UAS), such as drones, overcome line-of-sight limitations, offering aerial perspectives and bolstering digital communications. Subsurface surveillance tools, including ground-penetrating radar, enable exploration of underground spaces to uncover enemy hideouts or infiltration routes. Integration of artificial intelligence (AI) and automation facilitates predictive analysis, pattern recognition, and real-time decision-making, reducing human workloads and enhancing operational efficiency [486].

The role of surveillance in urban warfare underscores its indispensability in achieving mission success. By combining traditional intelligence methods with advanced technologies, military forces can navigate the complexities of urban settings effectively. The integration of these capabilities ensures that operations are strategically informed, responsive to dynamic conditions, and aligned with the overarching objective of minimizing unintended consequences while achieving tactical goals [486].

Digital Mapping, Real-Time Intelligence, and Augmented Reality for Navigation

Digital mapping is fundamental to urban warfare strategies, providing detailed and dynamic representations of the operational environment. Unlike traditional maps, digital maps integrate data from various sources, including satellite imagery and reconnaissance inputs, to create a comprehensive view of the urban landscape. This integration allows for the inclusion of critical infrastructure, transportation networks, and potential threat zones, which are essential for mission planning. The adaptability of digital maps to real-time changes is particularly valuable; they can reflect recent strikes, alterations in traffic patterns, and new obstacles such as rubble or barricades [25]. Furthermore, the incorporation of subsurface data, such as tunnels and sewer systems, enhances the utility of these maps, enabling military forces to plan multi-level operations effectively [487]

The ability of digital mapping to provide an up-to-date battlefield view is crucial for identifying safe routes, prioritizing targets, and minimizing civilian casualties. This capability is underscored by studies that demonstrate how digital mapping technologies can improve situational awareness and operational efficiency in urban warfare scenarios.

Digital mapping for urban warfare is a critical process that involves the creation, updating, and utilization of advanced geographic information systems (GIS) to provide military forces with comprehensive, dynamic, and actionable insights about the operational environment. This approach integrates diverse technologies and data sources to address the complexities of urban areas, which are characterized by dense populations, intricate infrastructures, and rapidly changing conditions. The objective is to support effective decision-making and navigation in these challenging settings.

The initial stage in developing digital maps involves the collection and integration of extensive data about the urban environment. High-resolution satellite imagery offers a broad view of the terrain, infrastructure, and changes within the landscape. Reconnaissance inputs, both aerial and ground-based, provide detailed information about specific zones, including streets, building interiors, and subterranean systems. Pre-existing geospatial databases contribute foundational geographical information, while Internet of Things (IoT) devices in urban areas— such as traffic cameras, environmental sensors, and communication networks— supply real-time updates on activities and conditions. Once gathered, this diverse data is processed and harmonized through advanced algorithms on a GIS platform, ensuring consistency and accuracy across various informational layers.

The construction of urban digital maps employs a layered approach, enabling users to focus on specific environmental aspects according to their operational needs. Infrastructure layers display buildings, roads, bridges, and essential utilities, while transportation networks highlight streets, railways, and transit routes, complete with real-time traffic patterns. Demographic layers detail population density, residential zones, and movement patterns, and subsurface features reveal tunnels, subways, and sewer systems crucial for urban operations. Threat zones indicate identified enemy positions, ambush sites, and areas with potential improvised explosive devices (IEDs). This layered approach allows commanders and troops to access targeted information relevant to their objectives.

Urban warfare is inherently dynamic, requiring digital maps to be continuously updated in real-time. Drones and reconnaissance aircraft provide live feeds of updated imagery and immediate identification of threats or environmental changes. Crowdsourced data, such as civilian reports and social media activity, highlights new hazards like roadblocks or damaged infrastructure. Artificial intelligence (AI) plays a pivotal role by analyzing incoming data, identifying patterns, and updating maps to ensure they remain current and actionable.

Advanced visualization tools enhance the utility of digital maps. Three-dimensional mapping offers a realistic view of terrain, enabling better understanding of building heights, rooftop access, and line-of-sight considerations. Heat maps emphasize areas of high activity or danger, such as enemy concentrations or civilian gatherings, while interactive interfaces allow users to zoom in on specific locations, annotate points of interest, and simulate various scenarios.

Integration with other operational systems ensures that digital maps provide seamless support to military operations. Command and control systems use these maps for troop coordination, mission planning, and resource allocation. Augmented reality devices offer navigation assistance and real-time updates directly to soldiers on the ground. Additionally, weapons systems incorporate map data for precise targeting, reducing collateral damage and enhancing mission effectiveness.

In urban warfare, digital mapping is indispensable for a range of applications. It aids navigation by helping troops find optimal routes, avoid hazards, and move efficiently through complex environments. Commanders rely on these maps for planning, identifying critical infrastructure, anticipating enemy actions, and formulating tactical strategies. By accurately mapping populated areas and vital facilities, digital maps also play a crucial role in protecting civilians and minimizing casualties. Post-operation, they facilitate damage assessment and guide recovery efforts by evaluating the impact on infrastructure.

Overall, digital mapping for urban warfare is a sophisticated process that integrates data collection, real-time updates, and advanced visualization to deliver actionable intelligence. This enables military forces to plan, navigate, and execute operations with greater precision and situational awareness, ultimately enhancing mission success while mitigating risks.

Real-time intelligence is another critical component that transforms static insights from digital maps into actionable information. This intelligence is sourced from various platforms, including drones and surveillance systems, allowing military units to respond swiftly to changing conditions. In urban warfare, where situations can evolve rapidly, the ability to process and analyse data streams continuously is vital. For example, drone surveillance can detect enemy ambushes, enabling ground forces to alter their strategies in real-time [25]. Additionally, monitoring civilian movements allows commanders to allocate resources effectively to protect vulnerable populations, thereby enhancing operational safety and efficiency [23].

The integration of real-time intelligence into command systems ensures that decisions are informed by the most current and accurate data, which is essential for maintaining operational effectiveness in the fluid dynamics of urban combat. This capability not only improves the speed of decision-making but also enhances the overall situational awareness of military forces engaged in urban warfare.

Augmented reality (AR) further enhances the capabilities provided by digital mapping and real-time intelligence by overlaying critical information onto the soldier's field of view. AR technologies, such as headsets and mobile applications, allow troops to access dynamic visual cues that guide their movements and enhance situational awareness [4]. For navigation, AR can project routes directly onto a soldier's view, highlighting safe paths and potential threats, which is particularly beneficial in the chaotic environments typical of urban warfare [488].

Moreover, AR facilitates tactical awareness by integrating data from multiple sources into a single interface, allowing soldiers to make informed decisions without needing to shift focus between different devices [4]. This integration is crucial for maintaining coordination among units during complex operations, as it enables commanders to share visual instructions and updates directly with their teams, thereby improving synchronization [487].

AR technology combines several sophisticated systems to function effectively in combat scenarios. The core of AR lies in its ability to superimpose digital elements—such as graphics, text, and symbols—onto the user's perception of the physical world. This is achieved through a combination of hardware and software working seamlessly to create a coherent and interactive experience.

1. Input Data Integration: AR systems rely on continuous input from a variety of sources. These include digital maps, real-time intelligence feeds, geospatial data, and sensor outputs from drones, surveillance cameras, and IoT devices. This input provides the raw data necessary for AR overlays, ensuring the visualizations are accurate and contextually relevant.

2. Localization and Orientation: Accurate geolocation and orientation are essential for AR to provide meaningful overlays. This is achieved using GPS, inertial measurement units (IMUs), and other location-tracking technologies. By pinpointing the soldier's position and perspective in relation to the surrounding environment, AR systems align digital information with the physical world.

3. Rendering and Display: The AR system processes the input data and converts it into 3D visualizations or annotations that are rendered in real time. This information is displayed through specialized AR devices such as smart glasses, helmet-mounted displays, or handheld devices. These displays project overlays directly into the soldier's line of sight, ensuring seamless access to critical data without diverting attention from the battlefield.

4. Interaction and Feedback: AR systems often include interactive features, allowing soldiers to engage with the digital overlays. For example, touch gestures on a handheld device, voice commands, or eye-tracking can be used to select, zoom, or query specific data points. Feedback loops ensure that the system adapts dynamically to user inputs and environmental changes.

One of the most impactful applications of AR is in navigation and pathfinding. By overlaying navigation cues directly onto the environment, AR systems guide soldiers through complex urban terrains with unparalleled precision. Visual elements such as arrows, waypoints, and highlighted routes appear in the soldier's field of view, aiding them in avoiding obstacles, identifying safe passages, or navigating efficiently to objectives. This capability is especially critical in urban environments characterized by narrow alleys, multi-level structures, and hidden threats, where traditional navigation tools may fall short.

AR also plays a pivotal role in threat identification and targeting. It enables the real-time detection and marking of threats, such as enemy positions, sniper locations, or improvised explosive devices (IEDs). These markers are superimposed onto the physical environment, ensuring soldiers can quickly recognize and prioritize threats without the need to consult separate maps or devices. This immediacy not only enhances operational efficiency but also reduces the cognitive load on soldiers in high-stress combat situations.

In terms of situational awareness, AR integrates real-time intelligence to provide a comprehensive understanding of the battlefield. Soldiers can access dynamic overlays that display troop movements, civilian populations, or areas under enemy control. This holistic view of the operational environment improves tactical coordination, enabling more informed decision-making while mitigating the risk of collateral damage. By presenting critical data in an intuitive and accessible manner, AR helps soldiers maintain a strategic advantage even in fast-changing conditions.

AR also fosters improved communication and collaboration among team members. Visual annotations and tactical updates shared through AR systems ensure seamless information flow within units. For example, a soldier can mark a threat or point of interest on their AR interface, making it instantly visible to their teammates. This shared awareness enhances coordination and response times, particularly in scenarios requiring synchronized actions across dispersed teams.

Lastly, AR has proven invaluable in military training and simulation. It provides soldiers with realistic combat scenarios in controlled environments by overlaying virtual enemies, hazards, and objectives onto real-world settings. These immersive simulations allow soldiers to practice navigating complex terrains, responding to threats, and executing missions, all without the risks associated with live exercises. Additionally, AR-based training adapts to individual learning needs, offering tailored experiences that prepare soldiers more effectively for real-world challenges. Through these applications, AR is not only transforming how military operations are conducted but also how soldiers are trained for the evolving demands of modern warfare.

The integration of augmented reality (AR) and mixed reality (MR) into military operations marks a significant evolution in how modern forces approach combat scenarios, particularly in urban and complex environments. The U.S. Army, NATO forces, and other advanced militaries are investing heavily in these technologies, aiming to enhance tactical awareness, situational comprehension, and target acquisition capabilities while streamlining the flow of critical information for command decision-making [489]. Among these efforts, the U.S. Army's Integrated Visual Augmentation System (IVAS) exemplifies a leading-edge approach to AR/MR deployment, combining advanced capabilities such as thermal imaging, holographic mapping, and integrated GPS [489].

The accelerated acquisition and fielding of AR/MR systems underscore the Army's commitment to leveraging commercial technology for rapid deployment. Through soldier-centred design (SCD), the process of developing and refining these systems has been condensed significantly. SCD emphasizes continuous feedback from end users, enabling iterative improvements that align technology with operational needs. While this streamlined approach brings innovation to the battlefield faster, it also introduces potential weaknesses due to the reduced timeline for testing and validation. Ensuring that these systems enhance operational effectiveness without introducing vulnerabilities requires a methodical approach to research and development [489].

One of the critical challenges facing AR/MR technology lies in its resilience and reliability under combat conditions. These systems are subjected to extreme environments, ranging from harsh weather to electromagnetic interference, which can compromise functionality. For example, soldiers rely on the integrity of data displayed in AR/MR systems, but data manipulation, signal spoofing, or disruptions in the electromagnetic spectrum could undermine trust in the technology. Effective deployment demands rigorous testing to ensure that systems perform consistently in diverse operational contexts, such as dense urban centres or remote environments with limited infrastructure [489].

Human factors also play a pivotal role in the successful implementation of AR/MR systems. Soldiers must be able to transition seamlessly between various tasks without being overwhelmed by information or distracted from their immediate surroundings. Effective system design requires prioritizing user-centric interfaces that present relevant and actionable data tailored to the user's role and level of expertise. For instance, while squad leaders might need a comprehensive operational picture, individual soldiers require information that supports their immediate tactical responsibilities without creating cognitive overload [489].

The risk of overdependence on AR/MR technology is another consideration. While these systems offer transformative capabilities, reliance on them must be balanced with the preservation of fundamental combat skills. Training programs must ensure that soldiers can perform critical tasks independently of AR/MR systems, maintaining operational readiness in scenarios where the technology becomes unavailable due to technical failure or adversarial actions. This balance is essential to prevent a degradation of core competencies and ensure adaptability in rapidly evolving combat situations [489].

The reliance on AR/MR systems introduces additional logistical challenges, particularly concerning energy consumption and connectivity. The battery life of current systems, such as the IVAS, may not suffice for extended missions, especially in austere environments where recharging or replacing batteries is difficult. Furthermore, the constant data streaming required for these systems creates detectable electromagnetic and infrared signatures, increasing the risk of adversary detection. Advanced adversaries are likely to exploit these vulnerabilities, emphasizing the need for robust countermeasures and stealth-focused design improvements [489].

Finally, the scalability of AR/MR technology across different echelons of command is a crucial factor in its effectiveness. While AR/MR systems offer significant benefits at the individual and squad levels, their full potential lies in their ability to integrate seamlessly into larger operational frameworks. Commanders must be equipped with aggregated and accurate real-time information to facilitate multi-domain operations and shorten decision-making cycles. Achieving this requires ensuring that network infrastructure can support the high bandwidth and low latency demands of AR/MR systems, even in contested environments [489].

The adoption of AR/MR systems represents a critical advancement in modern warfare, offering unprecedented capabilities for enhancing situational awareness and operational efficiency. However, the successful deployment of these technologies demands a comprehensive approach that addresses technical, human, and environmental considerations. By investing in rigorous testing, iterative development, and balanced training programs, military forces can harness the transformative potential of AR/MR systems while mitigating the risks and challenges they present. As conflicts evolve and near-peer adversaries continue to advance their capabilities, the careful integration of AR/MR technologies will be instrumental in maintaining a strategic and tactical edge on the battlefield [489].

Cyber Warfare and Electronic Warfare in Urban Combat Settings

Cyber Warfare

Cyber warfare represents a significant evolution in the landscape of conflict, characterized by the strategic use of cyberattacks to disrupt, damage, or manipulate the critical systems of adversaries. This form of warfare encompasses a range of tactics, including espionage, sabotage, propaganda, and economic disruption, which can have effects as profound as those of traditional military engagements. The complexity of defining cyberwarfare is underscored by ongoing debates among experts regarding its parameters and implications.

The concept of cyberwar remains a subject of intense debate among experts in international politics and computer security. Many argue that the term fails to accurately describe the activities occurring in cyberspace, which often align more closely with cybercrime, cyberespionage, or even cyberterrorism. These critics note that calling such activities "war" has significant political, legal, and military implications. For instance, espionage—whether conducted through traditional means or cyberspace—has rarely been considered an act of war. Alleged incidents of Chinese cyberespionage against nations such as India, Germany, and the United States illustrate this point. While these activities have caused tensions, they have not escalated to the level of war or irreparably damaged diplomatic relations [490].

Similarly, criminal activities carried out in cyberspace are typically addressed through law enforcement rather than military action. However, certain cases blur the lines, such as claims that Russian organized crime syndicates facilitated cyberattacks against Georgia in 2008 and supported attacks on Israeli websites in 2009, allegedly on behalf of Hamas or Hezbollah. By contrast, a cyberattack by one state against another targeting critical infrastructure—such as electrical grids, air traffic systems, or financial networks—might be considered an armed attack under international law if attribution is conclusively established [490].

Legal experts specializing in the laws of armed conflict often distinguish hostile cyber activities from traditional warfare, arguing that such actions serve as precursors or complements to conventional military operations rather than constituting a new form of warfare. For example, in conflicts like the Israeli-Hezbollah clash in 2006 or Russia's invasion of Georgia in 2008, cyberattacks were deployed both before and during the physical conflicts but did not directly instigate them. Similarly, the 2007 cyberattacks on Estonia occurred amid a broader political crisis involving the relocation of a Soviet war memorial, highlighting the connection between cyberattacks and geopolitical tensions [490].

Despite these nuances, many analysts believe that cyberwar will play an increasingly prominent role in future conflicts. It is anticipated that cyberattacks will often mark the opening phases of traditional wars, and their significance in hybrid warfare strategies is expected to grow. As the integration of cyberspace into military operations deepens, the prominence of cyberwarfare in both the preparation for and execution of conventional conflicts is likely to escalate further [490].

The practical manifestations of cyberwarfare are diverse and include espionage, sabotage, and propaganda. Espionage activities, such as the NSA's surveillance programs and the Office of Personnel Management breach attributed to Chinese actors, exemplify the digital infiltration tactics employed to steal sensitive information [491]. Sabotage is vividly illustrated by the Stuxnet worm, which targeted Iran's nuclear facilities, showcasing the potential for cyber tools to cause physical damage [491]. Additionally, propaganda in the digital realm has emerged as a critical component of cyberwarfare, with state and non-state actors utilizing social media to disseminate manipulated information and influence public opinion, thereby destabilizing societies [492].

The economic repercussions of cyberwarfare are also significant. High-profile ransomware attacks, such as WannaCry and NotPetya, have demonstrated the potential for cyber operations to disrupt global economies, affecting critical infrastructure and major corporations alike [493]. The interplay between cybercrime and cyberwarfare becomes particularly pronounced when financial motives intersect with strategic objectives, complicating the landscape of cybersecurity [494]. Furthermore, the vulnerability of infrastructure to cyberattacks raises alarms about national security, as evidenced by the 2015 Ukrainian grid attack attributed to Russian actors, which highlighted the potential for cascading failures in essential services [493].

Emerging threats in the realm of cyberwarfare, such as the concept of a "cyber Pearl Harbor," reflect fears of catastrophic attacks that could incapacitate nations. However, experts caution against sensationalizing these scenarios, advocating for a nuanced understanding of cyber risks [495]. The terms "unpeace" and "grey zone" illustrate the ambiguous nature of cyber actions, which often blur the lines between peace and conflict, further complicating traditional military paradigms [495].

The motivations for engaging in cyberwarfare are multifaceted, encompassing strategic, political, economic, and ideological dimensions. Nations and non-state actors increasingly view offensive cyber operations as a low-cost, high-reward alternative to traditional military engagements or diplomatic efforts. This perspective is supported by the notion that cyberwarfare allows for the weakening of adversaries, the enhancement of one's own strategic position, and the disruption of critical infrastructures without the need for physical confrontation [496, 497]. The strategic implications of cyber operations are profound, as they enable military forces to target and neutralize critical command-and-control systems, air defence networks, and weapon systems that heavily depend on digital infrastructure [498, 499].

The military domain has recognized cyberwarfare as an essential component of contemporary security strategies. General Keith B. Alexander, the first head of USCYBERCOM, emphasized the significance of cyberspace as a battlefield, highlighting the need for military forces to adapt to this new domain [500]. The challenges of attribution in cyber operations complicate the classification of certain actions as acts of war, which can lead to ambiguous responses and escalation dynamics [501]. Historical instances of cyberwarfare, such as the 2008 cyberattacks during the South Ossetia conflict and the ongoing cyber operations in the Russo-Ukrainian War, illustrate how cyber capabilities can be effectively integrated with military strategies to achieve tactical and operational advantages [496, 498, 502].

Moreover, the recruitment of personnel for cyberwarfare roles has become a priority across all branches of the U.S. military, reflecting the growing recognition of cyber operations as integral to modern warfare [498]. The increasing automation and interconnectivity of civilian infrastructure, including financial networks and telecommunications systems, present significant vulnerabilities that can be exploited in cyberwarfare [496-498]. Disrupting these critical systems can lead to cascading effects on national security and economic stability, further motivating state and non-state actors to engage in cyber operations [496, 498].

Figure 76: Cyber warfare operators serving with the 175th Cyberspace Operations Group of the Maryland Air National Guard at Warfield Air National Guard Base, Middle River, Md., monitor cyber attacks on the operations floor of the 275th Cyber Operations Squadron, known as the Hunter's Den, Dec. 2, 2017. Defense Visual Information Distribution Service, Public Domain, via Picryl.

In addition to state-sponsored cyberwarfare, hacktivism represents a politically motivated form of cyber activity that blurs the lines between activism and warfare. Groups like Anonymous utilize cyberattacks to draw attention to social and political issues, often resulting in significant disruptions [496, 497]. While hacktivism may not always align with traditional definitions of cyberwarfare, its politically charged nature and potential for disruption contribute to the broader discourse on cyber operations [496, 497].

Economic motivations also play a critical role in cyberwarfare. Ransomware attacks, for instance, have emerged as a lucrative avenue for generating revenue, allowing states to finance operations while undermining adversaries [496]. North Korea's reported generation of $2 billion through cyberattacks to fund its weapons programs exemplifies the dual-purpose nature of such operations, which combine economic gain with strategic disruption [496]. The private sector is not immune to these threats, as corporations face millions of cyberattacks daily, often aimed at stealing intellectual property or disrupting operations [496, 498].

Finally, the role of ethical hacking and research in understanding vulnerabilities and threats cannot be overlooked. Institutions and cybersecurity organizations engage in proactive measures to identify and mitigate risks, contributing to the overall resilience of cyber infrastructures [496-498]. As the global reliance on digital systems continues to grow, the motivations for engaging in cyberwarfare will likely evolve, underscoring the need for comprehensive strategies to address the complexities of this domain [496-498].

Cyber warfare encompasses a range of attack methods designed to destabilize or weaken adversaries in the digital and physical domains. These attacks target sensitive systems, critical infrastructure, and societal stability, often serving as a precursor or supplement to conventional military operations. Below is an in-depth exploration of the main types of cyber warfare attacks [503]:

Espionage: Espionage in cyber warfare involves covert operations aimed at extracting sensitive information from adversaries. This could include the theft of government secrets, trade negotiations, or defence strategies. Cyber espionage often employs sophisticated methods such as spear phishing attacks, where tailored emails trick individuals into revealing confidential information, or botnets, which are networks of compromised computers used to infiltrate secure systems. Unlike traditional espionage, cyber methods provide a high degree of anonymity and scalability, enabling attackers to collect vast amounts of intelligence with minimal risk of detection.

Sabotage: Sabotage is a disruptive strategy where attackers aim to damage, destroy, or compromise critical systems and information. Government agencies and private organizations must evaluate which assets are most sensitive and the risks if they are compromised. Sabotage can manifest through malicious insiders, such as disgruntled employees or individuals with loyalties to foreign states, who may leak information or disable key systems. Additionally, external threats, including malware or physical destruction facilitated through cyber access, can cripple an organization's operations, affecting national security, critical infrastructure, and societal functions.

Denial-of-Service (DoS) Attacks: Denial-of-Service (DoS) attacks are designed to overwhelm a system or website with a flood of fake requests, rendering it inaccessible to legitimate users. These attacks can paralyse critical operations, such as government services, financial institutions, or research facilities. In military contexts, DoS attacks might disrupt communication systems or deny personnel access to essential resources. The scalability of these attacks through distributed denial-of-service (DDoS) methods, which involve multiple devices in the assault, amplifies their impact, making them a potent weapon in cyber warfare.

Electrical Power Grid Attacks: The electrical power grid is a critical infrastructure vulnerable to cyber attacks. Disrupting the grid can have cascading effects, disabling other essential systems such as transportation, healthcare, and emergency services. Cyber attacks targeting the power grid can lead to widespread blackouts, interrupt communications, and render digital services unusable. These attacks not only cause economic losses but can also result in physical harm by crippling life-support systems or causing accidents due to infrastructure failures.

Propaganda Attacks: Propaganda in cyber warfare seeks to manipulate public perception, undermine trust in governments, and sow discord among citizens. These attacks use social media platforms, fake news websites, and other digital tools to spread misinformation, expose sensitive truths, or propagate divisive narratives. Propaganda campaigns aim to erode morale, foster mistrust in national institutions, and influence public opinion to align with the adversary's goals. In military settings, such tactics can weaken the resolve of troops or encourage defections.

Economic Disruption: Economic disruption attacks target the financial backbone of a country, such as banks, stock markets, and payment systems. By breaching these networks, attackers can steal funds, manipulate markets, or block access to essential financial services. Such disruptions undermine economic stability, causing panic and loss of trust among the populace. In a broader context, these attacks may aim to destabilize a nation's economy to weaken its global standing or cripple its ability to sustain military operations.

Surprise Attacks: Surprise attacks are large-scale, unforeseen cyber offensives akin to historic physical strikes like Pearl Harbor or the events of 9/11. These attacks aim to catch the target off guard, causing significant disruption and weakening defences. Often employed as part of hybrid warfare, surprise cyber attacks can pave the way for physical assaults by disorienting command structures, disrupting communication channels, and neutralizing critical systems.

Their effectiveness lies in their suddenness, scale, and ability to exploit vulnerabilities in a nation's preparedness.

The diverse range of cyber warfare tactics highlights the complexity and multifaceted nature of modern conflicts in the digital domain. From espionage and sabotage to economic disruption and surprise offensives, each type of attack is tailored to exploit specific vulnerabilities. Understanding these threats is crucial for governments, organizations, and individuals to build robust defences, ensuring resilience against increasingly sophisticated cyber adversaries [503].

Figure 77: Staff Sgt. Wiggin Bernadotte, a cyber warfare operator in the Washington Air National Guard's 262nd Cyberspace Operations Squadron, works with Capt. Benjamin Kolar, a cyberspace operations officer in the 262nd, on an electrical substation simulator on November 3, 2018. Defense Visual Information Distribution Service, Public Domain, via National Archives and Defense Visual Information Distribution Service.

Below are detailed accounts of several well-known cyber warfare operations, showcasing their tactics, impact, and underlying motivations [503].

Stuxnet Virus: The Stuxnet virus stands out as one of the most sophisticated and impactful cyber attacks in history. It specifically targeted Iran's nuclear program, aiming to derail its capability to enrich uranium for nuclear weapons development. Delivered via infected Universal Serial Bus (USB) drives, Stuxnet infiltrated Iran's supervisory control and data acquisition (SCADA) systems, which managed industrial processes within nuclear facilities. Once inside, the malware caused physical damage by altering the speed of centrifuges used for uranium enrichment, all while displaying normal operational readings to system operators. This stealthy sabotage set back Iran's nuclear ambitions significantly, underscoring the potential for cyber tools to achieve strategic objectives without direct military intervention.

Sony Pictures Hack: In 2014, Sony Pictures became the target of a high-profile cyber attack following the announcement of the film *The Interview*, a satirical portrayal of North Korea's leader, Kim Jong Un. The hack, attributed to North Korean government hackers, involved the release of sensitive employee information, unreleased films, and confidential emails. The attackers also deployed malware designed to erase data on Sony's networks. The FBI linked the attack to North Korea by identifying similarities in malware used in previous attacks, such as encryption methods and data deletion protocols. This operation highlighted how cyber attacks could be used as tools of geopolitical influence and retaliation.

Bronze Soldier Incident: In 2007, Estonia faced a series of significant cyber attacks following its decision to relocate a Soviet-era statue, the Bronze Soldier, from the centre of Tallinn to a military cemetery. This move sparked controversy among ethnic Russians in Estonia and Russia, leading to massive distributed denial-of-service (DDoS) attacks against Estonian government institutions, media outlets, and financial systems. These attacks paralysed key services and demonstrated the vulnerabilities of a heavily digitized society. The Bronze Soldier incident is considered one of the first instances of large-scale, politically motivated cyber warfare.

Fancy Bear Operation: Between 2014 and 2016, the Russian cybercrime group Fancy Bear, believed to have ties to Russian intelligence, launched a targeted cyber attack on Ukrainian rocket and artillery units. The operation involved the spread of malware through an infected Android application used by Ukraine's D-30 Howitzer artillery unit for managing targeting data. The malware, known as X-Agent spyware, provided detailed intelligence on the location and activities of Ukrainian forces. This attack resulted in the destruction of more than 80% of Ukraine's D-30 Howitzers, showcasing the devastating impact of cyber tools in conjunction with conventional warfare.

Enemies of Qatar Campaign: In 2018, Elliott Broidy, an American Republican fundraiser, accused the government of Qatar of orchestrating a cyber warfare campaign against him. Allegedly, Qatari operatives hacked and leaked his emails to undermine his influence and discredit him in Washington. The attack was reportedly part of a broader strategy targeting individuals and organizations perceived as adversaries of Qatar, including officials from Egypt, Saudi Arabia, the United Arab Emirates, and Bahrain. Over 1,200 individuals were reportedly affected by this campaign. This incident illustrates how cyber operations can be used for espionage, reputation damage, and strategic manipulation on the global stage.

These examples of cyber warfare operations highlight the evolving nature of modern conflict, where digital attacks serve as both standalone strategies and complements to traditional military actions. They demonstrate the diverse motivations behind cyber warfare, ranging from geopolitical retaliation to tactical battlefield advantages. These cases underscore the critical importance of cybersecurity in protecting national interests and maintaining global stability in an increasingly interconnected world [503].

Electronic Warfare

Electromagnetic warfare (EW), also referred to as electronic warfare, is a critical aspect of modern military operations that leverages the electromagnetic spectrum (EM spectrum) to achieve strategic advantages over adversaries. EW encompasses a range of activities aimed at controlling, disrupting, or exploiting the EM spectrum to deny adversaries access while ensuring its effective use by friendly forces. This warfare domain is not limited to traditional military platforms but extends to air, land, sea, and space, utilizing both crewed and uncrewed systems to conduct operations. The scope of EW includes targeting communication networks, radar systems, and various civilian and military technologies, thereby illustrating its expansive role in contemporary conflict scenarios [504, 505].

The electromagnetic environment (EME) is a subset of the EM spectrum that is particularly significant for military operations. It is characterized by the dense information flow that is crucial for transmitting and receiving operational data. As military forces increasingly rely on the EME for operational superiority, this dependence introduces vulnerabilities that EW seeks to exploit through countermeasures, attacks, or surveillance strategies. NATO recognizes the EME as a distinct operational space, akin to traditional domains such as air and sea, and categorizes EW activities into three core functions: electronic attack (EA), electronic protection (EP), and electronic warfare support (ES) [504, 505]. These activities are further enhanced by intelligence, surveillance, target acquisition, and reconnaissance (ISTAR) and signals intelligence (SIGINT), which collectively contribute to comprehensive situational awareness [504].

The primary activities within EW include jamming and counter-jamming, electro-optical and infrared countermeasures, spectrum management, and reconnaissance. Jamming involves disrupting enemy communications and radar systems, while counter-jamming refers to measures taken to protect friendly systems from such disruptions. Electro-optical and infrared countermeasures are employed to safeguard assets from targeting by optical and heat-seeking weapons. Spectrum management is crucial for optimizing the use of the EM spectrum to minimize interference, and reconnaissance activities focus on collecting and analysing EM signals to identify and prioritize battlefield threats [504, 505].

EW can be subdivided into three main categories: electronic attack (EA), electronic protection (EP), and electronic warfare support (ES). EA involves offensive strategies such as radar jamming and communication interference, which can significantly impact adversary operations. For instance, during the ongoing conflict in Ukraine, Ukrainian forces effectively employed EW to disrupt Russian drone operations, demonstrating the tactical importance of EA in modern warfare [504]. EP, on the other hand, focuses on protecting friendly forces from electronic attacks through measures such as stealth technologies and emission control. The U.S. Air Force, for example, utilizes EMP-resistant systems to ensure command and control resilience in high-intensity electromagnetic environments [504]. Finally, ES encompasses the detection and analysis of electromagnetic signals to provide actionable intelligence, enhancing situational awareness and target prioritization [504, 505].

Historically, EW has evolved from its early applications in World War II, where radar jamming played a pivotal role, to contemporary conflicts where advanced technologies are employed. The 2007 Israeli airstrike on a suspected Syrian nuclear facility exemplified the effective use of EW systems to neutralize enemy air defences, allowing for unimpeded operations [504]. The ongoing Russian invasion of Ukraine has further highlighted the tactical significance of EW, as extensive Russian EW systems have disrupted GPS signals and UAV operations, impacting Ukrainian military effectiveness [504, 505].

Despite its advantages, EW faces several challenges, including the increasing complexity of the EM spectrum, the risk of electronic fratricide, and the need for resilience against adversarial countermeasures. Future developments in EW are likely to focus on integrating artificial intelligence for real-time spectrum analysis, enhancing the resilience of EP systems, and reducing the detectability of EW platforms [504, 505]. The evolving nature of EW necessitates continuous innovation to maintain superiority in this critical aspect of modern warfare.

Electronic warfare (EW) has become a cornerstone of military operations in urban settings due to the unique challenges posed by these environments. The dense population, complex infrastructure, and proximity of civilian and military assets make traditional warfare approaches less practical. EW offers a sophisticated alternative by enabling forces to achieve tactical dominance without direct confrontation, which is particularly important in minimizing collateral damage in populated areas. Its ability to disrupt, deceive, and manipulate electromagnetic signals makes EW indispensable in modern urban combat scenarios [506].

One of the most critical functions of EW in urban warfare is its capacity to disrupt enemy communications. Urban combat typically involves adversaries that rely on wireless and radio communications to coordinate operations. By employing jamming techniques, EW can interfere with these communications, fracturing command structures and creating confusion among opposing forces. This disruption undermines the enemy's operational effectiveness and provides friendly forces with a strategic advantage [506].

EW also facilitates intelligence gathering in urban areas, where traditional surveillance methods often fall short due to the physical obstructions presented by buildings and underground facilities. By intercepting signals and leveraging advanced surveillance technologies, EW enables real-time data collection, offering enhanced situational awareness. This capability is crucial for understanding enemy movements, intentions, and vulnerabilities, thereby aiding tactical planning and execution [506].

EW technologies employed in urban warfare are tailored to address the unique challenges of densely populated environments. These technologies include jamming systems, electronic surveillance tools, and cyber operations, each of which plays a pivotal role in achieving mission success [506].

Jamming systems are fundamental to disrupting enemy communications and navigation signals. In urban combat, where effective communication is critical, jamming technologies create operational chaos by rendering enemy coordination efforts futile. These systems are designed to be mobile, enabling them to adapt to the fluid dynamics of urban battlefields. By neutralizing critical communication links, jamming systems allow friendly forces to operate with greater freedom and security [506].

Electronic surveillance technologies are equally vital in urban warfare, where physical reconnaissance is often restricted. These systems intercept enemy communications and analyse electromagnetic signals to extract actionable intelligence. The integration of unmanned aerial vehicles (UAVs) equipped with EW capabilities enhances surveillance efforts, providing aerial reconnaissance in areas that are otherwise difficult or dangerous for ground personnel to access [506].

Cyber operations are another critical aspect of EW in urban settings. By exploiting vulnerabilities in enemy networks, cyber operations can compromise information flow, disrupt command systems, and manipulate data. These actions degrade the enemy's operational capacity while safeguarding friendly forces. In a highly digital battlefield, cyber operations serve as a force multiplier, amplifying the impact of conventional tactics [506].

Figure 78: TPz FUCHS 1 of the Dutch army in electronic warfare configuration. rheinmetall-defence, CC BY-SA 4.0, via Wikimedia Commons.

The application of EW in urban warfare offers several tactical advantages that significantly enhance operational effectiveness. One of the most prominent benefits is the disruption of enemy communications. Urban combat often involves adversaries relying heavily on wireless networks for command and control. By deploying electronic countermeasures, military forces can incapacitate these communication systems, sowing disarray among enemy ranks and reducing their ability to execute coordinated actions [506].

Intelligence gathering is another key advantage of EW in urban settings. The intricate layouts of cities, combined with the challenges of civilian presence, make conventional reconnaissance methods less effective. EW tools enable the interception and analysis of enemy signals, offering critical insights into troop movements and operational plans. This real-time intelligence empowers commanders to make informed decisions and pre-empt enemy actions, thereby ensuring strategic superiority [506].

Furthermore, EW enhances the safety of ground forces by enabling remote neutralization of enemy capabilities. By disrupting radar, navigation systems, and electronic devices, EW minimizes the need for direct engagements, reducing the risk to personnel. This capability is particularly valuable in urban warfare, where the proximity of combatants and civilians increases the potential for escalation and unintended casualties [506].

In urban warfare, disrupting enemy communications is a cornerstone of EW strategy. Effective communication is vital for coordinating military actions, and its disruption can significantly impair the enemy's operational capabilities. EW employs several methods to achieve this, including jamming, spoofing, and signal intelligence (SIGINT) [506].

Jamming introduces interference into communication frequencies, rendering enemy transmissions unintelligible. This method disrupts command chains and creates confusion, preventing adversaries from executing coordinated manoeuvres. Spoofing, on the other hand, involves imitating legitimate signals to mislead the enemy. By transmitting false commands or information, spoofing can divert enemy resources and attention, further degrading their effectiveness [506].

SIGINT complements these techniques by actively monitoring enemy communications. By intercepting and analysing signals, military forces can gain valuable insights into enemy strategies, intentions, and locations. This intelligence allows for pre-emptive actions and ensures that commanders remain one step ahead in the operational landscape [506].

Urban environments present unique challenges for intelligence gathering due to their physical complexity and high population density. Buildings, underground facilities, and other urban structures create obstacles for traditional surveillance methods. EW addresses these challenges by leveraging advanced technologies to collect and analyse data in real time [506].

Signal interception systems are a primary tool for intelligence gathering in urban settings. These systems detect electronic emissions from enemy units, enabling the identification of their positions and movements. The data gathered can be used to construct a comprehensive picture of the battlefield, enhancing situational awareness and aiding in the formulation of effective strategies [506].

Unmanned aerial vehicles (UAVs) equipped with EW capabilities are also instrumental in urban intelligence operations. These drones can navigate dense urban landscapes, conducting surveillance in areas inaccessible to ground forces. Their ability to operate discreetly reduces the risk of detection, making them invaluable assets in high-stakes missions [506].

Through these methods, EW ensures that military forces can gather critical intelligence in even the most challenging urban environments. This capability not only enhances operational planning but also minimizes risks to personnel and civilian populations. As urban warfare continues to evolve, the role of EW in intelligence gathering will remain pivotal in achieving mission success [506].

Recent urban conflicts have underscored the transformative impact of EW, with prominent examples seen in the Middle East and Eastern Europe. These cases highlight the strategic applications of EW technologies in disrupting enemy operations, gathering intelligence, and supporting ground forces in urban settings [506].

Middle Eastern Urban Battles: The urban battles in cities like Mosul and Aleppo during conflicts in the Middle East vividly demonstrate the utility of EW. These densely populated cities posed significant challenges for traditional military tactics due to their intricate infrastructure and civilian presence. In response, combatants relied heavily on EW technologies to gain a strategic edge [506].

Jamming devices were deployed to disrupt enemy communications, crippling their ability to coordinate attacks and defences. Drones equipped with signal intelligence (SIGINT) capabilities provided real-time reconnaissance, offering aerial views of otherwise inaccessible areas. Cyber operations further targeted critical communication networks, undermining the adversary's operational capabilities. These EW applications not only hampered enemy coordination but also enhanced intelligence gathering, enabling more informed and precise tactical decisions [506].

Figure 79: A view of an electronic warfare (jammer) pods installed on an A-3 Skywarrior aircraft at the Pacific Missile Test Center. The U.S. National Archives, Public Domain, via National Archives and Defense Visual Information Distribution Service.

Urban Combat in Eastern Europe: The conflict in Ukraine offers a contemporary example of EW's role in urban warfare. Russian forces utilized sophisticated EW equipment to jam Ukrainian communications and disable drone operations, significantly affecting the latter's reconnaissance capabilities. This disruption was particularly impactful in urban areas like Mariupol, where situational awareness is crucial for effective operations [506].

The integration of electronic surveillance allowed forces to intercept and analyse enemy signals, providing valuable insights into troop movements and strategies. Such intelligence proved essential for navigating the complexities of urban battlefields and maintaining an operational advantage. The use of EW in Eastern Europe underscores its growing importance in modern conflicts, where technology increasingly dictates outcomes [506].

The integration of EW with ground forces has emerged as a critical factor in enhancing the effectiveness of urban military operations. This collaboration leverages EW capabilities to support traditional ground tactics, providing a cohesive strategy that maximizes both technological and human resources [506].

One key aspect of this integration is real-time information sharing between EW units and ground troops. By coordinating efforts, forces can disrupt enemy systems while simultaneously exploiting the resulting vulnerabilities. For instance, jamming enemy communications can create opportunities for ground units to manoeuvre undetected, increasing their operational security and effectiveness.

Joint training exercises are another vital component, ensuring that EW specialists and ground troops work seamlessly together. These exercises foster teamwork and synchronize tactics, allowing for smoother implementation of complex strategies. Mobile EW systems, designed to accompany ground forces, further enhance this integration by providing on-the-move capabilities tailored to the dynamic nature of urban combat [506].

The synergy between EW and ground operations strengthens mission outcomes by countering enemy capabilities while bolstering the efficiency and safety of ground forces. As urban warfare evolves, this integration will remain a cornerstone of modern military strategy.

The future of EW in urban warfare is set to be shaped by rapid technological advancements and evolving combat strategies. Emerging technologies like artificial intelligence (AI) and machine learning are poised to revolutionize EW by enabling automated signal analysis, predictive threat detection, and precision targeting. These capabilities will enhance the efficiency and effectiveness of EW systems, particularly in the complex and fast-paced environments of urban conflicts.

Miniaturized and mobile EW systems are another anticipated development, offering increased operational flexibility. These compact devices can be deployed quickly in dense urban terrains, providing real-time intelligence and disruption capabilities without compromising mobility. Such innovations will allow forces to adapt rapidly to shifting battlefield conditions, maintaining a tactical edge [506].

The integration of EW with cyber operations represents another significant trend. Joint operations between EW and cyber units will enable sophisticated cyber-physical attacks, simultaneously targeting digital and physical assets of the enemy. This approach will amplify the impact of electronic engagements, creating comprehensive disruptions that hinder adversary capabilities while protecting friendly assets.

The use of EW in urban warfare raises critical legal and ethical questions, particularly given the potential for unintended consequences in densely populated areas. International legal frameworks, such as the Geneva Conventions, mandate strict adherence to principles of distinction, proportionality, and necessity in military operations. These principles require forces to differentiate between combatants and civilians, minimize harm, and ensure that actions are justified by military necessity.

The unpredictability of EW, particularly in urban settings, poses significant ethical challenges. Disruptions to civilian infrastructure, such as hospitals or communication networks, can have severe humanitarian consequences. Balancing military objectives with the need to protect non-combatants is a delicate task that requires rigorous planning and execution [506].

Comprehensive training for military personnel is essential to address these legal and ethical concerns. This training should focus on the implications of EW tactics, ensuring that operations are conducted responsibly and in compliance with international law. By prioritizing the minimization of harm and respect for human rights, military forces can maintain the legitimacy of their actions while effectively leveraging the capabilities of EW in urban warfare [506].

Future Trends in Urban Warfare Technology

Future trends in urban warfare technology are increasingly shaped by advancements in precision targeting, unmanned vehicles, AI-integrated drones, loitering munitions, AI image recognition, and information warfare. These technologies are critical in enhancing military effectiveness while minimizing collateral damage and operational risks.

Precision Targeting is a significant trend in urban warfare, where smaller and more precise munitions are being developed to reduce collateral damage while increasing lethality. The integration of advanced targeting systems allows for more accurate strikes, which is essential in densely populated urban environments. This capability aligns with the military's shift towards minimizing civilian casualties while achieving strategic objectives [507]. The evolution of precision-guided munitions has been a focal point in military modernization efforts, emphasizing the importance of accuracy in urban combat scenarios [508].

Unmanned Vehicles are becoming increasingly versatile in urban warfare. These vehicles can perform various tasks, including resupply missions, reconnaissance, and even offensive operations. The deployment of unmanned aerial vehicles (UAVs) and ground vehicles enhances operational capabilities, allowing forces to conduct missions in high-risk areas without exposing personnel to danger [509]. The development of swarm technology, where multiple unmanned vehicles operate in coordination, further amplifies their effectiveness in complex urban environments [509].

AI-Integrated Drones represent a transformative shift in military operations. These drones can operate autonomously, making rapid decisions based on real-time data. Their integration with other systems allows for enhanced target acquisition and weapon selection, significantly improving operational efficiency [510]. The use of AI in drones also facilitates advanced reconnaissance capabilities, enabling forces to gather intelligence without direct human oversight [511]. The potential for AI to enhance decision-making processes in combat scenarios is a critical area of ongoing research and development [510].

Loitering Munitions are designed to reduce their detectability through minimized radar, visual, and thermal signatures. This stealth capability allows them to operate effectively in urban environments where traditional munitions might be easily intercepted [507]. The ability to loiter over a target area and strike at the optimal moment provides a tactical advantage, particularly in urban warfare where the dynamics of engagement can change rapidly [507].

AI Image Recognition technology is pivotal in urban warfare, enabling the identification of civilians and protected emblems amidst combat operations. This capability is crucial for adhering to international humanitarian laws and minimizing civilian casualties [510]. The application of AI in image recognition allows for the rapid processing of vast amounts of visual data, enhancing situational awareness for military commanders [510]. This technology can also assist in distinguishing between combatants and non-combatants, which is essential in urban settings where the lines can often blur [510].

Information Warfare is another critical aspect of modern urban combat. The ability to influence and inform operations through information manipulation can disrupt enemy command structures and undermine morale [508]. This form of warfare involves cutting off hostile forces from their strategic leadership and utilizing psychological operations to achieve military objectives without direct confrontation [508]. The integration of AI into information warfare strategies allows for more sophisticated approaches to data analysis and dissemination, enhancing the effectiveness of these operations [510].

In addition to these technological advancements, several other factors are shaping the future of urban warfare. The physical and human terrain of cities presents unique challenges and advantages for defenders, often mitigating the technological superiority of more powerful opponents [507]. Understanding the complexities of urban environments is essential for military planners as they adapt strategies to leverage these factors effectively.

The rapid rise in urbanization is another critical consideration, with projections indicating that by 2050, 68% of the world's population will reside in urban areas [507]. This demographic shift necessitates a re-evaluation of military strategies and technologies to address the complexities of urban warfare effectively.

Finally, the need for rapid adaptation and innovation in military forces is paramount. As technological advancements continue to evolve, military organizations must be agile in their approach to integrating new technologies and adapting to changing battlefield dynamics [508]. This adaptability is crucial for maintaining a technological edge in increasingly complex urban environments.

Chapter 10

COUNTER-INSURGENCY (COIN) AND ASYMMETRICAL URBAN WARFARE

Dealing with Unconventional Enemy Tactics in Urban Settings

Counter-Insurgency (COIN) and asymmetrical urban warfare represent some of the most complex and challenging forms of modern military conflict. These operations are defined by their focus on combating insurgent groups, which often operate within densely populated urban environments. Unlike conventional warfare, where opposing forces are clearly defined, COIN and asymmetrical warfare involve irregular tactics, blending military engagements with psychological, political, and social dimensions.

The urban setting amplifies these challenges. Cities are characterized by their dense populations, intricate infrastructure, and interwoven civilian-military dynamics. Insurgents often exploit this complexity, embedding themselves within civilian populations to complicate military responses and leverage the urban environment to their advantage. This creates a highly asymmetrical battlefield where conventional military power must adapt to unconventional threats.

COIN strategies aim not only to neutralize insurgent forces but also to address the underlying political and social grievances that fuel insurgencies. These operations emphasize a combination of military action, intelligence gathering, and civilian engagement to restore stability and deny insurgents the support of the local population.

Asymmetrical urban warfare further complicates these efforts. The irregular tactics employed by insurgents—such as guerrilla warfare, improvised explosive devices (IEDs), and cyber operations—require innovative and adaptive responses from conventional forces. Success in such conflicts often hinges on superior situational awareness, advanced technology, and a deep understanding of the sociopolitical landscape.

The convergence of COIN and asymmetrical urban warfare has led to the development of specialized strategies and technologies, such as electronic warfare, real-time intelligence systems, and urban-centric counter-insurgency doctrines. These approaches aim to mitigate the advantages of insurgents while minimizing collateral damage and maintaining legitimacy in the eyes of local populations and the international community.

Urban warfare presents a distinct set of challenges, particularly when addressing unconventional enemy tactics. The complexities of urban environments necessitate a re-evaluation of traditional military strategies, as insurgents and non-state actors exploit the dense terrain to negate conventional military advantages. This synthesis will explore the unique challenges posed by urban warfare and propose strategies for countering unconventional tactics.

The urban terrain itself is a significant challenge in urban warfare. The density and complexity of cities hinder long-range targeting capabilities and complicate the effectiveness of artillery and air support. The predictable nature of ground routes in urban areas makes them susceptible to ambushes, as insurgents can easily conceal themselves among civilian populations and structures [512]. This concealment allows guerrilla fighters to operate without large bases, thereby complicating the use of advanced military technologies such as aerial bombardment [513]. The urban landscape thus serves as a force multiplier for unconventional tactics, enabling insurgents to evade detection and stage surprise attacks [512].

Moreover, the information landscape in urban warfare has transformed with the rise of social media, which insurgents utilize to disseminate propaganda and shape public perception [408]. This manipulation of information complicates the operational environment for conventional forces, as they must navigate not only the physical terrain but also the psychological landscape shaped by these narratives. The challenge is further compounded by humanitarian concerns, as urban warfare often leads to significant civilian casualties and displacement, necessitating strategies that minimize harm to non-combatants while effectively targeting enemy combatants [514].

Urban warfare often involves unconventional tactics designed to exploit the unique challenges posed by densely populated city environments. These tactics are employed by insurgents, non-state actors, or adversaries to leverage the complexities of urban settings and challenge conventional military strategies. They make use of the structural intricacies, high civilian presence, and logistical challenges inherent in cities, creating significant operational and ethical dilemmas for military forces.

Improvised Explosive Devices (IEDs) are a prominent feature of unconventional urban warfare. These devices are ingeniously concealed in everyday objects, such as trash bins or vehicles, making detection exceptionally difficult. In cities, IEDs are often placed in high-traffic areas like marketplaces and transit hubs or along critical supply routes to disrupt operations and inflict casualties. For instance, during the Iraq War, insurgents frequently targeted coalition forces with IEDs on urban roads, while in Afghanistan, these devices were strategically placed near schools or within buildings to impede military actions and generate propaganda opportunities.

Sniper attacks exemplify the use of urban architecture for tactical advantage. Snipers utilize rooftops, windows, and other concealed positions to target opposing forces, disrupt troop movements, and demoralize soldiers. The Bosnian War's "Sniper Alley" in Sarajevo and ISIS's use of snipers in Mosul illustrate how such tactics exploit the multi-story structures of cities to create zones of denial and constant threats for advancing forces.

Guerrilla warfare tactics are another hallmark of urban combat, involving hit-and-run attacks, ambushes, and sabotage. Guerrilla fighters often rely on intimate knowledge of the city terrain to outmanoeuvre better-equipped forces. In Gaza, Hamas has employed these tactics effectively by attacking Israeli forces and retreating through densely packed neighbourhoods and tunnels. Similarly, the Viet Cong in Saigon demonstrated the effectiveness of guerrilla strategies during the Vietnam War, launching surprise attacks and blending seamlessly into civilian populations.

The use of human shields is a particularly challenging tactic in urban warfare, where insurgents embed themselves within civilian populations to deter military action. By positioning fighters and weaponry in schools, hospitals, or residential buildings, they create ethical and operational dilemmas for military forces. For example, ISIS and other groups in Raqqa frequently used civilians as shields during the Syrian Civil War, complicating military operations and raising humanitarian concerns.

Modern unconventional tactics also include cyber and electronic warfare. Adversaries disrupt communications, target critical infrastructure, or influence public opinion through cyberattacks. In Ukraine, Russian forces have used electronic jamming to impede Ukrainian drone operations, while in Hong Kong, protestors deployed encrypted communication apps to evade police surveillance during urban demonstrations.

Psychological warfare and propaganda campaigns are crucial tools in unconventional urban tactics. These campaigns aim to manipulate public opinion and erode the morale of military forces and civilians. ISIS, for example, used social media to instil fear and bolster its image during its occupation of cities like Mosul. Propaganda efforts during the Battle of Aleppo further illustrate how information campaigns can shape public perception and sway international opinion.

Ambushes in urban chokepoints are a classic example of leveraging the urban terrain for tactical advantage. Insurgents exploit narrow streets and alleyways to trap opposing forces. Notable examples include insurgents' ambushes in Fallujah during the Iraq War and Somali militia attacks during the Battle of Mogadishu, both of which relied on the maze-like structure of cities and high civilian presence to maximize impact.

Tunnels offer insurgents a covert means of transporting supplies, launching attacks, and evading capture. Hamas's tunnel networks in Gaza and the Viet Cong's underground routes in Saigon are prime examples of how such infrastructure can provide a tactical advantage in urban warfare.

The use of drones for surveillance and attack is another innovative urban tactic. ISIS's deployment of drones in Mosul to monitor coalition forces and drop explosives underscores their utility in modern urban conflicts. Similarly, modified consumer drones have been used by both sides in the Ukraine conflict for reconnaissance and tactical strikes.

Sabotaging infrastructure, such as water supplies, electricity grids, or transportation systems, is another tactic aimed at destabilizing urban life and straining resources. During the Syrian Civil War, both rebel and government forces targeted critical infrastructure to control urban populations, while in Iraq, insurgents frequently attacked oil pipelines and power grids to disrupt coalition operations.

Hostage-taking and kidnapping are also employed to gain leverage or provoke military responses. Chechen militants' hostage-taking during the Moscow theatre crisis and Al-Qaeda's kidnappings in Iraqi cities highlight how such tactics are used to manipulate public attention and negotiations.

Finally, vehicle-borne improvised explosive devices (VBIEDs) are a destructive urban tactic. Insurgents load vehicles with explosives to target checkpoints, military convoys, or crowded civilian areas. This method was widely used in Baghdad during the Iraq War and by ISIS in Mosul to slow advancing forces and create chaos.

These examples demonstrate how unconventional tactics in urban warfare are designed to exploit the inherent challenges of city environments. They highlight the necessity for military forces to adapt, integrating advanced technologies and innovative strategies to counter these evolving threats effectively.

To counter these unconventional tactics, military forces must adopt adaptive strategies. Defensive plans should focus on breaking apart attacking formations, separating mounted from dismounted forces, and limiting the enemy's ability to manoeuvre [515]. A dispersed defence strategy can leverage the concealment provided by urban terrain, establishing well-protected positions that deny insurgents freedom of movement [514]. Additionally, the use of loitering munitions can enhance operational effectiveness by reducing radar, visual, and thermal signatures, making it more difficult for insurgents to detect and counter military operations [516].

The use of improvised explosive devices (IEDs) and ambush tactics poses another significant threat in urban combat. Insurgents often deploy IEDs along critical supply routes or in crowded areas to disrupt movement and instil fear [513]. To mitigate these threats, military forces must maintain constant vigilance and employ advanced detection technologies, such as drones and ground-penetrating radar, to identify and neutralize threats before they can inflict harm [514].

Cyber operations represent another dimension of unconventional warfare, as insurgents increasingly leverage digital platforms to disrupt communication networks and manipulate information [517]. Countering these tactics requires robust cybersecurity measures and electronic warfare capabilities to detect and neutralize digital threats effectively [518].

Psychological warfare is also a critical component of unconventional tactics in urban settings. Insurgents utilize propaganda to undermine public trust in government and military forces, employing acts of terror to intimidate civilians [408]. Countering these tactics necessitates proactive information campaigns and community engagement to build public trust and support for military operations [408].

Key Concepts: Guerilla Tactics, Hit-and-Run, and Urban Insurgency

In urban warfare, guerrilla tactics, hit-and-run strategies, and urban insurgency form the backbone of unconventional combat, often employed by insurgents or non-state actors. These tactics leverage the complexities of urban environments, such as dense infrastructure and civilian populations, to neutralize the advantages of conventional military forces. Understanding these key concepts is essential for countering their effectiveness in modern conflicts.

Guerrilla tactics are a hallmark of asymmetrical warfare, focusing on mobility, surprise, and adaptability. In urban settings, these tactics capitalize on the city's intricate layouts, offering insurgents concealment and manoeuvrability. Unlike conventional forces, guerrilla fighters often lack significant firepower or resources, so they rely on exploiting their intimate knowledge of the terrain. Urban guerrilla operations typically include ambushes, sabotage, and covert attacks on high-value targets.

An ambush is a surprise attack launched from a concealed position to catch an enemy force or asset off guard. This tactic relies on meticulous planning, strategic positioning, and the element of surprise to maximize its impact. Urban environments, with their narrow streets, chokepoints, and dense infrastructure, are particularly conducive to ambushes. Attackers typically conduct thorough reconnaissance to identify enemy movement patterns, vulnerabilities, and potential escape routes. Once the target's predictable path is established, attackers position themselves in hidden locations, such as rooftops, alleys, or behind debris, ensuring maximum cover and visibility.

The ambush is triggered when the target enters the designated kill zone—a carefully chosen area where attackers can concentrate their firepower or detonate explosives. Techniques for initiating the ambush vary, ranging from tripwires and remote detonators to signals from lookout personnel. Execution involves a combination of small arms fire, explosives, and possibly heavier weapons like rocket-propelled grenades (RPGs). After delivering the blow, attackers swiftly withdraw to pre-planned escape routes to evade counterattacks, often employing decoys to confuse pursuers. Notable examples include insurgent ambushes on coalition convoys during the Iraq War, where roadside IEDs were combined with gunfire, and Somali militia attacks during the Battle of Mogadishu, where narrow streets and coordinated firepower overwhelmed U.S. forces.

Sabotage is another cornerstone of unconventional warfare, involving the deliberate disruption, destruction, or degradation of an adversary's infrastructure, equipment, or operations. This tactic is rooted in stealth and aims to weaken the opponent's capabilities without direct confrontation. Sabotage often begins with target selection, identifying critical assets whose compromise could cause widespread disruption, such as power grids, transportation networks, or communication systems. Attackers then infiltrate these locations, using disguises, forged identification, or insider collaboration to gain access.

Once inside, saboteurs deploy their tools—often explosives like C4 or improvised devices—to damage or destroy physical infrastructure. Cyber-sabotage is also prevalent, leveraging malware to disrupt digital systems. The act is executed stealthily to avoid detection, and the perpetrators escape without leaving evidence of their presence. Historical examples include World War II resistance fighters who targeted railway lines and factories in Nazi-occupied Europe and sabotage operations on power and water infrastructure during the Syrian Civil War to destabilize opposing forces.

Covert attacks on high-value targets focus on neutralizing key enemy personnel, assets, or facilities with minimal visibility and collateral damage. Such operations demand extensive reconnaissance to identify the target's routines, location, and security measures. Intelligence gathering may involve human intelligence (HUMINT), signal intelligence (SIGINT), and even surveillance drones. With detailed plans in place, including entry and exit strategies and weapon selection, attackers infiltrate the target area covertly, often using underground routes or disguises. Urban terrain, with its complex layout, enhances the feasibility of such infiltration.

The attack itself is swift and precise, employing tools such as sniper rifles, explosives, or sabotage equipment to achieve the objective. Noise and prolonged engagement are minimized to maintain stealth. Post-mission extraction relies on pre-planned routes and sometimes involves decoys or diversions to evade capture. A prominent example of this tactic is the 2020 assassination of Iranian nuclear scientist Mohsen Fakhrizadeh, which showcased meticulous planning and advanced surveillance. Similarly, U.S. special forces in Iraq conducted covert raids to eliminate high-value insurgent leaders based on actionable intelligence.

The tactical implications of these methods are profound. Ambushes disrupt enemy momentum, sabotage debilitates essential systems, and covert attacks decapitate leadership or neutralize critical assets. Together, these tactics form a triad of highly effective strategies in urban warfare, particularly against better-equipped conventional forces. Countering these methods demands robust intelligence, technological superiority, and adaptable operational strategies to mitigate their impact and safeguard mission success.

For example, during the Battle of Grozny in the Chechen Wars, insurgents used guerrilla tactics to challenge Russian forces. They employed the city's maze-like streets and densely packed buildings to mount ambushes, inflict significant casualties, and retreat before the enemy could regroup. By avoiding direct confrontations, guerrilla fighters effectively prolonged the conflict, exhausting their opponents and forcing them into costly engagements.

Guerrilla tactics also thrive in environments where the insurgents can blend into civilian populations. This approach complicates targeting for conventional forces, as distinguishing between combatants and civilians becomes difficult. Consequently, these tactics often result in prolonged conflicts and ethical dilemmas for military planners.

Hit-and-run is a central technique in guerrilla warfare, emphasizing speed and surprise. This strategy involves striking a target swiftly and retreating before the enemy can retaliate. The goal is to disrupt enemy operations, lower morale, and create chaos while minimizing the risk to the attacking force.

In urban insurgencies, hit-and-run tactics are especially effective due to the confined spaces and abundant hiding spots provided by cities. Insurgents can strike at military convoys, supply lines, or command centres and then disappear into the urban sprawl. For example, during the Vietnam War, the Viet Cong employed hit-and-run tactics in urban areas like Saigon, launching surprise attacks on American forces and swiftly retreating to underground tunnels or safe houses.

The success of hit-and-run strategies depends on precise planning and execution. Insurgents often conduct thorough reconnaissance to identify vulnerable targets and exploit gaps in enemy defences. They prioritize mobility and use lightweight weapons, ensuring they can escape quickly and evade pursuit. This strategy keeps conventional forces on edge, forcing them to expend resources guarding against potential attacks.

Urban insurgency represents a broader framework that incorporates guerrilla tactics and hit-and-run strategies to achieve political or ideological goals. Unlike rural insurgencies, which operate in less populated areas, urban insurgencies exploit cities' dense populations, infrastructure, and media visibility to exert pressure on governments or occupying forces. The urban setting offers insurgents proximity to political, economic, and military targets, enhancing their strategic impact.

Urban insurgencies often combine military actions with political and psychological warfare. Insurgents aim to erode public trust in the government or occupying forces by creating a sense of insecurity. This may involve coordinated attacks on infrastructure, such as power grids or transportation systems, to disrupt daily life and generate public discontent. For instance, during the Iraq War, insurgents targeted oil pipelines and electricity supplies in Baghdad to destabilize the government and undermine coalition efforts.

In addition to physical attacks, urban insurgencies leverage propaganda and media coverage to amplify their message. By staging dramatic attacks or high-profile kidnappings, insurgents draw attention to their cause, influencing both domestic and international perceptions. ISIS, for example, used urban insurgency tactics in Mosul, combining physical assaults with psychological operations to project power and attract recruits.

Countering guerrilla tactics, hit-and-run strategies, and urban insurgencies necessitates a multi-faceted approach that integrates conventional military capabilities with innovative strategies tailored to urban environments. Conventional forces must adapt to the complexities of urban warfare, which often involves a blend of guerrilla tactics and conventional military operations. For instance, the use of psychological warfare and information operations has been highlighted as essential in countering insurgent strategies effectively, as seen in the historical context of the Boer War where British forces employed a combination of scorched-earth policies and mobile columns to disrupt Boer guerrilla tactics [519]. This historical precedent underscores the importance of intelligence and adaptability in military operations.

Moreover, the integration of advanced technologies, such as augmented reality systems, can significantly enhance situational awareness for military personnel operating in urban environments. Research indicates that augmented reality can improve decision-making efficiency and reduce casualties by providing real-time information to combatants [4]. This technological leverage is crucial in urban settings where the risk of civilian casualties is high, necessitating precise targeting and operational restraint to minimize collateral damage while effectively neutralizing insurgent threats.

Engaging local communities is another critical component of counter-insurgency efforts. Collaborative strategies that involve local populations can help address the root causes of insurgencies and foster public support for military operations. The importance of community engagement is evident in various counter-insurgency campaigns, where building trust and cooperation with civilians has proven vital for operational success [520]. This approach aligns with findings from the Rhodesian War, where the insurgents' ability to sabotage key installations compelled the government to reconsider its military strategies, highlighting the need for a comprehensive understanding of local dynamics [521, 522].

Strategies to Counter Improvised Explosive Devices (IEDs) and Ambushes

Countering Improvised Explosive Devices (IEDs) and ambushes necessitates a multifaceted approach that integrates advanced technology, intelligence, tactical adaptation, and comprehensive training. Both tactics exploit the element of surprise and target vulnerabilities in conventional military operations. Effective counter-strategies are designed to mitigate risks, enhance situational awareness, and ensure rapid response to emerging threats.

The cornerstone of counter-IED and ambush strategies is robust intelligence gathering. Understanding the enemy's tactics, techniques, and procedures (TTPs) is crucial for anticipating potential threats. Intelligence agencies employ a combination of human intelligence (HUMINT), signal intelligence (SIGINT), and imagery intelligence (IMINT) to identify likely IED deployment zones and ambush sites. For instance, analysts focus on areas with restricted manoeuvrability, such as chokepoints and urban environments, where adversaries are likely to deploy these tactics. Furthermore, field units utilize threat analysis tools that leverage historical incident data and local terrain assessments to predict patterns and identify high-risk zones, thereby enabling precautionary measures such as route adjustments and increased vigilance.

Figure 80: A counter improvised explosive device route clearance convoy sits halted on a highway after and IED detonates in Ghazni province, Afghanistan Nov. 29, 2010. The convoy is halted while members of the CIED team search the area for additional explosives. Defense Visual Information Distribution Service, Public Domain, via Picryl.

To effectively counter IEDs, route clearance operations are vital. Specialized engineering units equipped with mine-resistant, ambush-protected (MRAP) vehicles and robotic systems are deployed to clear roads of potential explosive hazards. These units utilize ground-penetrating radar, metal detectors, and bomb-sniffing dogs to detect and neutralize threats before convoys proceed. Additionally, electronic countermeasures (ECMs) are employed to jam the signals of remotely detonated IEDs, rendering them ineffective. Protective measures such as convoy spacing, speed control, and staggered formations are implemented to minimize the impact of attacks, ensuring that even if an ambush occurs, its effects are mitigated.

Figure 81: WO2 Iain Martin pictured defusing an IED on the Bandi Barq Road in Gereshk. ResoluteSupportMedia, CC BY 2.0, via Flickr.

Extensive training is essential for troops to recognize potential threats and respond appropriately to ambushes and IEDs. Awareness programs educate soldiers on identifying indicators of IEDs, such as freshly disturbed earth or abandoned vehicles. Tactical training drills simulate ambush scenarios, teaching troops how to react under fire, establish defensive perimeters, and execute counterattacks effectively. Units are also trained in immediate-action drills, which focus on minimizing casualties and regaining control during an ambush, including rapid deployment of cover fire and coordinated manoeuvres to exit the kill zone.

Technology plays a pivotal role in countering IEDs and ambushes. Unmanned aerial vehicles (UAVs) provide real-time reconnaissance of convoy routes, allowing for the detection of potential threats from a safe distance. Surveillance drones equipped with thermal imaging can identify hidden attackers or unusual heat signatures associated with explosives. Moreover, artificial intelligence (AI) and machine learning algorithms analyse vast amounts of data to predict enemy behaviour and recommend safer routes, thereby enhancing decision-making in dynamic and high-stress environments.

In counterinsurgency operations, collaboration with local communities is crucial for identifying and neutralizing threats. Local populations often possess critical information regarding insurgent activities, including IED placements and planned ambushes. Building trust through community engagement programs encourages civilians to share intelligence with security forces. Additionally, information sharing among allied forces and agencies is essential for developing effective counter-strategies and staying ahead of evolving threats.

Despite preventative measures, some attacks may still succeed. Quick reaction forces (QRFs) are essential for responding to such incidents, trained to secure the area, provide medical aid, and extract personnel and equipment. Effective communication and coordination among units ensure that QRFs can respond rapidly to ambushes or IED attacks. Recovery operations also involve forensic analysis of explosive devices to gather intelligence on enemy methods and supply chains, which is vital for refining future countermeasures.

Flexibility and adaptability are key to countering the dynamic nature of IEDs and ambushes. Forces must continually vary their routes, timing, and methods of movement to avoid predictability, which insurgents often exploit. Commanders emphasize decentralized decision-making, allowing field units to respond swiftly to emerging threats without waiting for higher-level approval. By integrating intelligence, training, technology, and community collaboration, military forces can effectively counter IEDs and ambushes, thereby enhancing operational security and protecting personnel and civilians in conflict zones.

Training Troops for Rapid Adaptability in Urban Insurgency Contexts

Training troops for rapid adaptability in urban insurgency contexts is essential for operational effectiveness, particularly in environments marked by unpredictability and unconventional enemy tactics. The dynamic nature of urban warfare necessitates that soldiers possess not only technical skills but also the cognitive ability to assess, decide, and act swiftly in ever-changing scenarios. This multifaceted training approach is crucial for preparing military personnel to navigate the complexities of urban combat effectively.

Urban insurgencies demand heightened situational awareness, as threats can arise from diverse and unexpected directions, including multi-level buildings and dense civilian crowds. Training programs increasingly incorporate realistic simulations that replicate the chaos of urban combat, exposing soldiers to scenarios such as sudden ambushes and improvised explosive device (IED) threats. The integration of virtual reality (VR) and augmented reality (AR) technologies enhances this training by immersing soldiers in simulated environments where they must identify threats and make rapid decisions [395]. These technologies not only improve situational awareness but also allow for the practice of critical thinking in high-stress situations, which is vital for effective decision-making in combat [523].

In urban insurgency contexts, soldiers frequently face high-pressure situations requiring swift decision-making. Training programs focus on fostering cognitive resilience by simulating high-stress conditions that compel soldiers to analyse situations rapidly and prioritize threats effectively. This approach is supported by research indicating that decision-making drills, when integrated with physical exercises, can enhance soldiers' ability to perform under duress [81, 524]. The emphasis on cognitive readiness and operational adaptability is crucial, as soldiers must often make split-second decisions that can have significant consequences in combat scenarios [525].

Urban combat often involves small, decentralized units operating independently. Training emphasizes the mastery of small-unit tactics, including room clearing and securing high-value targets. Effective communication within teams is critical, and soldiers learn to use hand signals and radio communications to coordinate movements discreetly [88]. This training also fosters adaptability, as troops may need to shift between offensive and defensive postures based on real-time intelligence. The ability to operate cohesively in small units is essential for success in the unpredictable environments characteristic of urban insurgencies [526].

Interactions with civilian populations are frequent in urban insurgency contexts, making cultural sensitivity training vital. Soldiers are educated on local customs and social dynamics to distinguish between hostile insurgents and non-combatants, reducing the likelihood of alienating the local population [527]. This training not only minimizes collateral damage but also fosters trust and cooperation with civilians, which can yield valuable intelligence regarding insurgent activities [528]. Understanding the cultural landscape is thus integral to operational effectiveness in urban environments.

Modern urban warfare increasingly relies on advanced technology, necessitating that training programs equip soldiers with the skills to utilize these tools effectively. Soldiers learn to operate drones for reconnaissance, employ electronic warfare tools, and navigate using GPS systems in dense urban areas [529]. Moreover, training in cybersecurity measures is essential to protect communication networks from insurgent interference, highlighting the importance of technological proficiency in contemporary military operations [530].

The unpredictable nature of urban insurgencies requires soldiers to develop adaptive problem-solving skills. Training scenarios often present complex challenges where standard operating procedures may not apply, compelling soldiers to think creatively and leverage available resources [395]. This adaptability is crucial for navigating blocked routes or responding to sudden shifts in enemy behaviour, ensuring that troops can maintain operational effectiveness despite evolving threats [531].

The physical and mental demands of urban insurgency combat are substantial. Training programs focus on building resilience through rigorous physical conditioning and mental preparedness exercises. Soldiers engage in endurance drills and combat simulations that test their limits, while mental resilience is cultivated through stress inoculation training [81, 524]. This dual focus on physical and mental preparedness is vital for ensuring that soldiers can manage fear and fatigue during operations [88].

Training for urban insurgency contexts is an ongoing process that evolves based on lessons learned from real-world operations. After-action reviews (AARs) allow soldiers to analyse their performance, identify strengths and weaknesses, and refine their tactics [531]. This continuous feedback loop ensures that training programs remain dynamic, incorporating new techniques and technologies as they emerge, which is essential for maintaining operational readiness in complex environments [527].

Team cohesion is critical for effective operations in urban insurgencies. Training programs emphasize trust and communication within units, ensuring that soldiers can rely on one another in high-stakes situations [532]. Exercises designed to foster camaraderie and mutual support are integral to building a cohesive team capable of navigating the challenges of urban combat [524].

Urban insurgencies often involve asymmetric threats, such as guerrilla tactics and cyber warfare. Training addresses these challenges by exposing soldiers to unconventional scenarios where the enemy's approach deviates from traditional military norms [531]. Soldiers learn to anticipate and counter these threats, maintaining their adaptability in the face of unconventional tactics, which is essential for success in modern warfare [530].

Building Networks of Trust for Intelligence and Operations

In the realm of intelligence gathering and operational effectiveness, the establishment of robust networks of trust is paramount. These networks are essential for enhancing collaboration, facilitating the flow of accurate information, and ensuring the successful execution of complex missions. Trust, which is built on mutual respect, reliability, and shared objectives, serves as the cornerstone for these networks, particularly in high-stakes environments such as military operations, counterinsurgency efforts, and intelligence activities. The significance of trust in intelligence gathering cannot be overstated, as it fosters open communication and ensures the credibility of information. Intelligence networks often depend on human sources, technology, and collaborative partnerships, where the willingness of human sources, including informants and local contacts, to share valuable insights is heavily influenced by the trust they have in the information collectors [533].

Moreover, partnerships with allied organizations or nations necessitate a foundation of trust. Sharing intelligence entails significant risks; mishandled information can jeopardize operations or relationships. Trust ensures that shared intelligence is accurate, timely, and utilized appropriately, thereby maximizing its strategic value [534]. In counterinsurgency contexts, for instance, trust is crucial for effective collaboration among military personnel, local police, government officials, and community leaders. This collaboration is vital for resource coordination and minimizing misunderstandings that could jeopardize missions [533].

Operational collaboration is further enhanced through trust, which facilitates seamless interactions among diverse units and stakeholders. Modern missions often involve joint efforts between military forces, intelligence agencies, and civilian organizations. Building trust among these entities requires clear communication, shared goals, and a mutual understanding of each party's roles and capabilities [535]. For example, in urban insurgency scenarios, trust among military personnel and local authorities is essential for effective resource allocation and mission success [533].

To build networks of trust, several strategies can be employed. Consistent engagement is vital, as regular interactions foster familiarity and reliability, which are key components of trust. Engaging with local communities through meetings, shared initiatives, and cultural sensitivity demonstrates commitment and builds rapport [533]. Transparency also plays a crucial role; open communication about objectives, expectations, and limitations reinforces trustworthiness and minimizes suspicion [533]. Additionally, cultural competence—understanding and respecting cultural norms and values—can significantly enhance trust among local populations and stakeholders [533].

Establishing mutual benefits is another effective strategy for building trust. Networks that provide reciprocal advantages, such as offering security or developmental support in exchange for intelligence, create a win-win dynamic that strengthens relationships [533]. Furthermore, accountability is essential; upholding commitments and taking responsibility for actions solidifies trust, making collaborators more likely to rely on and contribute to the network [533].

However, building trust in volatile or hostile environments presents unique challenges. Distrust may arise from historical grievances, cultural differences, or fears of exploitation. Overcoming these barriers requires patience, consistent actions aligned with promises, and the involvement of credible intermediaries [533]. Maintaining trust is also an ongoing effort, as a single breach can undo years of relationship-building. Establishing clear protocols for handling sensitive information and addressing disputes promptly is crucial for preserving trust [533].

Technology plays a significant role in building and maintaining trust networks. Encrypted communication platforms, real-time data sharing systems, and secure databases enhance trust by ensuring the confidentiality and reliability of shared information. Furthermore, technology can provide transparency through tracking and reporting mechanisms, reinforcing confidence among collaborators [535].

Chapter 11

CASE STUDIES IN URBAN WARFARE

Analysis of Major Urban Battles in History and Modern Warfare

Historical Conflicts

U rban warfare has undergone significant evolution throughout history, particularly characterized by combat operations within densely populated cities. This form of warfare presents unique challenges and opportunities due to the complexities of urban environments, including dense infrastructure, civilian presence, and restricted mobility. Analysing major urban battles provides insights into strategic approaches, technological advancements, and lessons applicable to modern military operations.

The Battle of Stalingrad (1942-1943) serves as a pivotal example of urban warfare, where German and Soviet forces engaged in prolonged combat. The urban landscape of Stalingrad was marked by narrow streets and rubble-filled neighbourhoods, which severely limited the mobility of mechanized units [536]. The Soviet forces adeptly utilized the city's infrastructure, establishing defensive strongholds in factories and buildings, demonstrating the importance of urban terrain in combat [194]. The Soviets employed close-quarter combat and ambush tactics, while the German forces relied heavily on bombardment but faced logistical challenges and harsh winter conditions that hindered their advance [194, 536]. The outcomes of this battle underscored the significance of supply lines, urban defence strategies, and the psychological toll of urban combat, emphasizing the necessity for adaptability in dynamic environments [194, 536].

Similarly, the Battle of Hue (1968) during the Vietnam War illustrated the complexities of modern urban warfare. North Vietnamese forces occupied the city, prompting a counteroffensive from U.S. and South Vietnamese troops. The city's layout, featuring fortified positions and underground tunnels, complicated direct assaults and resulted in significant civilian casualties, raising ethical dilemmas regarding collateral damage [537]. The U.S. forces employed airstrikes and heavy artillery, which, while effective in some respects, led to extensive destruction and further civilian suffering [537]. The North Vietnamese utilized guerilla tactics and the urban infrastructure to mount a prolonged resistance, highlighting the difficulty of distinguishing combatants from civilians in urban settings [537]. This battle revealed the limitations of conventional firepower in densely populated areas, emphasizing the need for more nuanced strategies in urban combat [537].

The Battle of Mogadishu (1993), known as "Black Hawk Down," further exemplifies the challenges of urban combat in politically unstable environments. U.S. forces aimed to capture Somali warlord Mohamed Farrah Aidid's lieutenants but encountered heavy resistance from local militias [537]. The narrow streets and civilian interference created a chaotic battlefield, complicating the extraction of U.S. personnel [537]. Insurgent forces leveraged local knowledge and civilian populations for cover, while U.S. forces struggled with coordination amid the urban maze [537]. This battle highlighted the vulnerability of technologically advanced forces in dense urban environments and underscored the necessity for contingency planning and a comprehensive understanding of local dynamics [537].

In more contemporary contexts, the Second Battle of Fallujah (2004) and the Battle of Mosul (2016-2017) further illustrate the evolution of urban warfare. In Fallujah, insurgents fortified the city using improvised explosive devices (IEDs) and sniper positions, while coalition forces employed precision airstrikes and advanced reconnaissance technologies [537]. The battle emphasized the importance of combined arms operations and intelligence integration to minimize civilian casualties and maintain moral legitimacy [537]. The Battle of Mosul showcased the use of modern technology, such as drones and precision munitions, in urban combat, while also highlighting the critical need for post-conflict reconstruction to stabilize urban areas [537].

Current Conflicts

The current geopolitical landscape is characterized by a series of complex and overlapping conflicts, particularly in Gaza, Lebanon, Israel, Ukraine, and Syria. Each of these conflicts is shaped by distinct historical, political, and social contexts, yet they share common themes such as urban warfare, asymmetrical tactics, and significant international implications.

Gaza and the Israel-Palestine Conflict: The conflict between Israel and Hamas in Gaza is one of the most enduring and volatile disputes globally, characterized by recurrent cycles of violence, ceasefires, and humanitarian crises. Urban warfare is a defining feature of this conflict, as Gaza's densely populated areas complicate military operations. The embedding of Hamas operations within civilian infrastructure raises ethical dilemmas for Israeli forces, who must navigate the challenges of distinguishing combatants from civilians ([538, 539]. Hamas employs a variety of tactics, including the use of rockets, underground tunnels, and drones, to conduct both offensive and defensive operations. In response, Israel utilizes precision airstrikes, the Iron Dome missile defence system, and intelligence-led operations to mitigate threats [540].

The civilian impact of the conflict is profound, with heavy casualties and widespread displacement drawing international scrutiny. This humanitarian toll complicates Israel's strategic objectives and raises critical questions regarding military ethics and human rights [541]. Furthermore, the conflict has significant diplomatic implications, influencing regional relations and international discourse, particularly regarding the ongoing calls for a two-state solution [542, 543].

Lebanon and Israel: Lebanon remains a critical flashpoint in the Middle East, particularly due to the presence of Hezbollah, a powerful non-state actor allied with Iran. Hezbollah's military and political power in Lebanon enables it to engage in cross-border attacks on Israel, employing guerrilla tactics, precision missiles, and cyber capabilities [544, 545]. The integration of Hezbollah's operations within civilian areas complicates Israeli countermeasures and raises concerns about collateral damage [546]. The conflict between Israel and Hezbollah is deeply intertwined with broader regional tensions, particularly the Iran-Israel rivalry, and any escalation risks drawing in additional actors, including Syria and Iran [547, 548].

Ukraine and Russia: The ongoing war between Ukraine and Russia, which escalated with Russia's full-scale invasion in February 2022, has significantly reshaped European security dynamics and marked a resurgence of large-scale conventional warfare. Urban warfare has become a central feature of this conflict, with cities like Mariupol and Bakhmut witnessing devastating consequences for civilian populations [549, 550]. Russia employs heavy artillery and siege tactics, while Ukraine utilizes agile, decentralized defence strategies, often supported by Western military aid [551].

Electronic warfare (EW) plays a crucial role in this conflict, with both sides employing tactics to disrupt enemy communications and enhance their operational capabilities [552]. The asymmetrical nature of the conflict is evident as Ukraine leverages advanced military technology, including drones and precision weaponry, to counter Russia's numerical and firepower advantages [553, 554]. The war has led to significant civilian displacement and infrastructure destruction, raising allegations of war crimes against Russian forces and highlighting the global implications of the conflict on energy supplies and food security [555, 556].

Syria's Ongoing Civil War: The Syrian conflict, now in its second decade, involves multiple actors, including the Syrian government, opposition groups, ISIS remnants, and foreign powers like Russia, Iran, and Turkey. Urban destruction is rampant, with cities like Aleppo and Raqqa facing catastrophic devastation due to aerial bombardments [557, 558]. The conflict serves as a proxy war, with various regional and global powers backing different factions, complicating the humanitarian crisis that has resulted in one of the largest refugee crises in history.

The use of chemical weapons by the Assad regime has drawn international condemnation and limited military responses from the West, further complicating the geopolitical landscape. Asymmetrical tactics employed by non-state actors, including IEDs and urban guerrilla warfare, continue to pose significant challenges to superior military forces.

Across these conflicts, urban warfare emerges as a central feature, presenting tactical challenges such as civilian protection and logistical constraints. Asymmetrical tactics are prevalent among non-state actors like Hamas, Hezbollah, and Syrian opposition groups, who utilize unconventional methods to counter technologically superior adversaries. The integration of electronic and cyber warfare into modern conflicts is increasingly evident, as seen in Ukraine, where both sides employ electronic jamming and cyber operations.

The devastating impact on civilian populations is a common thread, with all these conflicts resulting in significant casualties, displacement, and infrastructure destruction, raising ethical and legal questions regarding military operations. The geopolitical implications of these conflicts extend beyond their immediate regions, influencing global energy and food supplies and shaping international diplomacy.

Urban warfare has a profound and multifaceted impact on civilian populations, often resulting in widespread suffering, displacement, and destruction. The dense and complex environments of cities amplify the consequences of conflict, making civilians not just collateral damage but often the central victims of urban combat. The effects of urban warfare are extensive and deeply affect civilian life.

One of the most immediate impacts is direct physical harm. Urban warfare frequently leads to high civilian casualties due to the use of heavy artillery, airstrikes, and indiscriminate weapons in densely populated areas. Civilians are often caught in the crossfire, and the proximity of combatants to residential zones increases the likelihood of injuries and deaths. For example, repeated conflicts between Hamas and Israel in Gaza have caused significant civilian casualties as airstrikes target areas where combatants operate near civilian infrastructure. Similarly, cities like Aleppo and Raqqa in Syria have been devastated by relentless bombardments, leading to catastrophic civilian tolls.

Another significant consequence of urban warfare is the destruction of infrastructure. Homes, schools, hospitals, water systems, and power grids are often destroyed during combat. This destruction has cascading effects on civilian well-being, resulting in shortages of basic necessities, reduced access to medical care, and prolonged hardship. In Mariupol, Ukraine, relentless shelling during the Russian invasion left the city without electricity, water, or heating, exacerbating civilian suffering. Similarly, the campaign to liberate Mosul, Iraq, from ISIS resulted in widespread destruction, rendering entire neighbourhoods uninhabitable.

Displacement and refugee crises are also common outcomes of urban warfare. Civilians are often forced to flee their homes to escape violence, resulting in massive displacement and overcrowded refugee camps. The Syrian Civil War, for instance, has displaced millions, with many fleeing urban centres such as Damascus and Aleppo. Refugees face significant challenges in accessing shelter, food, and security in host countries. In Gaza, repeated cycles of violence have displaced thousands, forcing them into overcrowded and inadequate shelters.

Urban warfare also inflicts immense psychological trauma on civilians. The experience of living under constant threat, witnessing violence, and losing loved ones leaves deep scars. Children, in particular, suffer long-term psychological effects, including anxiety, depression, and post-traumatic stress disorder (PTSD). In Aleppo, many children exposed to years of conflict exhibit high rates of PTSD and developmental delays, while survivors in Mariupol recount the terror of being trapped without food or water amid ongoing artillery strikes.

Combatants often exploit civilians as human shields, embedding themselves within residential areas, hospitals, or schools to deter attacks. This tactic creates moral and operational dilemmas for opposing forces, who must weigh military objectives against the need to minimize civilian casualties. Hamas in Gaza and Hezbollah in Lebanon have been accused of using civilian infrastructure to shield military operations, complicating responses from opposing forces. Similarly, ISIS fighters in Mosul used civilians as shields to delay coalition advances.

Urban warfare also disrupts local economies and social structures. The destruction of marketplaces, transportation networks, and workplaces results in unemployment, poverty, and food insecurity. Additionally, the breakdown of governance and law enforcement creates opportunities for crime and exploitation. In Raqqa, Syria, the battle against ISIS left the city's economy in ruins, with residents struggling to rebuild their lives amid destroyed infrastructure. In Ukraine, ongoing conflict has devastated urban industries, leaving civilians reliant on humanitarian aid.

Access to humanitarian aid is often obstructed during urban warfare. Active combat, destroyed roads, and blockades hinder the delivery of aid, leaving civilians in besieged cities facing acute shortages of food, medicine, and clean water. For example, besieged cities in Eastern Ukraine have struggled to receive aid, leaving residents reliant on dwindling supplies. Similarly, during the Syrian conflict, cities such as Homs and Aleppo were under siege for months, cutting off essential supplies and leading to famine-like conditions.

The effects of urban warfare extend far beyond the cessation of hostilities. Rebuilding destroyed cities takes years or even decades, leaving civilians in prolonged states of vulnerability. Economic and social recovery is slow, and psychological scars persist for generations. Post-war recovery in Fallujah, Iraq, has been marked by slow reconstruction efforts and ongoing poverty, hindering the return of displaced residents. In Syria, many urban areas remain in ruins, with millions of displaced civilians unable to return due to the lack of housing and services.

Mitigating the impact of urban warfare on civilians requires concerted efforts from governments, international organizations, and humanitarian groups. Strategies include adhering to international humanitarian law to protect civilians, establishing safe humanitarian corridors for aid delivery, prioritizing post-conflict reconstruction, and providing psychological support to affected populations. As conflicts increasingly occur in cities, addressing the humanitarian consequences and safeguarding civilian populations must remain a central focus for policymakers, military planners, and international organizations. Urban warfare profoundly disrupts civilian life, leaving lasting scars on individuals and communities, making it imperative to address its devastating impact comprehensively.

Comparative Study of Offensive and Defensive Operations

Offensive and defensive operations form the cornerstone of military strategy, each fulfilling distinct roles in achieving strategic objectives. These approaches, while contrasting in nature, are deeply interconnected, often requiring seamless integration for effective outcomes. A comprehensive understanding of their objectives, characteristics, and applications provides critical insight into their impact on the battlefield.

The primary goal of offensive operations is to seize the initiative, disrupt the enemy's plans, and secure territorial or strategic gains. These strategies are designed to force the enemy into a reactive stance, capturing key positions, neutralizing opposing forces, and reducing their capacity to retaliate. In contrast, defensive operations aim to preserve forces, maintain control over critical areas, and counter enemy offensives. Their central objective is to deny the adversary success, either by repelling attacks, delaying their progress, or absorbing their offensive momentum to create opportunities for counteraction.

Offensive operations are marked by proactivity, requiring aggressive tactics to push into enemy territory or positions. Success hinges on maintaining momentum and applying constant pressure to exploit weaknesses. Surprise and deception are often employed to catch opponents off-guard, while significant logistical resources are necessary to sustain the offensive. However, these operations come with high risks, as forces are exposed to counterattacks and logistical vulnerabilities. Defensive operations, on the other hand, are inherently reactive, leveraging fortifications, obstacles, and pre-prepared positions to slow and deplete attackers. These strategies require endurance and economy of force, often utilizing fewer resources by maximizing the advantages of terrain and pre-established defences. Defensive operations also prepare for counteroffensives, turning defensive advantages into opportunities for strategic gains once the enemy's momentum wanes.

The planning and execution of offensive operations rely heavily on precise intelligence gathering, manoeuvre warfare tactics, force concentration, and the synchronization of land, air, and naval assets. In contrast, defensive planning emphasizes terrain utilization, layered defences, disruption tactics, and the strategic positioning of reserves for counterattack potential. Both approaches benefit from advanced technologies, with offensive operations integrating precision-guided munitions, drones, and electronic warfare, while defensive strategies deploy anti-air and missile defence systems, surveillance technologies, and autonomous systems for perimeter security.

The advantages of offensive operations include securing the initiative, achieving decisive victories, and disrupting enemy cohesion. However, these gains often come at the cost of logistical strain, high casualties, and the risk of overextension. Defensive operations, while economical and capable of inflicting high attrition on attackers, face challenges such as resource depletion, difficulty transitioning to offense, and vulnerability to technological and psychological warfare aimed at eroding morale.

Historical case studies illustrate the dynamics of these strategies. Germany's Blitzkrieg during World War II demonstrated the effectiveness of fast, coordinated offensives combining tanks, air support, and infantry. Similarly, Operation Desert Storm showcased precision strikes and air superiority in achieving offensive objectives. On the defensive side, the Battle of Stalingrad exemplified the absorption and countering of enemy offensives, while the Siege of Leningrad highlighted the resilience and endurance required to repel prolonged attacks.

In modern warfare, the lines between offense and defence are increasingly blurred. Hybrid tactics and asymmetric threats often necessitate a fluid interplay between these strategies. For instance, Ukraine's defensive operations against Russian advances have utilized urban environments and international support to transition into successful counteroffensives. In the Middle East, insurgent groups have employed unconventional offensives, such as ambushes and improvised explosive devices, against well-fortified defensive positions.

Ultimately, offensive and defensive operations represent complementary facets of military strategy, each with unique strengths and limitations. The effective integration of these approaches, along with well-timed transitions between offense and defence, remains pivotal in achieving strategic success on the battlefield. The dynamic interplay between these operations underscores their critical role in shaping the outcomes of conflicts across diverse scenarios.

Key Takeaways and Strategies That Emerged from Urban Battle Experiences

Urban battles, drawn from both historical and contemporary conflicts, have provided critical insights into the unique challenges and opportunities that cities present as battlefields. These experiences underscore the need for tailored strategies, advanced technologies, and nuanced approaches to urban warfare, where the dense population, complex infrastructure, and unpredictable dynamics shape every aspect of military operations.

One of the key takeaways from urban battle experiences is the importance of adaptability. Urban combat often deviates from conventional battlefield norms, requiring forces to adjust quickly to unexpected challenges. The intricate layouts of cities, combined with the presence of civilians and non-combatants, demand real-time decision-making and flexibility in tactics. Successful operations rely on the ability to shift strategies based on the evolving battlefield, blending offensive and defensive actions seamlessly to maintain momentum while minimizing collateral damage.

Another critical lesson is the necessity of intelligence superiority. Urban settings are dense with information, much of it hidden or obscured by the environment. Accurate, timely intelligence is crucial for understanding the terrain, identifying enemy positions, and anticipating insurgent tactics. Technologies such as drones, surveillance systems, and signals intelligence play a pivotal role in collecting and analysing data, enabling commanders to make informed decisions that can significantly influence the course of urban engagements.

The experiences of urban warfare also highlight the role of civilian considerations. In cities, the battlefield is inherently intertwined with the lives of ordinary people. Protecting civilians, minimizing casualties, and preserving critical infrastructure are not only ethical imperatives but also strategic necessities. Winning the support of the local population can provide invaluable advantages, including intelligence-sharing and reduced resistance, while mishandling civilian relations can lead to long-term instability and hostility.

Logistical challenges in urban warfare are another recurring theme. Supplying forces in densely built areas, where access routes are often blocked or contested, requires innovative solutions. The destruction of infrastructure, such as bridges and roads, complicates the movement of troops and supplies, making logistics a critical component of operational planning. Effective strategies include pre-positioning resources, using smaller, more mobile supply units, and leveraging local assets where feasible.

Technology has emerged as both a tool and a battleground in urban conflicts. From the use of drones for reconnaissance to electronic warfare that disrupts enemy communications, advanced technologies have transformed urban combat. However, adversaries have also exploited technology, employing cyberattacks, IEDs, and commercial drones for their purposes. These developments emphasize the need for forces to maintain technological superiority while preparing for the creative and adaptive use of technology by opponents.

The experiences of urban battles also stress the value of training troops for urban-specific scenarios. Traditional combat training is often insufficient for the complexities of city warfare. Soldiers must be skilled in navigating tight spaces, conducting close-quarters combat, and adapting to the unpredictability of urban environments. Simulated urban training environments, including mock cities, have proven invaluable for preparing troops for these challenges.

Finally, the necessity of integrating humanitarian considerations into military strategies has become increasingly apparent. Urban warfare often results in significant humanitarian crises, with displaced populations, damaged infrastructure, and strained resources. Coordinating with humanitarian organizations, establishing safe corridors for civilians, and prioritizing post-conflict reconstruction are essential elements of a comprehensive urban warfare strategy.

In conclusion, urban battle experiences underscore the complexity and multifaceted nature of city-based conflicts. Success in urban warfare requires a combination of adaptability, intelligence, civilian engagement, logistical innovation, technological prowess, specialized training, and humanitarian sensitivity. As urbanization continues to grow worldwide, these lessons and strategies will remain vital for addressing the challenges of future urban conflicts.

Chapter 12

PREPARING FOR URBAN WARFARE: TRAINING AND SIMULATION

Training Methods for Troops in Urban Warfare Environments

Training troops for urban warfare environments is indeed a complex and multifaceted process that necessitates specialized methods to prepare soldiers for the unique challenges posed by urban combat scenarios. Urban warfare presents distinctive obstacles such as dense infrastructure, civilian presence, limited visibility, and unpredictable combat scenarios, which require tailored training approaches that address both technical and psychological demands to ensure operational effectiveness and minimize risks [17, 23].

One of the primary methods for preparing soldiers for urban combat is the establishment of training facilities that simulate urban environments. These facilities are designed to replicate various urban terrains, including buildings with multiple levels, rooftops, and access points. They often incorporate civilian role-players, booby traps, and the use of paint ammunition to create realistic combat scenarios [23]. Such environments allow soldiers to practice navigation, tactical manoeuvres, and close-quarters combat, which are essential skills in urban warfare. The realistic conditions help troops adapt to the sensory overload typical of urban combat, including noise and confined spaces, while also allowing commanders to refine coordination and decision-making processes under pressure [17, 23].

Figure 82: Embarked security team Sailors assigned to Coastal Riverine Group 1 (CRG 1), Det. Guam, prepare to enter a structure during tactical team movement and urban warfare training in Barrigada, Guam. U.S. NAVY, Public Domain, via Picryl.

Wargames also play a crucial role in urban warfare training. Custom-designed scenarios enable larger groups to engage in simulated combat, where individuals can assume various roles, including enemy combatants, local populations, and host nation forces. This approach not only enhances tactical understanding but also fosters an appreciation for the complexities of urban operations, including the necessity of maintaining civilian safety and infrastructure [23]. Furthermore, preliminary training in urban combat skills, urban marksmanship, and physical fitness is essential. Soldiers must master individual movement techniques, battle drills, and weapon positioning specific to urban environments to enhance their effectiveness in combat situations [23].

Figure 83: Georgian soldiers from the 1st Infantry Brigade's NATO Response Force company keep a lookout during an urban warfare training exercise with the 173rd Airborne Brigade and the Georgian 1st Infantry Brigade as part of Exercise Noble Partner here May 17. Noble Partner is a combined U.S. Army Europe-Georgian army exercise designed to increase interoperability between Georgia's contribution to the NATO Response Force and allied militaries. U.S. Army Europe Images from Wiesbaden, Germany, CC BY 2.0, via Wikimedia Commons.

The development of urban training estates, such as the British Army's "Afghan village" and the French Army's urban training areas, exemplifies the commitment to realistic training environments. These estates provide soldiers with opportunities to practice urban operations in settings that closely resemble real-world scenarios, thereby improving their readiness for actual combat [23]. Additionally, the incorporation of masked combat tactics, which emphasize concealment and deception, is vital for minimizing exposure while manoeuvring in urban settings [23].

Moreover, training for urban warfare must encompass various other considerations. Soldiers should practice urban movement extensively, both inside and outside buildings, and develop skills for hasty obstacle reduction and breaching operations [23]. Training for urban defence is equally important, as soldiers must be prepared for defensive operations in addition to offensive manoeuvres. Current survival doctrines often overlook urban environments, necessitating a re-evaluation of survival training to include urban-specific strategies [23]. Effective communication is paramount, particularly for new infantry soldiers, as urban combat scenarios often require coordinated efforts among multiple units and supporting elements [23].

Incorporating advanced technological tools such as virtual reality (VR) and augmented reality (AR) systems into training programs is increasingly common. These technologies allow soldiers to engage in immersive training experiences that simulate urban combat scenarios, enabling them to practice tactical operations and interact with virtual civilians [23]. The adaptability of VR and AR technologies allows for tailored scenarios that reflect specific mission objectives, providing a safe environment for soldiers to learn and make mistakes without real-world consequences [23].

Finally, addressing the psychological resilience of troops is a critical aspect of urban warfare training. The stress associated with urban combat, including exposure to civilian suffering and close-quarters engagements, can significantly impact mental health. Training programs often include stress inoculation exercises to prepare soldiers for high-stress scenarios, alongside access to mental health support and counselling [106, 559]. The emphasis on psychological resilience is vital for maintaining operational effectiveness and ensuring that soldiers can perform under pressure [106, 559].

Importance of Urban Warfare Simulation Facilities and Mock Cities

Urban warfare simulation facilities and mock cities are critical components of modern military training, offering realistic and controlled environments to prepare troops for the complexities of urban combat. These facilities replicate the physical, operational, and psychological challenges of urban warfare, allowing forces to develop the skills and strategies necessary to navigate densely populated, structurally complex, and dynamic urban settings effectively.

One of the primary advantages of urban warfare simulation facilities is their ability to replicate real-world environments. Modern mock cities are designed with features such as multi-level buildings, narrow alleyways, open plazas, marketplaces, and underground tunnels. These replicas mimic the spatial dynamics and obstacles found in urban combat zones, enabling troops to practice movement, positioning, and coordination in settings similar to those they will encounter in actual operations. Such facilities often include functional elements like public transportation systems, industrial zones, and residential neighbourhoods to create varied scenarios.

These simulations also allow for the integration of advanced technologies, such as virtual and augmented reality, to enhance training. By incorporating these technologies, mock cities can simulate dynamic combat scenarios, such as sudden ambushes, civilian evacuations, or infrastructure collapses, in ways that are safe yet challenging for trainees. Troops can experience diverse combat conditions, including day and night operations, adverse weather, and cyber or electronic warfare elements, which are often encountered in urban settings.

Urban warfare simulation facilities are invaluable for refining the coordination of combined arms operations. Urban combat often requires close collaboration between infantry, armour, air support, and intelligence units. Mock cities provide a platform to rehearse these joint operations, allowing teams to test communication systems, tactical synchronization, and rapid decision-making in high-pressure environments. This training ensures that all units understand their roles and can operate seamlessly during real combat scenarios.

Another critical aspect of these facilities is their role in improving soldiers' psychological preparedness. Urban warfare is notoriously stressful, with troops facing risks from sniper fire, IEDs, and civilian entanglements. Mock cities expose soldiers to high-intensity simulations that replicate the chaos and unpredictability of urban battles, helping them develop resilience and decision-making skills under duress. This training is essential for mitigating the cognitive overload and emotional strain associated with real-world combat.

For commanders, urban warfare simulation facilities offer opportunities to test and refine strategies without the immediate risks of live operations. Commanders can evaluate the effectiveness of tactics, adapt to evolving urban combat trends, and identify weaknesses in planning and execution. The use of mock cities for large-scale drills allows for the assessment of command and control systems, logistical support, and rules of engagement in urban warfare contexts.

In addition to their military applications, these facilities are also critical for training allied forces, peacekeepers, and law enforcement agencies. International collaboration in mock cities fosters interoperability among allied militaries and equips peacekeeping units with the skills needed for stability operations in urban areas. Law enforcement can use these facilities for counter-terrorism and riot control training, further extending their utility beyond traditional military purposes.

Role of Virtual Reality (VR) And Artificial Intelligence in Urban Warfare Training

Virtual Reality (VR) and Artificial Intelligence (AI) have emerged as transformative tools in urban warfare training, revolutionizing how military forces prepare for the complexities of combat in dense urban environments. These technologies offer unparalleled realism, adaptability, and precision, addressing many of the challenges inherent in traditional training methods and significantly enhancing combat readiness.

Virtual Reality provides an immersive training environment that replicates urban combat scenarios with high fidelity. Unlike conventional mock cities or physical training grounds, VR allows trainees to engage in diverse, customizable environments, including multi-story buildings, crowded marketplaces, underground tunnels, and industrial zones. These simulations can be programmed to include varying levels of difficulty, environmental conditions, and mission objectives, offering a scalable and dynamic training experience.

One of VR's significant advantages is its ability to create realistic, high-pressure situations without the risks associated with live training exercises. Soldiers can practice room-clearing techniques, hostage rescues, or counter-IED operations in a controlled digital environment. The ability to pause, rewind, and review these simulations enables trainees to analyse their performance, learn from mistakes, and refine their tactics.

VR also enhances teamwork and coordination in urban warfare training. Multiplayer VR platforms allow entire units to participate in simulations, practicing communication and synchronization in scenarios that mimic real-world combat. By simulating scenarios like ambushes, civilian evacuations, or infrastructure collapses, VR ensures that soldiers are prepared to operate cohesively under the stress and unpredictability of urban combat.

Artificial Intelligence plays a complementary and critical role in augmenting VR simulations and other training methodologies. AI-driven virtual adversaries can mimic the behaviour of insurgents, terrorists, or conventional forces, providing realistic and adaptive opposition. These AI entities respond dynamically to the trainees' actions, forcing soldiers to adapt their strategies in real-time, mirroring the unpredictability of actual combat.

AI also enhances the realism of urban environments by simulating civilian behaviour, traffic patterns, and the effects of environmental factors like fire, smoke, or flooding. This adds layers of complexity to the training, teaching soldiers to make split-second decisions while navigating ethical dilemmas and minimizing collateral damage.

Another significant application of AI in urban warfare training is data analysis and performance feedback. AI algorithms can track individual and unit performance metrics during training, such as reaction times, accuracy, communication efficiency, and adherence to mission objectives. This data-driven feedback provides actionable insights, enabling instructors and trainees to identify strengths and weaknesses, adjust training regimens, and continuously improve.

The integration of VR and AI offers a synergistic approach to urban warfare training. For instance, VR environments powered by AI can generate unpredictable scenarios, ensuring that no two training sessions are identical. This variability prevents trainees from becoming overly reliant on scripted exercises, promoting adaptability and critical thinking.

Additionally, VR and AI integration facilitates the development of custom training programs tailored to specific mission requirements or geographic locations. AI can analyse real-world data from past conflicts to recreate scenarios reflective of actual urban combat experiences, such as the battles in Mosul, Aleppo, or Mariupol. This specificity ensures that soldiers are not only trained generically but are also equipped for the unique challenges of their assigned missions.

The use of VR and AI in urban warfare training offers significant advantages over traditional methods. These technologies reduce costs associated with building and maintaining physical training facilities, minimize the environmental impact of live exercises, and eliminate the risks of injuries during training. They also provide a level of flexibility and scalability that physical environments cannot match, allowing for training in diverse urban settings and under varied operational conditions.

As VR and AI technologies continue to advance, their role in urban warfare training is expected to expand. Emerging developments include integrating haptic feedback devices to simulate physical sensations, such as recoil or impact, and using AI-powered virtual instructors to guide trainees through complex scenarios. Advances in AI-driven predictive analytics may also allow for the simulation of future combat scenarios based on emerging trends and intelligence.

Virtual Reality and Artificial Intelligence are reshaping the landscape of urban warfare training, offering immersive, adaptive, and data-driven methodologies that prepare soldiers for the challenges of modern combat. By combining the strengths of these technologies, militaries can enhance the effectiveness, safety, and efficiency of their training programs, ensuring readiness for the complexities of urban warfare.

Preparing Soldiers Psychologically and Physically for Urban Combat

Understanding the complex interplay between the brain and body during combat is a critical focus within military science. Combat environments exert significant psychological and physiological pressures on soldiers, with each combat type presenting unique challenges that influence performance and well-being.

Investigating these responses allows for the development of tailored strategies to prepare soldiers for diverse battle scenarios [560].

Symmetric combat is characterized by engagements between adversaries with comparable military capabilities and resources. These battles are defined by their intensity, as both sides employ advanced technology, skilled personnel, and tactical acumen in a bid to outmanoeuvre each other. The equilibrium in power amplifies the psychological strain on soldiers, as the outcomes hinge heavily on strategic precision and adaptability [560].

Soldiers in symmetric combat often experience heightened stress due to the unpredictability of evenly matched confrontations. This acute stress response manifests through elevated levels of adrenaline and cortisol, enhancing immediate performance but risking long-term health impacts if not managed. On the battlefield, soldiers must maintain high levels of vigilance and resilience to navigate the uncertainty inherent in such encounters. From a tactical perspective, symmetric combat necessitates mastery of conventional warfare techniques, such as coordinated firepower and mobility. Rigorous training programs that incorporate advanced simulation technologies are essential for preparing soldiers to respond effectively to the high-stakes scenarios typical of symmetric engagements [560].

Leadership plays a pivotal role in symmetric combat, requiring a balance of tactical expertise and morale maintenance under intense conditions. Effective communication and decision-making are paramount, with leaders tasked to sustain operational efficiency while safeguarding the mental well-being of their units. Furthermore, symmetric combat often drives innovation in military technology, as competing forces strive for superiority through advancements in weaponry, communication systems, and defensive measures. The psychological support for soldiers engaged in symmetric combat is equally critical, addressing the risks of anxiety, PTSD, and combat-related mental health issues through counselling and resilience-building initiatives [560].

Asymmetric combat involves engagements between forces with stark disparities in strength, resources, or tactics. The more powerful side often relies on conventional warfare methods, while the weaker side employs guerrilla tactics, ambushes, and improvisation to exploit vulnerabilities. This imbalance introduces unique stressors, as soldiers confront unpredictable threats that demand rapid adaptation and heightened situational awareness [560].

Psychologically, asymmetric combat imposes enduring uncertainty and fear. Soldiers face constant risks from unconventional attacks such as improvised explosive devices (IEDs) and hidden ambushes, leading to chronic stress and anxiety. Prolonged exposure to these unpredictable conditions can erode mental resilience, necessitating robust psychological support systems to mitigate the long-term effects [560].

From a physiological standpoint, the sporadic nature of asymmetric engagements results in fluctuating stress hormone levels, affecting cognitive function and overall health. Soldiers require training in flexibility and adaptability, mastering counterinsurgency tactics and survival skills to navigate the dynamic challenges of asymmetric combat effectively. Advanced technologies, such as surveillance drones and reconnaissance tools, are vital in asymmetric warfare, providing critical intelligence to counter unconventional strategies. Leaders in asymmetric scenarios must exhibit exceptional agility, making swift decisions in ambiguous and resource-constrained environments while fostering unit cohesion and morale [560].

Ethical and legal considerations in asymmetric warfare add another layer of complexity. Upholding international norms and rules of engagement becomes challenging against adversaries who may exploit civilian populations or infrastructure. This dynamic underscores the need for continuous dialogue on military ethics and legal frameworks to address the evolving challenges of asymmetric conflict [560].

Combat in specialized environments, including close-quarters, subterranean, and urban settings, demands tailored approaches to address their distinct challenges. These scenarios introduce physical and psychological hurdles that test soldiers' preparedness and adaptability [560].

Close-Quarters Combat: Close-quarters combat (CQC) occurs in extremely confined spaces, requiring intense physical fitness and advanced combat skills. Soldiers must master rapid reflexes, agility, and decision-making under extreme stress. Physiological responses, such as adrenaline surges and elevated heart rates, enhance performance but can also lead to fatigue. Rigorous training programs that combine martial arts, tactical simulations, and psychological conditioning are critical for equipping soldiers to navigate the intensity of CQC scenarios [560].

Subterranean Warfare: Subterranean warfare involves combat in tunnels and underground facilities, presenting unique challenges such as confined spaces, poor visibility, and limited communication. Soldiers must overcome disorientation, claustrophobia, and environmental stressors like high humidity and low air quality. Specialized training in navigation, communication, and small-unit tactics is essential for operational effectiveness in these settings. Simulations replicating subterranean conditions prepare soldiers to maintain situational awareness and mental resilience in these demanding environments [560].

Urban Combat: Urban warfare, particularly in regions like the Middle East, combines the complexities of dense terrains, civilian populations, and unconventional tactics. Soldiers must navigate narrow streets, towering buildings, and urban chokepoints while distinguishing between combatants and non-combatants. Advanced surveillance technologies, such as UAVs and remote sensing tools, enhance intelligence gathering and situational awareness. Effective training regimens incorporating building clearance, urban sniping, and close-quarters engagements are indispensable for preparing soldiers for urban operations. Additionally, cultural sensitivity and collaboration with local populations are crucial for mission success in urban settings [560].

Understanding the psychological and physiological dimensions of symmetric, asymmetric, and specialized combat environments informs the development of comprehensive military training and operational strategies. Emphasizing mental resilience, advanced tactical skills, and technological integration ensures that soldiers are prepared to meet the challenges of modern warfare. Recognizing the unique demands of each combat type allows military forces to adapt their approaches, safeguarding the well-being and effectiveness of soldiers in the complex and dynamic landscapes of contemporary conflict [560].

The Physical Demands of Urban Warfare: The Need for Rigorous Preparation

Urban warfare imposes unparalleled physical demands on soldiers, requiring a level of fitness that surpasses what is needed in other combat environments. The complex terrain, confined spaces, and vertical challenges inherent in urban settings make physical preparation a cornerstone of military readiness. The events of 9/11 serve as a stark reminder of these demands, with firefighters ascending 110 flights of stairs while carrying up to 75 pounds of gear. This extraordinary feat, often replicated in memorial workouts, underscores the gruelling physical challenges soldiers face in urban combat, where such exertion is compounded by the stressors of warfare and sustained over days or even weeks [561].

Current military fitness programs, such as the Army Combat Fitness Test (ACFT), provide a baseline for physical readiness but fall short of fully preparing soldiers for the unique demands of urban combat. While the ACFT includes events designed to mimic combat tasks, such as lifting casualties and sprinting with gear, it cannot encompass the full spectrum of challenges posed by urban warfare. Recent changes to the ACFT, including the removal of the leg tuck and a reduction in deadlift standards for female soldiers, further dilute its relevance to urban combat scenarios. These alterations inadvertently diminish the preparation of soldiers for tasks like scaling walls or carrying comrades over rugged terrain [561].

The Army's Holistic Health and Fitness (H2F) program, outlined in FM 7-22, emphasizes a triad of strength, speed, and endurance, providing a valuable framework for enhancing physical fitness. However, the lack of specificity in identifying occupational tasks limits its applicability to urban warfare. To bridge this gap, units and individual soldiers must go beyond ACFT standards, incorporating tailored physical training programs that address the unique rigors of urban combat [561].

Urban warfare demands both strength and cardiovascular endurance. Soldiers must be capable of lifting, pushing, and carrying heavy loads, including casualties and equipment, while maintaining the agility and speed to navigate dangerous urban terrain. A soldier with exceptional endurance but insufficient strength to carry a wounded comrade is operationally ineffective. Training programs should focus on functional strength, emphasizing exercises like squats, deadlifts, and carries that replicate combat tasks. Hypertrophy training, which balances muscular strength with endurance, is particularly effective, aiming for repeated heavy lifts rather than maximum loads [561].

Cardiovascular endurance remains critical, as the movement to and within the battlefield often involves climbing, sprinting, and prolonged exertion under load. Urban environments exacerbate these challenges, with soldiers navigating multi-story buildings, broken ground, and steep gradients. Sprint training, combined with strength exercises targeting the hamstrings and glutes, is essential for preparing soldiers for high-stakes movements across open kill zones. Such training not only builds physical resilience but also enhances the mental fortitude required to persevere under stress [561].

Close-quarters combat in urban settings demands unparalleled physical strength and a warfighting mindset. Soldiers must be prepared for hand-to-hand encounters, requiring not only brute force but also the psychological resilience to overcome fear and fatigue. Regular training in combatives, such as wrestling, mixed martial arts (MMA), and the Army combatives program, builds the physical and mental stamina needed for these intense confrontations. This type of training mirrors the full-body exertion and unpredictability of combat, providing a level of preparation unmatched by traditional cardio or weightlifting routines [561].

The evacuation of casualties in urban warfare presents a unique set of challenges. Soldiers must transport injured comrades across obstructed and hazardous terrain, often while under fire. This requires exceptional anaerobic endurance, grip strength, and coordination. Litter carries and buddy carries become exponentially more difficult in the confined and treacherous spaces of urban environments. Training programs must include these tasks as core components, preparing soldiers for the physical and mental strain of moving casualties under combat conditions [561].

Rigorous physical training also serves to enhance mental toughness, a crucial attribute for urban combat. Tough workouts not only release endorphins and improve self-esteem but also foster unit cohesion and individual resilience. Soldiers who train together in challenging conditions build trust and camaraderie, creating a supportive environment that encourages personal growth. This shared experience strengthens the collective will to persevere in the face of adversity, a quality essential for success in urban warfare [561].

Urban combat is one of the most physically demanding forms of warfare, requiring soldiers to prepare for unparalleled levels of strain and complexity. Current fitness standards provide a starting point but must be augmented with rigorous, combat-specific training programs that address the unique challenges of urban environments. By prioritizing strength, endurance, and psychological resilience, military forces can ensure that soldiers are not only ready to meet these challenges but also capable of thriving in the demanding and unpredictable conditions of urban warfare. Early investment in such preparation is essential to securing victory in future conflicts [561].

Psychophysiology across Military Branches of Service

The psychophysiological responses of military personnel differ significantly across branches of service, shaped by the specific demands, roles, and operational environments inherent to each division. Understanding these differences is critical for enhancing psychological resilience, operational readiness, and the development of tailored support systems that meet the unique challenges faced by members of each branch [560].

Army personnel, often engaged in ground combat, encounter a distinct set of psychophysiological challenges. Their roles demand extended periods of heightened alertness, strenuous physical exertion, and direct exposure to life-threatening situations. The mental strain associated with these high-stress environments can impair cognitive functions and decision-making abilities, necessitating robust mental resilience. Physically, soldiers face the demands of carrying heavy equipment, enduring long marches, and operating under extreme fatigue, which collectively impact their endurance, overall health, and cognitive performance. The cumulative effect of these stressors highlights the need for comprehensive training programs that address both physical and psychological readiness [560].

Air Force personnel, particularly pilots, face unique psychophysiological stressors tied to their specialized roles. High G-forces during flight, the need for acute visual and sensory perception, and the mental demands of managing complex aircraft systems create a distinct array of challenges. Prolonged flights, aerial combat, and surveillance missions add layers of mental strain that can affect neuropsychological health. Research underscores the importance of understanding these effects to develop interventions that mitigate stress and optimize performance. Tailored training in stress management, alongside strategies for maintaining focus under extreme physical and cognitive pressures, is essential for this branch [560].

Naval personnel, including those on submarines, operate in environments characterized by prolonged isolation, confinement, and the absence of natural light. These conditions disrupt circadian rhythms, elevate stress levels, and can adversely affect mental health. Submarine crews, in particular, contend with unique challenges such as underwater confinement and limited personal space, which exacerbate fatigue and psychological strain. Prolonged deployments at sea also contribute to stress, necessitating mental health support systems that address these specific conditions. Training programs that enhance adaptability to isolation and include psychological support mechanisms are critical for sustaining operational readiness in naval environments [560].

Marine Corps personnel navigate stressors from both naval and ground combat, particularly in the context of amphibious and expeditionary warfare. The dynamic and often unpredictable nature of amphibious operations amplifies psychological stress, requiring Marines to adapt quickly to changing scenarios. Both combat and non-combat stressors, such as extended deployments and the uncertainty of missions, contribute to cumulative mental health challenges. Research emphasizes the importance of resilience training and psychological support tailored to the dual demands of naval and ground environments. Such measures ensure that Marines maintain mental and physical readiness for their diverse operational roles [560].

Special operations forces, including Navy SEALs and Army Special Forces, endure the most extreme physical and psychological challenges within the military. Rigorous training prepares them for high-risk missions that demand exceptional mental toughness, emotional regulation, and physical endurance. Covert operations in hostile environments add layers of stress, while intense training regimens impose significant physiological strain. Research highlights the necessity of specialized health monitoring and support systems to address these unique challenges. Strategies for maintaining psychological resilience and managing the effects of operational stress are vital for sustaining the performance and well-being of these elite units [560].

The psychophysiological variability across military branches underscores the complexity of military operations and the need for nuanced strategies to support personnel effectively. Each branch faces distinct challenges, from the physical endurance required for ground combat to the mental strain of aerial missions or the psychological toll of isolation at sea. Tailored approaches that integrate specialized training, mental health support, and operational readiness programs are essential for optimizing performance and safeguarding the well-being of military personnel [560].

Military environments impose unique and intense stressors on soldiers, resulting in the manifestation of acute and chronic stress, each with distinct psychophysiological implications. Acute stress, characterized by immediate and high-intensity situations, triggers the body's "fight or flight" response, releasing adrenaline and cortisol. This hormonal surge heightens heart rate, increases alertness, and boosts energy levels, enabling soldiers to respond swiftly to immediate threats. Research indicates that acute stress can temporarily enhance short-term memory and cognitive functions, allowing for quick, potentially life-saving decisions. However, this heightened state is unsustainable and, if prolonged, can lead to fatigue and cognitive overload, diminishing operational effectiveness [560].

Conversely, chronic stress develops gradually due to prolonged exposure to challenging conditions, such as lengthy deployments or continuous combat. Unlike the rapid onset of acute stress, chronic stress maintains a persistent state of heightened alertness, causing sustained release of stress hormones. This disruption to normal bodily functions can lead to sleep disturbances, weakened immune response, and increased cardiovascular risk. Psychologically, chronic stress is closely linked to conditions such as depression, anxiety, and post-traumatic stress disorder (PTSD), profoundly affecting mental health and long-term resilience [560].

The interplay between acute and chronic stress in military contexts adds complexity to soldiers' psychophysiological responses. Acute stress can provide short-term benefits, but repeated exposure to acute stressors without adequate recovery transitions soldiers into a chronic stress state. This prolonged stress compromises cognitive functions, operational readiness, and overall health. Soldiers often remain in a heightened state of alertness even in the absence of immediate threats, a phenomenon that exacerbates mental and physical fatigue over time [560].

Effectively managing both acute and chronic stress is vital for preserving soldiers' well-being and operational capabilities. Training programs and support systems that emphasize stress resilience and management are pivotal. Evidence supports the effectiveness of interventions such as cognitive-behavioural techniques, mindfulness, and relaxation training in mitigating the impacts of chronic stress. These practices enhance soldiers' coping strategies, enabling them to navigate the challenges of acute stress while reducing the likelihood of transitioning into chronic stress states [560].

Comprehensive mental health support systems, including access to counselling and psychological resources, play an integral role in addressing chronic stress. Encouraging a supportive environment within military units fosters camaraderie and shared resilience, mitigating feelings of isolation that often accompany chronic stress. Additionally, leadership training emphasizing the recognition of stress indicators and proactive intervention further ensures the well-being of personnel under prolonged stress [560].

Psychological training in military settings emphasizes the critical interplay between mental fortitude and physical readiness. Resilience training, a cornerstone of psychological preparation, is designed to enhance soldiers' capacity to withstand, recover from, and thrive amidst stressors. Studies on veterans demonstrate that resilience training significantly improves adaptability and mental health, underscoring its transformative impact [560].

Stress inoculation training (SIT) is another pivotal component, exposing soldiers to controlled stressors in a secure environment to build coping skills for real-world scenarios. This method has proven effective in reducing anxiety and enhancing performance under stress, equipping soldiers for high-pressure environments. Additionally, mindfulness training, exemplified by programs like Mind Fitness Training (MFT), enhances cognitive resilience, emotional regulation, and situational awareness, offering a holistic approach to managing stress [560].

Cognitive-behavioural therapy (CBT), tailored for military contexts, addresses combat-related mental health challenges, including PTSD and anxiety. CBT's application has yielded significant symptom reduction, improving psychological well-being and operational focus. Emerging technologies such as virtual reality (VR) add an innovative dimension to psychological training, replicating combat scenarios for exposure therapy and enhancing training authenticity [560].

Understanding and addressing the distinct impacts of acute and chronic stress are essential for fostering the well-being and effectiveness of military personnel. While acute stress can enhance immediate performance, the risks associated with chronic stress necessitate proactive management through tailored training programs and comprehensive mental health support systems. The integration of advanced psychological training techniques, such as resilience building, mindfulness, CBT, and VR, reflects a progressive approach to preparing soldiers for the complex demands of modern military life. By embracing these strategies, military organizations can safeguard the long-term health and operational readiness of their personnel, ensuring success in the face of evolving challenges [560].

Real-World Military Training Applications

The evolution of military training from traditional drills to advanced, real-world applications marks a pivotal shift in preparing soldiers for the complexities of modern warfare. This transformation goes beyond physical readiness and tactical proficiency, emphasizing realistic, immersive, and multifaceted training programs that closely replicate authentic combat scenarios. These innovations bridge the gap between controlled training environments and the unpredictable nature of actual combat, ensuring soldiers are equipped to face diverse challenges in the field [560].

A cornerstone of this modern approach is the use of advanced simulation systems, incorporating technologies like Virtual Reality (VR) and Augmented Reality (AR). These systems immerse soldiers in environments that mirror real-world combat scenarios, enabling them to refine their decision-making, navigation, and tactical skills in a controlled and repeatable setting. VR simulations recreate specific geographical terrains and combat situations, providing soldiers with targeted preparation for their deployment regions. Research consistently highlights the benefits of these simulations, showing significant improvements in spatial awareness, reaction times, and cognitive performance under stress. By immersing soldiers in dynamic, technology-driven scenarios, these systems transform how critical skills are acquired and honed [560].

Another groundbreaking development in military training is the integration of live, virtual, and constructive (LVC) environments. LVC training combines real-world exercises, virtual simulations, and computer-generated forces to create comprehensive scenarios that mimic the unpredictability of combat. This approach allows soldiers to practice managing civilian interactions, responding to unforeseen events, and making rapid decisions in dynamic situations. Studies underscore the effectiveness of LVC training in enhancing situational awareness and adaptability—key competencies for modern military operations. This holistic methodology ensures soldiers are well-prepared for the multifaceted challenges of contemporary combat environments [560].

Physical and psychological resilience training also plays a critical role in modern military readiness. Programs emphasizing endurance, strength, and agility prepare soldiers for the physical demands of combat, while stress inoculation and mental toughness exercises fortify them against the psychological pressures of warfare. This dual approach integrates physical resilience with psychological adaptability, fostering a robust foundation for handling the stresses of military life. Research supports the efficacy of comprehensive resilience training, citing improvements in performance, reduced injury rates, and enhanced mental health outcomes among personnel. By addressing both physical and mental dimensions, these programs provide a holistic framework for soldier preparedness [560].

Wearable technology and biometrics have further revolutionized military training by enabling real-time monitoring of soldiers' physiological responses during exercises. Metrics such as heart rate, stress levels, and fatigue are tracked, offering instructors actionable insights to tailor training scenarios dynamically. This data-driven approach ensures that exercises replicate the physical and mental demands of combat while minimizing the risk of injury. Studies validate the impact of wearable technology, showcasing its ability to optimize training regimes and enhance soldier readiness for the complexities of modern operations [560].

Cultural and linguistic training has also become essential in preparing soldiers for global missions. Understanding local customs and languages is crucial for effective communication and fostering positive relationships with local populations. Training initiatives that include language courses and immersive cultural exercises have proven effective in augmenting soldiers' ability to operate in diverse environments. By building linguistic proficiency and cultural awareness, soldiers not only improve mission effectiveness but also contribute to diplomacy and cooperation, vital in today's interconnected world [560].

Finally, integrating unmanned systems and robotics into training reflects the rapid technological advancements shaping modern warfare. Soldiers engage in hands-on exercises with drones, uncrewed vehicles, and robotic systems, gaining proficiency in their use for surveillance, logistics, and combat support. These technologies enhance operational capabilities, and training soldiers to navigate their complexities ensures they are prepared for the dynamic demands of contemporary battlefields. Familiarity with unmanned systems fosters adaptability and maximizes mission success in technologically advanced operational environments [560].

In conclusion, the shift from traditional drill-based training to cutting-edge real-world applications represents a transformative evolution in military preparedness. By incorporating advanced simulations, LVC environments, resilience training, wearable technology, cultural immersion, and robotics, modern military training equips soldiers with the skills, experience, and adaptability required to navigate the complexities of contemporary warfare. These innovations not only enhance operational effectiveness but also ensure the holistic development of soldiers as they confront the multifaceted challenges of modern combat [560].

Exercises, Drills, and Scenarios to Improve Tactical Readiness

Tactical readiness is a critical component of military operations, particularly in urban environments where the complexities of combat demand precision, adaptability, and seamless coordination. Exercises, drills, and scenarios play an integral role in preparing military forces for the challenges of urban warfare, ensuring that both individuals and units are equipped to operate effectively under high-pressure conditions. These training methods are meticulously designed to simulate real-world combat situations, fostering skills that are essential for success in the field.

Military exercises are comprehensive training programs that simulate large-scale operations involving multiple units and integrated components such as infantry, artillery, air support, and intelligence. These exercises are conducted in controlled environments, which may include live-fire ranges, urban mock cities, or digital simulators. They are designed to test operational plans, enhance inter-unit coordination, and prepare forces for specific mission objectives.

Urban combat exercises often incorporate elements such as building clearance operations, civilian evacuations, counter-insurgency tactics, and logistics management. For example, a multi-day urban warfare exercise might involve securing a high-value target in a densely populated area, managing supply lines under fire, or coordinating with allied forces to neutralize threats. These exercises not only refine tactical skills but also provide commanders with insights into the strengths and weaknesses of their strategies, enabling iterative improvements.

Drills focus on the repetition of specific tasks or manoeuvres to build muscle memory and ensure that actions can be performed instinctively under stress. They are a fundamental component of tactical readiness, emphasizing precision, discipline, and consistency. Drills are particularly effective for individual and small-unit training, allowing soldiers to master essential skills before integrating them into larger exercises.

In urban warfare, drills may include breaching and entering techniques, room clearing, establishing defensive perimeters, and operating in low-visibility conditions. For example, a team might practice entering and securing a building repeatedly until every member can execute their role flawlessly. These drills are often conducted with increasing levels of complexity, such as adding live ammunition, simulated enemy combatants, or civilian actors to test decision-making and situational awareness.

Scenarios are dynamic training modules that replicate specific combat situations, challenging soldiers to apply their skills in realistic and unpredictable environments. Scenario-based training bridges the gap between theoretical knowledge and practical application, fostering critical thinking and adaptability.

Urban warfare scenarios are particularly diverse, ranging from counter-terrorism operations in high-rise buildings to humanitarian missions in conflict zones. For instance, a scenario might simulate a hostage rescue operation where soldiers must navigate narrow corridors, deal with booby traps, and negotiate with armed adversaries while minimizing collateral damage. Another scenario could involve responding to an ambush in a marketplace, requiring rapid tactical adjustments and coordinated team movements to neutralize threats and secure the area.

Modern exercises, drills, and scenarios increasingly incorporate advanced technologies such as virtual reality (VR), augmented reality (AR), and artificial intelligence (AI). These tools enhance the realism and adaptability of training environments. For example, VR simulations can recreate urban combat scenarios with high fidelity, while AI-driven adversaries respond dynamically to the actions of trainees, ensuring that no two sessions are identical.

Additionally, wearable devices and sensors can track physiological and performance metrics during drills and scenarios, providing data-driven feedback to improve training outcomes. For instance, heat maps generated from movement data can identify areas where soldiers hesitate or bottlenecks occur, allowing instructors to address these issues in subsequent training sessions.

A key focus of exercises, drills, and scenarios is fostering teamwork and interoperability. Urban combat requires seamless coordination among soldiers, units, and even allied forces. Training modules often emphasize communication protocols, role clarity, and mutual support. For example, drills might simulate the synchronization of sniper teams and assault squads during a building clearance or the integration of aerial drones with ground operations for real-time intelligence sharing.

Scenarios that involve multi-unit operations, such as securing a city block or conducting joint counter-insurgency missions, test the ability of forces to operate cohesively under complex and evolving conditions. These scenarios are particularly valuable for building trust and understanding among different units and branches of the military.

As the nature of urban combat evolves, so too must the exercises, drills, and scenarios used to prepare military forces. Increasingly, training programs address hybrid threats, including cyberattacks, electronic warfare, and the use of drones by adversaries. Scenarios now often include elements like GPS jamming, infrastructure sabotage, and misinformation campaigns to simulate the multifaceted challenges of modern warfare.

Moreover, urban combat training is adapting to include humanitarian and peacekeeping scenarios, reflecting the diverse roles that military forces play in contemporary conflicts. These scenarios emphasize the protection of civilians, adherence to rules of engagement, and collaboration with non-military actors.

Conclusion: The Evolving Future of Urban Warfare

Urban warfare remains one of the most challenging and multifaceted domains of modern military conflict, requiring comprehensive strategies and an adaptive approach to ensure success while minimizing harm. This discussion has highlighted several key insights and strategies central to mastering urban combat. The dense, unpredictable environment of cities demands a robust blend of physical preparedness, psychological resilience, technological integration, and culturally informed operations. Critical lessons have emerged from past and ongoing conflicts, emphasizing the importance of dynamic strategies and the ability to balance military objectives with the imperative of civilian protection.

The evolving nature of urban warfare underscores the transformative role of technological advancements. Tools such as virtual reality for immersive training, unmanned systems for reconnaissance and support, and augmented reality for situational awareness have redefined how soldiers prepare for and engage in combat. These technologies not only enhance precision and efficiency but also provide invaluable support in reducing collateral damage and ensuring ethical conduct in urban settings.

Adaptive tactics and continuous learning are vital in this dynamic landscape. The unpredictability of urban combat necessitates flexible strategies that account for both conventional and unconventional threats. Lessons learned from operations in cities like Aleppo, Fallujah, and Mariupol have demonstrated the importance of integrating real-time intelligence, fostering unit cohesion, and training soldiers to operate effectively under physical and psychological stress. Moreover, the emphasis on resilience training and mental health support has proven essential in maintaining operational readiness and safeguarding the well-being of personnel.

Looking to the future, urban warfare presents significant implications for military forces and policymakers. As urbanization continues to expand, conflicts are increasingly likely to unfold in cities, necessitating the prioritization of urban-specific doctrines and resources. Policymakers must invest in advanced training facilities, innovative technologies, and robust frameworks for international cooperation to address the unique challenges of urban conflict. Furthermore, ethical considerations, including civilian protection and compliance with international humanitarian law, must remain at the forefront of military planning and execution.

Ultimately, urban warfare is not merely a test of military might but also a measure of strategic ingenuity and moral responsibility. By embracing adaptive tactics, fostering continuous learning, and leveraging technological advancements, military forces can navigate the complexities of urban combat effectively. At the same time, the commitment to protecting civilian populations and mitigating the long-term impacts of conflict must guide every decision. In doing so, the future of urban warfare can evolve toward greater precision, humanity, and accountability, ensuring that military operations align with the broader goal of achieving sustainable peace and stability.

REFERENCES

1. Graham, S., *When Life Itself Is War: On the Urbanization of Military and Security Doctrine.* International Journal of Urban and Regional Research, 2011. **36**(1): p. 136-155.

2. Jurišić, D., *The Challenges of Military Operations in Urban Areas.* Vojno Delo, 2022. **74**(2): p. 75-94.

3. King, A., *Close Quarters Battle.* Armed Forces & Society, 2015. **42**(2): p. 276-300.

4. Xiong, Y., et al., *Survey on Urban Warfare Augmented Reality.* Isprs International Journal of Geo-Information, 2018. **7**(2): p. 46.

5. Bużantowicz, W. and P. Turek, *Autonomous Combat Support Vehicles in Urban Operations: Tactical and Technical Determinants.* Advances in Military Technology, 2020. **15**(1): p. 97-114.

6. Sánchez-Molina, J., J.J. Robles-Pérez, and V.J. Clemente-Suárez, *Effect of Parachute Jump in the Psychophysiological Response of Soldiers in Urban Combat.* Journal of Medical Systems, 2017. **41**(6).

7. Aguilera, J.F.T., V.F. Elias, and V.J. Clemente-Suárez, *Autonomic and Cortical Response of Soldiers in Different Combat Scenarios.* BMJ Military Health, 2020. **167**(3): p. 172-176.

8. Spanu, G., *Urban Military Geographies: New Directions in the (Re)production of Space, Militarism, and the Urban.* Geography Compass, 2023. **17**(12).

9. Liu, P.C., et al., *Effects of Female Body Size and Age and Male Mating Status on Male Combat in <scp><i>Anastatus Disparis</I></Scp> (Hymenoptera: Eupelmidae).* Ecological Entomology, 2020. **45**(5): p. 1071-1079.

10. Muñoz-Reyes, J.A., et al., *Fighting Ability Influences Mate Value In Late Adolescent Men.* Personality and Individual Differences, 2015. **80**: p. 46-50.

11. Fang, W., *Fighting Fish and Two-Stack Sortable Permutations.* 2017.

12. Liu, M.-S., et al., *Chromosomal Variations of Lycoris Species Revealed by FISH With rDNAs and Centromeric Histone H3 Variant Associated DNAs.* Plos One, 2021. **16**(9): p. e0258028.

13. Cui, X.-y. and J.-q. Diao, *Design of Human-Computer Interactive Fire Extinguishing Training System Based on Virtual Reality Technology.* 2020: p. 124-136.

14. Taglieri, I., et al., *Effects of Flaxseed Cake Fortification on Bread Shelf Life, and Its Possible Use as Feed for <scp><i>Tenebrio Molitor</I></Scp> Larvae in a Circular Economy: Preliminary Results.* Journal of the Science of Food and Agriculture, 2021. **102**(4): p. 1736-1743.

15. Liu, P. and D. Hao, *Effect of Variation in Objective Resource Value on Extreme Male Combat in a Quasi-Gregarious Species, Anastatus Disparis.* BMC Ecology, 2019. **19**(1).

16. Zuiderhoek, A., *The Ancient City.* 2016.

17. Ljungkvist, K., *A New Horizon in Urban Warfare in Ukraine?* Scandinavian Journal of Military Studies, 2022. **5**(1): p. 91-98.

18. Graham, S., *Cities and the 'War on Terror'.* International Journal of Urban and Regional Research, 2006. **30**(2): p. 255-276.

19. Coward, M., *Network-Centric Violence, Critical Infrastructure and the Urbanization of Security.* Security Dialogue, 2009. **40**(4-5): p. 399-418.

20. Zeitoun, M., et al., *Urban Warfare Ecology: A Study of Water Supply in Basrah.* International Journal of Urban and Regional Research, 2017. **41**(6): p. 904-925.

21. Hägerdal, N., *Starvation as Siege Tactics: Urban Warfare in Syria.* Studies in Conflict and Terrorism, 2020. **46**(7): p. 1241-1262.

22. Caro, C.J.V., *Unpacking the History of Urban Warfare and its Challenges in Gaza.* 2023, Stimson.

23. Massingham, E., E. Almila, and M. Piret, *War in Cities: Why the Protection of the Natural Environment Matters Even When Fighting in Urban Areas, and What Can Be Done to Ensure Protection.* International Review of the Red Cross, 2023. **105**(924): p. 1313-1336.

24. Khorram-Manesh, A., et al., *Review of Military Casualties in Modern Conflicts—The Re-Emergence of Casualties From Armored Warfare.* Military Medicine, 2021. **187**(3-4): p. e313-e321.

25. Bin Inam, Z. and S. Rauf, *Understanding Urban Warfare and its Manifestation in Russia-Ukraine War.* Strategic Studies, 2023. **43**(2): p. 97-115.

26. Fleming, C., et al., *Ethnic Economic Inequality and Fatalities From Terrorism.* Journal of Interpersonal Violence, 2020. **37**(11-12): p. NP9089-NP9114.

27. Slav, M., et al., *Deprivation, Instability, and Propensity to Attack: How Urbanization Influences Terrorism.* International Interactions, 2021. **47**(6): p. 1100-1130.

28. Bilukha, O., M. Brennan, and M.C. Anderson, *The Lasting Legacy of War: Epidemiology of Injuries From Landmines and Unexploded Ordnance in*

Afghanistan, 2002–2006. Prehospital and Disaster Medicine, 2008. **23**(6): p. 493-499.

29. Khan, I.D., *Grenade Explosion Leading to Penetrative Brain Trauma and Demise of Three Children During High Altitude Counter Insurgency Operations.* Journal of Archives in Military Medicine, 2016. **4**(4).

30. Bilukha, O., et al., *Seen but Not Heard: Injuries and Deaths From Landmines and Unexploded Ordnance in Chechnya, 1994–2005.* Prehospital and Disaster Medicine, 2007. **22**(6): p. 507-512.

31. Sampaio, A., *Before and After Urban Warfare: Conflict Prevention and Transitions in Cities.* International Review of the Red Cross, 2016. **98**(901): p. 71-95.

32. Olson, K.R., *Review and Analysis: United States Cluster Munitions and Unexploded Ordnance Left in Laos After the Second Indochina War.* Open Journal of Soil Science, 2023. **13**(08): p. 355-369.

33. Lin, E., *How War Changes Land: Soil Fertility, Unexploded Bombs, and the Underdevelopment of Cambodia.* American Journal of Political Science, 2020. **66**(1): p. 222-237.

34. Ispas, L. and P. Tudorache, *Mission Command &Amp; Manoeuvrist Approach – Basic Principles of Nato Urban Operations.* International Conference Knowledge-Based Organization, 2019. **25**(1): p. 91-96.

35. Widodo, P., G.R. Somantri, and A.H.S. Ghafur, *Anticipative Urban Warfare Strategies on the Neo'post-Truth Era in Indonesia.* Journal of Law and Sustainable Development, 2024. **12**(1): p. e1610.

36. Jensen, B., H. Breitenbauch, and B. Valeriano, *Complex Terrain: Megacities and the Changing Character of Urban Operations.* 2019.

37. Tortonesi, M., K. Wrona, and N. Suri, *Secured Distributed Processing and Dissemination of Information in Smart City Environments.* Ieee Internet of Things Magazine, 2019. **2**(2): p. 38-43.

38. Wan, M., et al., *Unmanned Aerial Vehicle Video-Based Target Tracking Algorithm Using Sparse Representation.* Ieee Internet of Things Journal, 2019. **6**(6): p. 9689-9706.

39. Vincent, B., *War in Cities: The Spectre of Total War.* International Review of the Red Cross, 2016. **98**(901): p. 1-11.

40. Rahman, A.H., et al., *Design Configuration of a Generation Next Main Battle Tank for Future Combat.* Defence Science Journal, 2017. **67**(4): p. 343.

41. Öğünç, G.İ., *The Effectiveness of Armoured Vehicles in Urban Warfare Conditions.* Defence Science Journal, 2021. **71**(1): p. 25-33.

42. Öztürk, A., et al., *Injury Mechanisms and Injury Severity Scores as Determinants of Urban Terrorism-Related Thoracoabdominal Injuries.* Turkish Journal of Surgery, 2022. **38**(1): p. 67-73.

43. Desch, M.C., *Soldiers in Cities: Military Operations on Urban Terrain.* 2001.

44. Tas, A., *Urban Destruction and Redevelopment as Counterinsurgency: The Void, the Limbo and New Face of Sur.* The Commentaries, 2023. **3**(1): p. 25-50.

45. Emah, S.-P. and J.A. Ekah, *Small Wars and the War Crime Dilemma: The "Weaponization" of the Laws of War in Non-Conventional Warfare.* Greener Journal of Social Sciences, 2017. **7**(4): p. 045-053.

46. Maisaia, V., *Small Armaments Non-Proliferation and Terrorism Security Dilemma Smuggling in Aegis of the Asymmetric Warfare.* 2023.

47. Kittana, A.M.G. and B.D. Meulder, *Architecture as an Agency of Resilience in Urban Armed Conflicts.* International Journal of Architectural Research Archnet-Ijar, 2019. **13**(3): p. 698-717.

48. Graham, S., *The Urban 'Battlespace'.* Theory Culture & Society, 2009. **26**(7-8): p. 278-288.

49. Nedal, D.K., M. Stewart, and M. Weintraub, *Urban Concentration and Civil War.* Journal of Conflict Resolution, 2019. **64**(6): p. 1146-1171.

50. Mousavi, B., et al., *Quality of Life in Caregivers of Severely Disabled War Survivors.* Rehabilitation Nursing, 2015. **40**(3): p. 139-147.

51. John-Hopkins, M., *Regulating the Conduct of Urban Warfare: Lessons From Contemporary Asymmetric Armed Conflicts.* International Review of the Red Cross, 2010. **92**(878): p. 469-493.

52. Zehfuss, M., *Targeting: Precision and the Production of Ethics.* European Journal of International Relations, 2010. **17**(3): p. 543-566.

53. Millard-Ball, A., *The Width and Value of Residential Streets.* Journal of the American Planning Association, 2021. **88**(1): p. 30-43.

54. Gitelman, V., F. Pesahov, and R. Carmel, *Speed Perception by Drivers as Dependent on Urban Street Design; A Case-Study.* Transactions on Transport Sciences, 2020. **11**(2): p. 5-18.

55. Álías, F., J.C. Socoró, and R.M. Alsina-Pagès, *WASN-Based Day–Night Characterization of Urban Anomalous Noise Events in Narrow and Wide Streets.* Sensors, 2020. **20**(17): p. 4760.

56. Lee, H. and S.-N. Kim, *Perceived Safety and Pedestrian Performance in Pedestrian Priority Streets (PPSs) in Seoul, Korea: A Virtual Reality Experiment and Trace Mapping.* International Journal of Environmental Research and Public Health, 2021. **18**(5): p. 2501.

57. Li, M. and J. Pan, *Assessment of Influence Mechanisms of Built Environment on Street Vitality Using Multisource Spatial Data: A Case Study in Qingdao, China.* Sustainability, 2023. **15**(2): p. 1518.

58. Hu, F., et al., *Urban Function as a New Perspective for Adaptive Street Quality Assessment.* Sustainability, 2020. **12**(4): p. 1296.

59. Lynn, T. and C.M. Wood, *Smart Streets as a Cyber-Physical Social Platform: A Conceptual Framework.* Sensors, 2023. **23**(3): p. 1399.

60. Lovasi, G.S., et al., *Body Mass Index, Safety Hazards, and Neighborhood Attractiveness.* American Journal of Preventive Medicine, 2012. **43**(4): p. 378-384.

61. Wu, W., X. Niu, and M. Ли, *Influence of Built Environment on Street Vitality: A Case Study of West Nanjing Road in Shanghai Based on Mobile Location Data.* Sustainability, 2021. **13**(4): p. 1840.

62. Majer, S. and A. Sołowczuk, *Traffic Calming Measures and Their Slowing Effect on the Pedestrian Refuge Approach Sections (Case Study—on Urban Streets in Poland).* 2023.

63. Groves, P.D., *Shadow Matching: A New GNSS Positioning Technique for Urban Canyons.* Journal of Navigation, 2011. **64**(3): p. 417-430.

64. Chen, Z. and K.E. Haynes, *Regional Impact of Public Transportation Infrastructure.* Economic Development Quarterly, 2015. **29**(3): p. 275-291.

65. Kang, B.S., K.-M. Seo, and T.G. Kim, *Communication Analysis of Network-Centric Warfare via Transformation of System of Systems Model Into Integrated System Model Using Neural Network.* Complexity, 2018. **2018**(1).

66. Lu, T., et al., *Research on Dynamic Evolution Model and Method of Communication Network Based on Real War Game.* Entropy, 2021. **23**(4): p. 487.

67. Anderson, M.G., *Decided among the Cities: The Past, Present, and Future of War in Urban Environments.* Military Review, 2023. **103**(3): p. 25-33.

68. Williams, M. and M. Cheng, *The Future Of Urban Warfare.* 2023, Irregular Warefare Initiative.

69. Guha-Sapir, D., et al., *Patterns of Civilian and Child Deaths Due to War-Related Violence in Syria: A Comparative Analysis From the Violation Documentation Center Dataset, 2011–16.* The Lancet Global Health, 2018. **6**(1): p. e103-e110.

70. Lin, T., H. Lin, and M. Hu, *Three-Dimensional Visibility Analysis and Visual Quality Computation for Urban Open Spaces Aided by Google SketchUp and WebGIS.* Environment and Planning B Urban Analytics and City Science, 2015. **44**(4): p. 618-646.

71. Gartzke, E. and J.I. Walsh, *The Drawbacks of Drones: The Effects of UAVs on Escalation and Instability in Pakistan.* Journal of Peace Research, 2022. **59**(4): p. 463-477.

72. Melzer, N., *The Principle of Distinction Between Civilians and Combatants.* 2014: p. 296-331.

73. Heße, M., *Logistics: Situating Flows in a Spatial Context.* Geography Compass, 2020. **14**(7).

74. Gauchat, G., et al., *The Military Metropolis: Defense Dependence in U.S. Metropolitan Areas.* City and Community, 2011. **10**(1): p. 25-48.

75. Gohari, S., et al., *Prevailing Approaches and Practices of Citizen Participation in Smart City Projects: Lessons From Trondheim, Norway.* Infrastructures, 2020. **5**(4): p. 36.

76. Lugova, H., et al., *Civil-Military Coordination of Public Health Response to Urban Disasters in Malaysia.* Ukrainian Journal of Military Medicine, 2020. **1**(2): p. 35-38.

77. García-Ayllón Veintimilla, S., J.L. Miralles-Garcia, and B. Sowińska-Świerkosz, *Editorial: Challenges in Sustainable Urban Planning and Territorial Management for the XXI Century.* Frontiers in Environmental Science, 2023. **11**.

78. Takács, Á. and T. Haidegger, *Infrastructural Requirements and Regulatory Challenges of a Sustainable Urban Air Mobility Ecosystem.* Buildings, 2022. **12**(6): p. 747.

79. Bauder, H., *Sanctuary Cities: Policies and Practices in International Perspective.* International Migration, 2016. **55**(2): p. 174-187.

80. Silva, M.R.S.d., et al., *Effect of Alveolar Recruitment Maneuver on Pulse Pressure Variation: A Case Report.* International Journal of Health Science, 2022. **2**(35): p. 2-6.

81. Stergiou, M., et al., *Psychophysiological Responses in Soldiers During Close Combat: Implications for Occupational Health and Fitness in Tactical Populations.* Healthcare, 2023. **12**(1): p. 82.

82. Tornero-Aguilera, J.F., V.E. Fernández-Elías, and V.J. Clemente-Suárez, *Ready for Combat, Psychophysiological Modifications in a Close-Quarter Combat Intervention After an Experimental Operative High-Intensity Interval Training.* The Journal of Strength and Conditioning Research, 2020. **36**(3): p. 732-737.

83. Clemente-Suárez, V.J., *New Training Program for the New Requirements of Combat of Tactical Athletes.* Sustainability, 2022. **14**(3): p. 1216.

84. Head, J., et al., *Prior Mental Fatigue Impairs Marksmanship Decision Performance.* Frontiers in Physiology, 2017. **8**.

85. Hassan, S.A.M. and S.A.M. Ali, *Limitations On Warfare Methods: A Brief Examination Under International Humanitarian Law.* Journal of Advances in Humanities Research, 2023. **2**(4): p. 1-19.

86. Britt, T.W., et al., *Morale as a Moderator of the Combat Exposure-PTSD Symptom Relationship.* Journal of Traumatic Stress, 2013. **26**(1): p. 94-101.

87. Renshaw, K.D., *Deployment Experiences and Postdeployment PTSD Symptoms in National Guard/Reserve Service Members Serving in Operations Enduring Freedom and Iraqi Freedom.* Journal of Traumatic Stress, 2010. **23**(6): p. 815-818.

88. Sundin, J., et al., *Mental Health Among Commando, Airborne and Other UK Infantry Personnel.* Occupational Medicine, 2010. **60**(7): p. 552-559.

89. Osório, C., et al., *Combat Exposure and Posttraumatic Stress Disorder Among Portuguese Special Operation Forces Deployed in Afghanistan.* Military Psychology, 2013. **25**(1): p. 70-81.

90. Alexander, D. and S. Klein, *Combat-Related Disorders: A Persistent Chimera.* Journal of the Royal Army Medical Corps, 2008. **154**(2): p. 96-101.

91. Vogt, D., et al., *Deployment Stressors and Posttraumatic Stress Symptomatology: Comparing Active Duty and National Guard/Reserve Personnel From Gulf War I.* Journal of Traumatic Stress, 2008. **21**(1): p. 66-74.

92. Miller, N.L., L.G. Shattuck, and P. Matsangas, *Sleep and Fatigue Issues in Continuous Operations: A Survey of U.S. Army Officers.* Behavioral Sleep Medicine, 2011. **9**(1): p. 53-65.

93. Shattuck, N.L. and S. Brown, *Wounded in Action: What the Sleep Community Can Learn From Sleep Disorders of US Military Service Members.* Sleep, 2013.

94. Clemente-Suárez, V.J. and J.J. Robles-Pérez, *Respuesta Psico-Fisiológica De Soldados en Combate Urbano.* Anales De Psicología, 2013. **29**(2).

95. Smith-MacDonald, L., et al., *Transitioning Fractured Identities: A Grounded Theory of Veterans' Experiences of Operational Stress Injuries.* Traumatology an International Journal, 2020. **26**(2): p. 235-245.

96. Drescher, K.D., et al., *An Exploration of the Viability and Usefulness of the Construct of Moral Injury in War Veterans.* Traumatology an International Journal, 2011. **17**(1): p. 8-13.

97. Held, P., et al., *Using Prolonged Exposure and Cognitive Processing Therapy to Treat Veterans With Moral Injury-Based PTSD: Two Case Examples.* Cognitive and Behavioral Practice, 2018. **25**(3): p. 377-390.

98. Nichter, B., et al., *Moral Injury and Suicidal Behavior Among US Combat Veterans: Results From the 2019–2020 National Health and Resilience in Veterans Study.* Depression and Anxiety, 2021. **38**(6): p. 606-614.

99. Schwartz, G., E. Halperin, and Y. Levi-Belz, *Moral Injury and Suicide Ideation Among Combat Veterans: The Role of Trauma-Related Shame and Collective Hatred.* Journal of Interpersonal Violence, 2021. **37**(15-16): p. NP13952-NP13977.

100. Gutierrez, I.A., S.W. Krauss, and A.B. Adler, *Leadership in Moral Awareness: Initial Evidence From U.S. Army Soldiers Returning From Deployment.* Psychological Trauma Theory Research Practice and Policy, 2023.

101. Phelps, A., et al., *Addressing Moral Injury in the Military.* BMJ Military Health, 2022. **170**(1): p. 51-55.

102. Marigliano, R., L.H.X. Ng, and K.M. Carley, *Analyzing Digital Propaganda and Conflict Rhetoric: A Study on Russia's Bot-Driven Campaigns and Counter-Narratives During the Ukraine Crisis.* 2024.

103. Cao, F., et al., *The Impact of Resilience on the Mental Health of Military Personnel During the COVID-19 Pandemic: Coping Styles and Regulatory Focus.* Frontiers in Public Health, 2023. **11**.

104. Kim, P.Y., et al., *Stigma, Negative Attitudes About Treatment, and Utilization of Mental Health Care Among Soldiers.* Military Psychology, 2011. **23**(1): p. 65-81.

105. Shen, Y.C., J. Arkes, and P.B. Lester, *Association Between Baseline Psychological Attributes and Mental Health Outcomes After Soldiers Returned From Deployment.* BMC Psychology, 2017. **5**(1).

106. Nindl, B.C., et al., *Perspectives on Resilience for Military Readiness and Preparedness: Report of an International Military Physiology Roundtable.* Journal of Science and Medicine in Sport, 2018. **21**(11): p. 1116-1124.

107. Kanapeckaitė, R. and D. Bagdžiūnienė, *Relationships Between Team Characteristics and Soldiers' Organizational Commitment and Well-Being: The Mediating Role of Psychological Resilience.* Frontiers in Psychology, 2024. **15**.

108. Smith, R.T. and G. True, *Warring Identities.* Society and Mental Health, 2014. **4**(2): p. 147-161.

109. Mallen, M.J., et al., *Providing Coordinated Care to Veterans of Iraq and Afghanistan Wars With Complex Psychological and Social Issues in a Department of Veterans Affairs Medical Center: Formation of Seamless Transition Committee.* Professional Psychology Research and Practice, 2014. **45**(6): p. 410-415.

110. Thomas, K.H., et al., *Mental Health Needs of Military and Veteran Women: An Assessment Conducted by the Service Women's Action Network.* Traumatology an International Journal, 2018. **24**(2): p. 104-112.

111. Worthen, M.E. and J. Ahern, *The Causes, Course, and Consequences of Anger Problems in Veterans Returning to Civilian Life.* Journal of Loss and Trauma, 2014. **19**(4): p. 355-363.

112. Gordon, N. and N. Perugini, *The Politics of Human Shielding: On the Resignification of Space and the Constitution of Civilians as Shields in*

Liberal Wars. Environment and Planning D Society and Space, 2015. **34**(1): p. 168-187.

113.   McDermott, R., et al., *International Law Applicable to Urban Conflict and Disaster*. Disaster Prevention and Management an International Journal, 2017. **26**(5): p. 553-564.

114.   Darcy, S. and J. Reynolds, *An Enduring Occupation: The Status of the Gaza Strip From the Perspective of International Humanitarian Law*. Journal of Conflict and Security Law, 2010. **15**(2): p. 211-243.

115.   Vautravers, A., *Military Operations in Urban Areas*. International Review of the Red Cross, 2010. **92**(878): p. 437-452.

116.   Epps, V., *Civilian Casualties in Modern Warfare: The Death of the Collateral Damage Rule*. SSRN Electronic Journal, 2011.

117.   Dalmolen, S., et al., *Trusted Data Sharing in Federated and Dynamic Mission Contexts: Improving Communication Flexibility With Emerging Data Control Architectures and Concepts*. Ieee Communications Magazine, 2021. **59**(8): p. 66-72.

118.   Fisentzou, I., *Blurred Lines: Social Media in Armed Conflict*. Legal Information Management, 2019. **19**(01): p. 65-67.

119.   Sowers, J.L., E. Weinthal, and N. Zawahri, *Targeting Environmental Infrastructures, International Law, and Civilians in the New Middle Eastern Wars*. Security Dialogue, 2017. **48**(5): p. 410-430.

120.   Burkle, F.M., *Revisiting the Battle of Solferino: The Worsening Plight of Civilian Casualties in War and Conflict*. Disaster Medicine and Public Health Preparedness, 2019. **13**(5-6): p. 837-841.

121.   Shortland, N., H. Sari, and E. Nader, *Recounting the Dead: An Analysis of ISAF Caused Civilian Casualties in Afghanistan*. Armed Forces & Society, 2017. **45**(1): p. 122-139.

122.   Milham, L., et al., *Squad-Level Training for Tactical Combat Casualty Care: Instructional Approach and Technology Assessment*. The Journal of Defense Modeling and Simulation Applications Methodology Technology, 2016. **14**(4): p. 345-360.

123.   Condra, L.N., et al., *The Effect of Civilian Casualties in Afghanistan and Iraq*. 2010.

124.   Condra, L.N. and J.N. Shapiro, *Who Takes the Blame? The Strategic Effects of Collateral Damage*. American Journal of Political Science, 2011. **56**(1): p. 167-187.

125.   Johns, R. and G.A.M. Davies, *Civilian Casualties and Public Support for Military Action: Experimental Evidence*. Journal of Conflict Resolution, 2017. **63**(1): p. 251-281.

126. Baum, M. and T. Groeling, *Reality Asserts Itself: Public Opinion on Iraq and the Elasticity of Reality.* International Organization, 2010. **64**(3): p. 443-479.

127. Levy, Y., *How Casualty Sensitivity Affects Civilian Control: The Israeli Experience.* International Studies Perspectives, 2011. **12**(1): p. 68-88.

128. Stodola, P., et al., *Collective Perception Using UAVs: Autonomous Aerial Reconnaissance in a Complex Urban Environment.* Sensors, 2020. **20**(10): p. 2926.

129. Tayarani, M., M. Tayarani, and M. Dehmollaian, *Two-dimensional Synthetic Aperture Imaging of Targets Behind Reinforced Concrete Walls.* Iet Radar Sonar & Navigation, 2022. **16**(11): p. 1851-1860.

130. Guo, Y., et al., *Collision-Free 4D Dynamic Path Planning for Multiple UAVs Based on Dynamic Priority RRT* and Artificial Potential Field.* Drones, 2023. **7**(3): p. 180.

131. Sánchez-Lopera, J. and J.L. Lerma, *Classification of Lidar Bare-Earth Points, Buildings, Vegetation, and Small Objects Based on Region Growing and Angular Classifier.* International Journal of Remote Sensing, 2014. **35**(19): p. 6955-6972.

132. Soldovieri, F., R. Solimene, and F. Ahmad, *Experimental Validation of a Microwave Tomographic Approach for Through-the-Wall Radar Imaging.* 2010.

133. Medina, S.A.O., et al., *Localization and Mapping Approximation for Autonomous Ground Platforms, Implementing SLAM Algorithms.* 2014: p. 1-5.

134. Durhin, N., *Protecting Civilians in Urban Areas: A Military Perspective on the Application of International Humanitarian Law.* International Review of the Red Cross, 2016. **98**(901): p. 177-199.

135. Hicks, M.H.-R., et al., *Violent Deaths of Iraqi Civilians, 2003–2008: Analysis by Perpetrator, Weapon, Time, and Location.* Plos Medicine, 2011. **8**(2): p. e1000415.

136. Ruiz, J., *The Civilianization of War.* 2018.

137. Khorram-Manesh, A., et al., *Estimating the Number of Civilian Casualties in Modern Armed Conflicts–A Systematic Review.* Frontiers in Public Health, 2021. **9**.

138. Ayoub, H.H., H. Chemaitelly, and L.J. Abu-Raddad, *Mortality Impact of the Israel-Gaza Conflicts (2008-2023): A Comparative Analysis of Civilians Versus Combatants.* 2023.

139. Clayton, G. and A. Thomson, *Civilianizing Civil Conflict: Civilian Defense Militias and the Logic of Violence in Intrastate Conflict.* International Studies Quarterly, 2016. **60**(3): p. 499-510.

140.   Wood, R.M., *Rebel Capability and Strategic Violence Against Civilians.*
       Journal of Peace Research, 2010. **47**(5): p. 601-614.

141.   Goniewicz, K., M. Goniewicz, and D. Lasota, *Armed Forces Operation in
       the Scope of the Civilian Health Protection During Peacekeeping and
       Stabilization Missions: A Short Review.* Safety & Defense, 2018. **4**: p. 27-
       30.

142.   Garbett, C., *The Concept of the Civilian: Legal Recognition, Adjudication
       and the Trials of International Criminal Justice.* International Journal of
       Law in Context, 2012. **8**(4): p. 469-486.

143.   Hicks, M.H.-R., et al., *Global Comparison of Warring Groups in 2002–
       2007: Fatalities From Targeting Civilians vs. Fighting Battles.* Plos One,
       2011. **6**(9): p. e23976.

144.   Barter, S.J., *Unarmed Forces: Civilian Strategy in Violent Conflicts.* Peace
       &Amp Change, 2012. **37**(4): p. 544-571.

145.   Baines, E. and E. Paddon, *'This Is How We Survived': Civilian Agency and
       Humanitarian Protection.* Security Dialogue, 2012. **43**(3): p. 231-247.

146.   Patel, R. and F.M. Burkle, *Rapid Urbanization and the Growing Threat of
       Violence and Conflict: A 21st Century Crisis.* Prehospital
       and Disaster Medicine, 2012. **27**(2): p. 194-197.

147.   Hultman, L., J.D. Kathman, and M. Shannon, *United Nations
       Peacekeeping and Civilian Protection in Civil War.* American Journal of
       Political Science, 2013. **57**(4): p. 875-891.

148.   Williams, P.D., *Protection, Resilience and Empowerment: United Nations
       Peacekeeping and Violence Against Civilians in Contemporary War
       Zones.* Politics, 2013. **33**(4): p. 287-298.

149.   Krause, J. and E.M. Kamler, *Ceasefires and Civilian Protection Monitoring
       in Myanmar.* Global Studies Quarterly, 2022. **2**(1).

150.   Jose, B. and P.A. Medie, *Understanding Why and How Civilians Resort to
       Self-Protection in Armed Conflict.* International Studies Review, 2015: p.
       n/a-n/a.

151.   Suarez, C., *'Living Between Two Lions': Civilian Protection Strategies
       During Armed Violence in the Eastern Democratic Republic of the Congo.*
       Journal of Peacebuilding & Development, 2017. **12**(3): p. 54-67.

152.   Wirtz, J.J., *The Sources and Methods of Intelligence Studies.* 2010: p. 59-
       69.

153.   Meng, X., L. Nie, and J. Song, *Research on Urban Anti-Terrorism
       Intelligence Perception System From the Perspective of Internet of Things
       Application.* International Journal of Electrical Engineering Education,
       2019. **58**(2): p. 248-257.

154.   Maharani, T.D., et al., *Separatist and Terrorist Movements in Papua: The
       Challenges of Social Disaster Management and the Important Role of*

Human-Made Disaster Intelligence. Jurnal Pertahanan Media Informasi TTG Kajian & Strategi Pertahanan Yang Mengedepankan Identity Nasionalism & Integrity, 2023. **9**(3): p. 443-457.

155. Jensen, T.W., *National Responses to Transnational Terrorism.* Journal of Conflict Resolution, 2014. **60**(3): p. 530-554.

156. Dash, B. and P.K. Sharma, *Role of Artificial Intelligence in Smart Cities for Information Gathering and Dissemination (A Review).* Academic Journal of Research and Scientific Publishing, 2022. **4**(39): p. 58-75.

157. Oleszkiewicz, S., P.A. Granhag, and S.C. Montecinos, *The Scharff-Technique: Eliciting Intelligence From Human Sources.* Law and Human Behavior, 2014. **38**(5): p. 478-489.

158. Yiğitcanlar, T., et al., *Contributions and Risks of Artificial Intelligence (AI) in Building Smarter Cities: Insights From a Systematic Review of the Literature.* Energies, 2020. **13**(6): p. 1473.

159. Li, X., *Suitability Evaluation Method of Urban and Rural Spatial Planning Based on Artificial Intelligence.* Journal of Intelligent Systems, 2022. **31**(1): p. 245-259.

160. Campaña, I., et al., *Air Tracking and Monitoring for Unmanned Aircraft Traffic Management.* 2019: p. 1-9.

161. Besada, J.A., et al., *Drone Flight Planning for Safe Urban Operations: UTM Requirements and Tools.* 2019: p. 924-930.

162. Ni, D., G. Yu, and S. Rathinam, *Unmanned Aircraft System and Its Applications in Transportation.* Journal of Advanced Transportation, 2017. **2017**: p. 1-2.

163. Raharja, S. and T. Sugawara, *An Extension of Particle Swarm Optimization to Identify Multiple Peaks Using Re-Diversification in Static and Dynamic Environments.* International Journal of Smart Computing and Artificial Intelligence, 2023. **7**(2): p. 1.

164. Sawarkar, A., et al., *HMD Vision-Based Teleoperating UGV and UAV for Hostile Environment Using Deep Learning.* 2016.

165. Srinivasan*, K., et al., *Aerial and Under-Water Dronal Communication: Potentials, Issues and Vulnerabilities.* International Journal of Innovative Technology and Exploring Engineering, 2019. **9**(1): p. 3874-3885.

166. Ji, P., et al., *Tele-aiming Control Design for Reconnaissance Robot Using a Strong Tracking Multi-model Extended Super-twisting Observer.* Iet Control Theory and Applications, 2022. **17**(6): p. 696-712.

167. Turnage, D.M., *Localization and Mapping of Unknown Locations With Unmanned Ground Vehicles.* 2019.

168. Sebastian, B. and P. Ben-Tzvi, *Active Disturbance Rejection Control for Handling Slip in Tracked Vehicle Locomotion.* Journal of Mechanisms and Robotics, 2019. **11**(2).

169. Liu, D., *The Research About Collaboration Techniques for Aerial and Ground Mobile Robots*. 2015.
170. Maistrenko, O., et al., *Devising a Procedure for Justifying the Choice of Reconnaissance-Firing Systems*. Eastern-European Journal of Enterprise Technologies, 2021. **1**(3 (109)): p. 60-71.
171. Cho, S., et al., *Priority Determination to Apply Artificial Intelligence Technology in Military Intelligence Areas*. Electronics, 2020. **9**(12): p. 2187.
172. Drozd, J., et al., *Effectiveness Evaluation of Aerial Reconnaissance in Battalion Force Protection Operation Using the Constructive Simulation*. The Journal of Defense Modeling and Simulation Applications Methodology Technology, 2021. **20**(2): p. 181-196.
173. Song, Z., et al., *An Intelligent Mission Planning Model for the Air Strike Operations Against Islands Based on Neural Network and Simulation*. Discrete Dynamics in Nature and Society, 2022. **2022**(1).
174. Atyabi, A., S. MahmoudZadeh, and S. Nefti-Meziani, *Current Advancements on Autonomous Mission Planning and Management Systems: An AUV and UAV Perspective*. Annual Reviews in Control, 2018. **46**: p. 196-215.
175. Yang, L., et al., *Research on Innovation of Urban Combat Equipment Support Model Based on Smart City and Artificial Intelligence*. Journal of Physics Conference Series, 2021. **1732**(1): p. 012048.
176. Greipl, A.R., *Artificial Intelligence in Urban Warfare: Opportunities to Enhance the Protection of Civilians?* The Military Law and the Law of War Review, 2023. **61**(2): p. 191-211.
177. Dostri, O. and K. Michael, *The Role of Human Terrain and Cultural Intelligence in Contemporary Hybrid and Urban Warfare*. The International Journal of Intelligence Security and Public Affairs, 2019. **21**(1): p. 84-102.
178. Liu, W., et al., *Special Issue on Intelligent Urban Computing With Big Data*. Machine Vision and Applications, 2017. **28**(7): p. 675-677.
179. Tornero-Aguilera, J.F., et al., *Optimising Combat Readiness: Practical Strategies for Integrating Physiological and Psychological Resilience in Soldier Training*. Healthcare, 2024. **12**(12): p. 1160.
180. López-Rodríguez, G. and D. Montoya-Roldan, *Challenges and Prospects in Urban Warfare: An Analytical Framework*. Contemporary Military Challenges, 2024. **26**(2): p. 35-47.
181. Bierman, A. and R. Kelty, *The Threat of War and Psychological Distress Among Civilians Working in Iraq and Afghanistan*. Social Psychology Quarterly, 2014. **77**(1): p. 27-53.

182.   McKenzie, F.D., et al., *Integrating Crowd-Behavior Modeling Into Military Simulation Using Game Technology.* Simulation & Gaming, 2007. **39**(1): p. 10-38.

183.   Obinwa, C., *Civil-Military Coordination in Peacebuilding: The Obstacle in Somalia.* SSRN Electronic Journal, 2019.

184.   Beal, H.L., *Military Foreign Humanitarian Assistance and Disaster Relief (FHA/FDR) Evolution: Lessons Learned for Civilian Emergency Management Response and Recovery Operations.* International Journal of Mass Emergencies & Disasters, 2015. **33**(2): p. 273-309.

185.   Warren, R., *Situating the City and September 11th: Military Urban Doctrine, 'Pop–up' Armies and Spatial Chess.* International Journal of Urban and Regional Research, 2002. **26**(3): p. 614-619.

186.   Комісаров, О., et al., *Formation of the Civil-Military Cooperation During the Settlement of the Armed Conflict in the South-East of Ukraine.* Path of Science, 2018. **4**(7): p. 2009-2015.

187.   Kaniewski, P., et al., *Heterogeneous Wireless Sensor Networks Enabled Situational Awareness Enhancement for Armed Forces Operating in an Urban Environment.* 2023: p. 1-8.

188.   Pawgasame, W., *A Survey in Adaptive Hybrid Wireless Sensor Network for Military Operations.* 2016: p. 78-83.

189.   Sun, Z.F., X. Ma, and D.X. Sun, *Construction of the Air Offensive Operation Battlefield Support System Based on the Internet of Things Technology.* Advanced Materials Research, 2013. **834-836**: p. 1873-1876.

190.   United State Marine Corps, *Urban Operations II Offensive and Defensive Operations B4R5379 Student Handout.* n.d., Marine Corps Training Command, Camp Barret, Virginia: United State Marine Corps.

191.   Dincecco, M. and M.G. Onorato, *Military Conflict and the Rise of Urban Europe.* Journal of Economic Growth, 2016. **21**(3): p. 259-282.

192.   Bekesiene, S. and D. Leliūnas, *Assessment of Soldiers' Resilience to Cognitive Attacks of Russian Hybrid Warfare.* 2023: p. 683-683.

193.   Sharma, B.K.O.P., *Military Psychology: Origin, Evolution, and Future Prospects.* International Journal of Research Publication and Reviews, 2024. **5**(5): p. 7184-7189.

194.   Brady, M., *Improvisation Versus Rigid Command and Control at Stalingrad.* Journal of Management History, 2011. **17**(1): p. 27-49.

195.   Clemente-Suárez, V.J., P. Ruisoto, and J.J. Robles-Pérez, *Psychophysiological Response to Acute -high-stress Combat Situations in Professional Soldiers.* Stress and Health, 2017. **34**(2): p. 247-252.

196.   Lampton, D.R., B.R. Clark, and B.W. Knerr, *Urban Combat: The Ultimate Extreme Environment.* Journal of Human Performance in Extreme Environments, 2003. **7**(2).

197.   Kiselev, I.V., *Difficulties of Liberation: Battles for Krasnodar in Early 1943.* Nasledie Vekov, 2020(2(22)).

198.   Lee, Y. and T. Lee, *Network-Based Metric for Measuring Combat Effectiveness.* Defence Science Journal, 2014. **64**(2): p. 115-122.

199.   Mishra, N., R.K. Singh, and S.K. Yadav, *Detection of DDoS Vulnerability in Cloud Computing Using the Perplexed Bayes Classifier.* Computational Intelligence and Neuroscience, 2022. **2022**: p. 1-13.

200.   Tabish, N. and T. Chaur-Luh, *Maritime Autonomous Surface Ships: A Review of Cybersecurity Challenges, Countermeasures, and Future Perspectives.* Ieee Access, 2024. **12**: p. 17114-17136.

201.   Raghavender, K.V. and P. Premchand, *Network Anomaly Detection for Protecting Web Services From the Application Layer Bandwidth Flooding Attack.* International Journal of Engineering & Technology, 2018. **7**(2): p. 907.

202.   Ovsyannikova, Y., et al., *The Nature of Combat Stress Development During Military Operations and Psychotherapy in Extreme Situations.* The Journal of Nervous and Mental Disease, 2024. **212**(5): p. 270-277.

203.   Pinto, F.C.L., H.P. Neiva, and R. Ferraz, *Theoretical Basis of Technical-Tactical Behavior and Its Application in Ultimate Full Contact Training.* The Open Sports Sciences Journal, 2021. **14**(1): p. 9-13.

204.   Cohen-Blankshtain, G. and O. Rotem-Mindali, *Key Research Themes on ICT and Sustainable Urban Mobility.* International Journal of Sustainable Transportation, 2013. **10**(1): p. 9-17.

205.   Krabben, K., D. Orth, and J.v.d. Kamp, *Combat as an Interpersonal Synergy: An Ecological Dynamics Approach to Combat Sports.* Sports Medicine, 2019. **49**(12): p. 1825-1836.

206.   Maksymenko, S., et al., *Psychodiagnostic Toolkit of Combat Stress for Aviation Specialists.* E3s Web of Conferences, 2021. **258**: p. 10001.

207.   Wen, K., et al., *An Integrative Model for Measuring Combat Readiness.* International Journal of Advanced Research, 2017. **5**(7): p. 1449-1462.

208.   Chen, G., et al., *Application and Development of Computer Technology in Modern Military Logistics Construction.* Academic Journal of Business & Management, 2022. **4**(15).

209.   Yassin, M. and E. Rachid, *A Survey of Positioning Techniques and Location Based Services in Wireless Networks.* 2015: p. 1-5.

210.   Vey, Q., et al., *POUCET: A Multi-Technology Indoor Positioning Solution for Firefighters and Soldiers.* 2021.

211.   Adey, P., et al., *Blurred Lines: Intimacy, Mobility, and the Social Military.* Critical Military Studies, 2016. **2**(1-2): p. 7-24.

212.   Markov, D., *Use of Artillery Fire Support Assets in the Attrition Approach in the Russia-Ukraine Conflict.* Environment Technology Resources

Proceedings of the International Scientific and Practical Conference, 2024. **4**: p. 178-182.

213. Zha, Q., et al., *Study on the Dynamic Modeling and the Correction Method of the Self-Propelled Artillery.* 2017.

214. He, F.H., *Intelligent Video Surveillance Technology in Intelligent Transportation.* Journal of Advanced Transportation, 2020. **2020**: p. 1-10.

215. Prasetiawan, E., et al., *Distinction Principle in International Humanitarian Law Related to Civilian Objects and Military Objects.* 2018.

216. Bouchet-Saulnier, F., *International Law for Healthcare Workers.* 2019: p. 173-194.

217. Khawaja, I.A., *Invisible Warfare: The Psychological Impact of Biological Weapons in the Context of International Humanitarian Law.* 2023. **2**(1): p. 15-25.

218. Martin, C., *A Means-Methods Paradox and the Legality of Drone Strikes in Armed Conflict.* The International Journal of Human Rights, 2015. **19**(2): p. 142-175.

219. Аракелян, М.Р., Х.Н. Бехруз, and L. Yarova, *Prohibited Means and Methods of Armed Conflicts.* Revista Amazonia Investiga, 2020. **9**(26): p. 349-355.

220. Wren, S.M., et al., *A Consensus Framework for the Humanitarian Surgical Response to Armed Conflict in 21st Century Warfare.* Jama Surgery, 2020. **155**(2): p. 114.

221. Reid-Henry, S. and O.J. Sending, *The "Humanitarianization" of Urban Violence.* Environment and Urbanization, 2014. **26**(2): p. 427-442.

222. Eyal, H., L. Samimian-Darash, and N. Davidovitch, *Humanitarian Aid, Security and Ethics.* Journal of Extreme Anthropology, 2020. **4**(1): p. 135-156.

223. Büscher, K. and K. Vlassenroot, *Humanitarian Presence and Urban Development: New Opportunities and Contrasts in Goma, DRC.* Disasters, 2010. **34**(s2).

224. Vartanian, O., et al., *Blast in Context: The Neuropsychological and Neurocognitive Effects of Long-Term Occupational Exposure to Repeated Low-Level Explosives on Canadian Armed Forces' Breaching Instructors and Range Staff.* Frontiers in Neurology, 2020. **11**.

225. Ytterbøl, C., D. Collins, and A.C. MacPherson, *Shooter Ready? Integrating Mental Skills Training in an Advanced Sniper Course.* Frontiers in Psychology, 2023. **14**.

226. Mišković, M., et al., *Method for Direct Localization of Multiple Impulse Acoustic Sources in Outdoor Environment.* Electronics, 2022. **11**(16): p. 2509.

227.   Baksheev, A.I., et al., *The Transition of Insurgent-Guerrilla Movements to Radical Terrorist Methods of Struggle: Retrospective Features.* Cuestiones Políticas, 2021. **39**(71): p. 822-832.

228.   Andrade, E.R., et al., *Human and Environmental Bias Affecting Risk Perception in Military Radiological and Nuclear Operations.* Defence Science Journal, 2023. **73**(1): p. 20-28.

229.   Boix, C., *Civil Wars and Guerrilla Warfare in the Contemporary World: Toward a Joint Theory of Motivations and Opportunities.* 2008: p. 197-218.

230.   Joes, A.J., *Urban Guerrilla Warfare.* 2007.

231.   Thunholm, P. and L. Henåker, *A Tentative Model on Effective Army Combat Tactics.* Comparative Strategy, 2020. **39**(5): p. 490-504.

232.   Wang, Y., Y. Li, and C. Lu, *Evaluating the Effects of Logistics Center Location: An Analytical Framework for Sustainable Urban Logistics.* Sustainability, 2023. **15**(4): p. 3091.

233.   Russo, F. and A. Comi, *Urban Freight Transport Planning Towards Green Goals: Synthetic Environmental Evidence From Tested Results.* Sustainability, 2016. **8**(4): p. 381.

234.   Feuer, A., *Environmental Warfare Tactics in Irregular Conflicts.* Perspectives on Politics, 2022. **21**(2): p. 533-549.

235.   Johnson, B. and W.A. Treadway, *Artificial Intelligence — An Enabler of Naval Tactical Decision Superiority.* Ai Magazine, 2019. **40**(1): p. 63-78.

236.   Mao, X.H., Y.H. Lee, and S.H. Ting, *Study of Path Loss for Ground Based Communication in Military UHF Band.* 2012.

237.   Nair, B. and S.M.S. Bhanu, *Task Scheduling in Fog Node Within the Tactical Cloud.* Defence Science Journal, 2022. **72**(1): p. 49-55.

238.   Klyszcz, I.U.K., *Chechnya's Paradiplomacy 2000–2020: The Emergence and Evolution of External Relations of a Reincorporated Territory.* Nationalities Papers, 2022. **51**(6): p. 1397-1413.

239.   O'Connor, S., et al., *Examining the Origin of Fortifications in East Timor: Social and Environmental Factors.* The Journal of Island and Coastal Archaeology, 2012. **7**(2): p. 200-218.

240.   Schapper, A., *Build the Wall!* Indonesia and the Malay World, 2019. **47**(138): p. 220-251.

241.   Lenticchia, E. and E. Coïsson, *The Use of Gis for the Application of the Phenomenological Approach to the Seismic Risk Analysis: The Case of the Italian Fortified Architecture.* The International Archives of the Photogrammetry Remote Sensing and Spatial Information Sciences, 2017. **XLII-5/W1**: p. 39-46.

242.   Shepperson, M., *An Architectural Analysis of the Sealand Building at Tell Khaiber, Southern Iraq.* Iraq, 2020. **82**: p. 207-226.

243. Schapper, A., *Historical and Linguistic Perspectives on Fortified Settlements in Southeastern Wallacea: Far Eastern Timor in the Context of Southern Maluku.* 2020: p. 221-246.

244. Munzi, M., G. Schirru, and I. Tantillo, *<i>Centenarium</I>*. Libyan Studies, 2014. **45**: p. 49-64.

245. Miguelez, C.V., et al., *Guidelines for Renewal and Securitization of a Critical Infrastructure Based on IoT Networks.* Smart Cities, 2023. **6**(2): p. 728-743.

246. Larsson, J., et al., *Physiological Demands and Characteristics of Movement During Simulated Combat.* Military Medicine, 2022. **188**(11-12): p. 3496-3505.

247. Taiwo, O.J., *Urban Growth During Civilian and Military Administrations in Osogbo, Nigeria.* Indonesian Journal of Geography, 2018. **50**(1): p. 1.

248. Aşık, M.B. and M. Bınar, *Retrospective Analyses of High-Energy Explosive Device Related Injuries of Ear and Auricular Region: Experiences in Operative Field Hospital Emergency Room.* Turkish Journal of Trauma and Emergency Surgery, 2018.

249. Beall, J.A., *The United States Army and Urban Combat in the Nineteenth Century.* War in History, 2009. **16**(2): p. 157-188.

250. Carter, D.B., *A Blessing or a Curse? State Support for Terrorist Groups.* International Organization, 2012. **66**(1): p. 129-151.

251. Salehyan, I., K.S. Gleditsch, and D. Cunningham, *Explaining External Support for Insurgent Groups.* International Organization, 2011. **65**(4): p. 709-744.

252. Moghadam, A. and M. Wyss, *The Political Power of Proxies: Why Nonstate Actors Use Local Surrogates.* International Security, 2020. **44**(4): p. 119-157.

253. Chertoff, M., P. Bury, and D. Richterova, *Bytes Not Waves: Information Communication Technologies, Global Jihadism and Counterterrorism.* International Affairs, 2020. **96**(5): p. 1305-1325.

254. Qiu, X., *State Support for Rebels and Interstate Bargaining.* American Journal of Political Science, 2022. **66**(4): p. 993-1007.

255. Collins, S., *State-Sponsored Terrorism: In Decline, Yet Still a Potent Threat.* Politics &Amp Policy, 2014. **42**(1): p. 131-159.

256. Porch, D., *Book Review: <i>Invisible Armies: An Epic History of Guerrilla Warfare From Ancient Times to the Present</I> by Max Boot.* War in History, 2013. **20**(3): p. 412-414.

257. Dowling, S., et al., *Insects as Chemical Sensors: Detection of Chemical Warfare Agent Simulants and Hydrolysis Products in the Blow Fly Using LC-MS/MS.* Environmental Science & Technology, 2022. **56**(6): p. 3535-3543.

258.   Schussler, L., F.M. Burkle, and S.M. Wren, *Protecting Surgeons and Patients During Wars and Armed Conflicts*. Jama Surgery, 2019. **154**(8): p. 683.

259.   Uwa, O.G. and M. Dada, *The Geneva Convention on Laws of War and the Sudan Armed Conflict*. International Journal of Social Service and Research, 2023. **3**(7): p. 1606-1623.

260.   Bakhsh, F., *Scope of the Application of International Humanitarian Law (Ihl) in Situations of Non-International Armed Conflicts: Extended Applicability of Common ARTICLE.3*. Journal of Law & Social Studies, 2019. **1**(2): p. 99-105.

261.   Salman, H.A.K., N. Shahrul Mizan Ismail, and N. Rohaida Nordin, *Prisoners of War: Classification and Legal Protection Under International Humanitarian Law*. Uum Journal of Legal Studies, 2023. **14**(2): p. 677-708.

262.   Henckaerts, J.-M., *Bringing the Commentaries on the Geneva Conventions and Their Additional Protocols Into the Twenty-First Century*. International Review of the Red Cross, 2012. **94**(888): p. 1551-1555.

263.   Wallace, G.P.R., *Regulating Conflict: Historical Legacies and State Commitment to the Laws of War1*. Foreign Policy Analysis, 2011. **8**(2): p. 151-172.

264.   Pfanner, T., *Various Mechanisms and Approaches for Implementing International Humanitarian Law and Protecting and Assisting War Victims*. International Review of the Red Cross, 2009. **91**(874): p. 279-328.

265.   Nevers, R.d., *The Geneva Conventions and New Wars*. Political Science Quarterly, 2006. **121**(3): p. 369-395.

266.   Iman, K.F., et al., *Comparative Analysis of a Multi-Layered Weapon System for City Air Defense in the Modern Warfare*. International Journal of Humanities Education and Social Sciences (Ijhess), 2023. **3**(3).

267.   Bueno-Suárez, C. and D. Coq-Huelva, *Sustaining What Is Unsustainable: A Review of Urban Sprawl and Urban Socio-Environmental Policies in North America and Western Europe*. Sustainability, 2020. **12**(11): p. 4445.

268.   Toet, A. and M.A. Hogervorst, *Urban Camouflage Assessment Through Visual Search and Computational Saliency*. Optical Engineering, 2012. **52**(4): p. 041103.

269.   Hogervorst, M.A., A. Toet, and P.A. Jacobs, *Design and Evaluation of (Urban) Camouflage*. 2010.

270.   Bell, J.B., *Stalingrad: 1942–1943 the Strategy of a Siege*. 2017: p. 123-160.

271.   Coldwell, P., *A Visual Response to the Siege of Sarajevo*. Journal of Visual Art Practice, 2019. **18**(2): p. 145-159.

272. Lichtenwald, T.G., *Terrorist Use of Smuggling Tunnels.* International Journal of Criminology and Sociology, 2013. **2**: p. 210-226.

273. Rais, W., *Defending Israelis or Suppressing Palestinian Self-Determination? An Analysis of Operation Protective Edge Using the Two-Factor Test.* Fiu Law Review, 2015. **11**(1).

274. Slesinger, I., *A Cartography of the Unknowable: Technology, Territory and Subterranean Agencies in Israel's Management of the Gaza Tunnels.* Geopolitics, 2018. **25**(1): p. 17-42.

275. U.S. Army, *INFANTRY RIFLE PLATOON AND SQUAD.* 2001.

276. The University of Akron, *Introduction to Tactics I.* 2011.

277. Kolkwitz, M., E. Luotonen, and S. Huuhka, *How Changes in Urban Morphology Translate Into Urban Metabolisms of Building Stocks: A Framework for Spatiotemporal Material Flow Analysis and a Case Study.* Environment and Planning B Urban Analytics and City Science, 2022. **50**(6): p. 1559-1576.

278. Bernardini, G. and T.M. Ferreira, *Simulating to Evaluate, Manage and Improve Earthquake Resilience in Historical City Centers: Application to an Emergency Simulation-Based Method to the Historic Centre of Coimbra.* The International Archives of the Photogrammetry Remote Sensing and Spatial Information Sciences, 2020. **XLIV-M-1-2020**: p. 651-657.

279. Cui, W., et al., *Co-Benefits Analysis of Buildings Based on Different Renewal Strategies: The Emergy-Lca Approach.* International Journal of Environmental Research and Public Health, 2021. **18**(2): p. 592.

280. Futcher, J., et al., *Creating Sustainable Cities One Building at a Time: Towards an Integrated Urban Design Framework.* Cities, 2017. **66**: p. 63-71.

281. Yuan, J., et al., *Effect of Highly Reflective Building Envelopes on Outdoor Environment Temperature and Indoor Thermal Loads Using CFD and Numerical Analysis.* E3s Web of Conferences, 2019. **111**: p. 06031.

282. Kim, D.H., H.-S. Cha, and S. Jiang, *The Prediction of Fire Disaster Using BIM-Based Visualization for Expediting the Management Process.* Sustainability, 2023. **15**(4): p. 3719.

283. Santamouris, M., *Heat Mitigation in Cities: A Catalyst for Building Energy Saving.* 2023.

284. Yuan, J., K. Emura, and C. Farnham, *Highly Reflective Roofing Sheets Installed on a School Building to Mitigate the Urban Heat Island Effect in Osaka.* Sustainability, 2016. **8**(6): p. 514.

285. Qiao, Z., et al., *Scale Effects of the Relationships Between 3D Building Morphology and Urban Heat Island: A Case Study of Provincial Capital Cities of Mainland China.* Complexity, 2020. **2020**: p. 1-12.

286.   Lin, Z., et al., *Genetic Algorithm-Based Building Geometric Opening Configurations Optimization for Enhancing Ventilation Performance in the High-Density Urban District*. 2023.

287.   Jagani, C. and U. Passe, *Simulation-Based Sensitivity Analysis of Future Climate Scenario Impact on Residential Weatherization Initiatives in the US Midwest*. 2017.

288.   Hong, T., et al., *Ten Questions on Urban Building Energy Modeling*. Building and Environment, 2020. **168**: p. 106508.

289.   Bal, P.M. and P. Smit, *The Older the Better!* Career Development International, 2012. **17**(1): p. 6-24.

290.   Bal, P.M., D.S. Chiaburu, and I. Diaz, *Does Psychological Contract Breach Decrease Proactive Behaviors? The Moderating Effect of Emotion Regulation*. Group & Organization Management, 2011. **36**(6): p. 722-758.

291.   Abdalla, M.d.J., et al., *COVID-19 and Unpaid Leave: Impacts of Psychological Contract Breach on Organizational Distrust and Turnover Intention: Mediating Role of Emotional Exhaustion*. Tourism Management Perspectives, 2021. **39**: p. 100854.

292.   Putro, A.N.S., et al., *Enhancing Security and Reliability of Information Systems Through Blockchain Technology: A Case Study on Impacts and Potential*. 2023. **1**(01): p. 35-43.

293.   Aslam, M., et al., *Getting Smarter About Smart Cities: Improving Data Security and Privacy Through Compliance*. Sensors, 2022. **22**(23): p. 9338.

294.   Narusaka, Y., et al., *Cytological and Molecular Analyses of Non-host Resistance of <i>Arabidopsis Thaliana</i> to <i>Alternaria Alternata</i>*. Molecular Plant Pathology, 2005. **6**(6): p. 615-627.

295.   Wang, X.T., *Self-framing of Risky Choice*. Journal of Behavioral Decision Making, 2003. **17**(1): p. 1-16.

296.   Dhillon, G. and J. Backhouse, *Current Directions in IS Security Research: Towards Socio-organizational Perspectives*. Information Systems Journal, 2001. **11**(2): p. 127-153.

297.   Ghimire, S., J. Ehrlich, and S. Sanders, *Measuring Individual Worker Output in a Complementary Team Setting: Does Regularized Adjusted Plus Minus Isolate Individual NBA Player Contributions?* Plos One, 2020. **15**(8): p. e0237920.

298.   Huang, Q., et al., *Critical Success Factors Affecting Implementation of Cloud ERP Systems: A Systematic Literature Review With Future Research Possibilities*. 2021.

299.   Derick Musundi Kesa, N., *Ensuring Resilience: Integrating IT Disaster Recovery Planning and Business Continuity for Sustainable Information*

Technology Operations. World Journal of Advanced Research and Reviews, 2023. **18**(3): p. 970-992.

300. Shields, R. and A. Sandoval-Hernández, *Mixed Signals: Cognitive Skills, Qualifications and Earnings in an International Comparative Perspective.* Oxford Review of Education, 2020. **46**(1): p. 111-128.

301. Guo, Y., C. Wang, and X. Chen, *Functional or Financial Remedies? The Effectiveness of Recovery Strategies After a Data Breach.* Journal of Enterprise Information Management, 2023. **37**(1): p. 148-169.

302. Zhu, T., X. Li, and W. Zhang, *Applying Markov Decision Processes to Evaluate Ransomware Data Theft Risks.* 2023.

303. Hollinger, G.A. and S. Singh, *Towards Experimental Analysis of Challenge Scenarios in Robotics.* 2014: p. 909-921.

304. Hvozdiuk, V. and N. Morhun, *Application of the European Court of Human Rights Practices by the Investigator During the Search.* Naukovij Visnik Nacìonal'noï Akademìï Vnutrìšnìh Sprav, 2024. **29**(2): p. 57-66.

305. Winston, C.A., *Truth Commissions as Tactical Concessions: The Curious Case of Idi Amin.* The International Journal of Human Rights, 2020. **25**(2): p. 251-273.

306. Olejníček, A., et al., *Determinants of Military Robotics Proliferation.* Advances in Military Technology, 2018. **13**(1): p. 71-86.

307. Ranjan, R., S. Lee, and J. Kye, *Design of Tactical Multipurpose All–Terrain Mobile Robot.* International Journal of Membrane Science and Technology, 2023. **10**(2): p. 2224-2237.

308. Santoso, F. and A. Finn, *Trusted Operations of a Military Ground Robot in the Face of Man-in-the-Middle Cyberattacks Using Deep Learning Convolutional Neural Networks: Real-Time Experimental Outcomes.* Ieee Transactions on Dependable and Secure Computing, 2024. **21**(4): p. 2273-2284.

309. Muda, N.R.S., et al., *Design and Construction of a Remotely Controlled Multy-Tasking Chain-Wheel Combat Robot.* Eduvest - Journal of Universal Studies, 2024. **4**(3): p. 723-740.

310. Zhao, J. and Q. Wang, *Design and Implementation of an Intelligent Moving Target Robot System for Shooting Training.* International Journal of Information Technologies and Systems Approach, 2023. **16**(2): p. 1-19.

311. Zhang, E., et al., *Obstacle Capability of an Air-Ground Amphibious Reconnaissance Robot With a Planetary Wheel-Leg Type Structure.* Applied Bionics and Biomechanics, 2021. **2021**: p. 1-12.

312. Kumar, P.S., et al., *ROSMOD: A Toolsuite for Modeling, Generating, Deploying, and Managing Distributed Real-Time Component-Based Software Using ROS.* Electronics, 2016. **5**(3): p. 53.

313.   Kapitonov, A., et al., *Robotic Services for New Paradigm Smart Cities Based on Decentralized Technologies.* Ledger, 2019.

314.   Adeniyi, A.O. and A.O. Akanle, *United Nations' Policy on Weapons of Warfare and the Implication on Global Peace and Security.* International Journal of Multidisciplinary Research and Analysis, 2024. **07**(01).

315.   Rosert, E. and F. Sauer, *Prohibiting Autonomous Weapons: Put Human Dignity First.* Global Policy, 2019. **10**(3): p. 370-375.

316.   Asaro, P., *On Banning Autonomous Weapon Systems: Human Rights, Automation, and the Dehumanization of Lethal Decision-Making.* International Review of the Red Cross, 2012. **94**(886): p. 687-709.

317.   Gunawan, Y., M.H. Aulawi, and A.R. Ramadhan, *Command Responsibility of Autonomous Weapons Systems Under International Humanitarian Law.* Jurnal Cita Hukum, 2019. **7**(3): p. 351-368.

318.   Garcia, D., *Killer Robots: Why the <scp>US</Scp> Should Lead the Ban.* Global Policy, 2015. **6**(1): p. 57-63.

319.   Masood, M. and M.A. Baig, *Potential Impact of Lethal Autonomous Weapon Systems on Strategic Stability and Nuclear Deterrence in South Asia.* Margalla Papers, 2023. **27**(2): p. 27-43.

320.   Egeland, K., *Lethal Autonomous Weapon Systems Under International Humanitarian Law.* Nordic Journal of International Law, 2016. **85**(2): p. 89-118.

321.   Дремлюга, Р., *General Legal Limits of the Application of the Lethal Autonomous Weapons Systems Within the Purview of International Humanitarian Law.* Journal of Politics and Law, 2020. **13**(2): p. 115.

322.   Kozyulin, V., *Lethal Autonomous Weapons Systems: Problems of Current International Legal Regulation and Prospects for Resolving Them.* International Affairs, 2019. **65**(002): p. 86-97.

323.   Maqbool, A. and A. Anwar, *Warfare and Machines: An in-Depth Study of Autonomous Weapons in the Context of International Humanitarian Law.* 2023. **2**(1): p. 01-14.

324.   Davison, N., *A Legal Perspective: Autonomous Weapon Systems Under International Humanitarian Law.* 2018: p. 5-18.

325.   Foy, J., *Autonomous Weapons Systems: Taking the Human Out of International Humanitarian Law.* SSRN Electronic Journal, 2013.

326.   Nasu, H., *The UN Security Council's Responsibility and the "Responsibility to Protect".* Max Planck Yearbook of United Nations Law Online, 2011. **15**(1): p. 377-418.

327.   Altmann, J. and F. Sauer, *Autonomous Weapon Systems and Strategic Stability.* Survival, 2017. **59**(5): p. 117-142.

328.   Belikova, K.M. and M.A. Akhmadova, *Development of Russian and International Legal Regulation of the Use of Lethal Autonomous Weapon*

Systems Equipped With Artificial Intelligence. Laplage Em Revista, 2021. **7**(Extra-C): p. 259-272.

329. Mao, X.H., et al., *Propagation Over the Outdoor-to-Indoor Channel in Military UHF Band.* 2013.

330. Caverley, J.D. and T.S. Sechser, *Military Technology and the Duration of Civil Conflict.* International Studies Quarterly, 2017. **61**(3): p. 704-720.

331. Fedotenko, K., *Cyber Warfare as Part of Information Warfare of Russia Against Ukraine Since the Beginning of the 2022 Russian Invasion.* Věda a Perspektivy, 2023(8(27)).

332. Karaman, O. and Y. Yurkiv, *Deformation of Personality as a Consequence of a Hybrid Warfare.* Postmodern Openings, 2020. **11**(1): p. 42-56.

333. Rusnáková, S., *Russian New Art of Hybrid Warfare in Ukraine.* Slovak Journal of Political Sciences, 2017. **17**(3-4): p. 343-380.

334. Veebel, V., *Is the European Migration Crisis Caused by Russian Hybrid Warfare?* Journal of Politics and Law, 2020. **13**(2): p. 44.

335. Parakhonskyi, B. and G. Yavorska, *The War as a Result of the Fragile Peace: The Conceptual Logic of War.* Strategic Panorama, 2022: p. 12-24.

336. Rammohan, S., et al., *Prediction of Abrasive Waterjet Machining Parameters of Military-Grade Armor Steel by Semi-Empirical and Regression Models.* Materials, 2022. **15**(12): p. 4368.

337. Hoffenson, S., S. Arepally, and P.Y. Papalambros, *A Multi-Objective Optimization Framework for Assessing Military Ground Vehicle Design for Safety.* The Journal of Defense Modeling and Simulation Applications Methodology Technology, 2013. **11**(1): p. 33-46.

338. Valladares, H. and A. Tovar, *Multilevel Design of Sandwich Composite Armors for Blast Mitigation Using Bayesian Optimization and Non-Uniform Rational B-Splines.* Sae International Journal of Advances and Current Practices in Mobility, 2021. **3**(4): p. 2146-2158.

339. Fan, J., et al., *Design of the Supporting Structure of Protective Plates.* 2023: p. 17.

340. Choi, E.-H., et al., *Optimum Suspension Unit Design for Enhancing the Mobility of Wheeled Armored Vehicles.* Journal of Mechanical Science and Technology, 2010. **24**(1): p. 323-330.

341. Mohsen, M., et al., *Investigation of the Ride Response of a Multi-Wheeled Combat Vehicle in Pitch-Bounce Plane.* International Conference on Aerospace Sciences and Aviation Technology, 2015. **16**(AEROSPACE SCIENCES): p. 1-13.

342. Aparow, V.R., et al., *Identification of an Optimum Control Algorithm to Reject Unwanted Yaw Effect on Wheeled Armored Vehicle Due to the Recoil Force.* Advances in Mechanical Engineering, 2017. **9**(1).

343. Radovanović, M., et al., *Analysis of the Development of Five Generation of Anti-Armor Missile Systems.* Scientific Technical Review, 2023. **73**(1): p. 26-37.

344. Grujičić, M., et al., *The Effect of Up-armoring of the High-mobility Multi-purpose Wheeled Vehicle (HMMWV) on the Off-road Vehicle Performance.* Multidiscipline Modeling in Materials and Structures, 2010. **6**(2): p. 229-256.

345. Tepanosyan, G., et al., *Machine Learning-Based Modeling of Air Temperature in the Complex Environment of Yerevan City, Armenia.* Remote Sensing, 2023. **15**(11): p. 2795.

346. Lin, H., et al., *VGEs as a New Platform for Urban Modeling and Simulation.* Sustainability, 2022. **14**(13): p. 7980.

347. Fang, C., et al., *Estimating the Impact of Urbanization on Air Quality in China Using Spatial Regression Models.* Sustainability, 2015. **7**(11): p. 15570-15592.

348. Bueno, B., et al., *The Urban Weather Generator.* Journal of Building Performance Simulation, 2012. **6**(4): p. 269-281.

349. Barbosa, P.R., et al., *Controlled Tracking in Urban Terrain: Closing the Loop.* Asian Journal of Control, 2019. **21**(4): p. 1630-1643.

350. Yang, L., J.A. Smith, and D. Niyogi, *Urban Impacts on Extreme Monsoon Rainfall and Flooding in Complex Terrain.* Geophysical Research Letters, 2019. **46**(11): p. 5918-5927.

351. Freitag, B., U.S. Nair, and D. Niyogi, *Urban Modification of Convection and Rainfall in Complex Terrain.* Geophysical Research Letters, 2018. **45**(5): p. 2507-2515.

352. Ren, Q., et al., *Influence of Urban Spatial Structure on the Spatial Distribution of Gaseous Pollutants.* Atmosphere, 2023. **14**(8): p. 1231.

353. Ku, C.-A., *Exploring the Spatial and Temporal Relationship Between Air Quality and Urban Land-Use Patterns Based on an Integrated Method.* Sustainability, 2020. **12**(7): p. 2964.

354. Giyasov, B. and I. Giyasova, *Environmental Analysis of Cities With Hot Climate and Rugged Terrain.* E3s Web of Conferences, 2019. **135**: p. 03034.

355. Weger, M. and B. Heinold, *Air Pollution Trapping in the Dresden Basin From Gray-Zone Scale Urban Modeling.* Atmospheric Chemistry and Physics, 2023. **23**(21): p. 13769-13790.

356. Weger, M. and B. Heinold, *Air Pollution Trapping in the Dresden Basin Induced by Natural and Urban Topography.* 2023.

357. Zhan, C., *Land Use and Anthropogenic Heat Modulate Ozone by Meteorology: A Perspective From the Yangtze River Delta Region.* Atmospheric Chemistry and Physics, 2022. **22**(2): p. 1351-1371.

358.   Weger, M., et al., *On the Application and Grid-Size Sensitivity of the Urban Dispersion Model CAIRDIO v2.0 Under Real City Weather Conditions.* Geoscientific Model Development, 2022. **15**(8): p. 3315-3345.

359.   Wang, L., et al., *Analysis on Urban Densification Dynamics and Future Modes in Southeastern Wisconsin, USA.* Plos One, 2019. **14**(3): p. e0211964.

360.   Burger, M., M. Gubler, and S. Brönnimann, *Modeling the Intra-Urban Nocturnal Summertime Air Temperature Fields at a Daily Basis in a City With Complex Topography.* Plos Climate, 2022. **1**(12): p. e0000089.

361.   Theeuwes, N., et al., *A Diagnostic Equation for the Daily Maximum Urban Heat Island Effect for Cities in Northwestern Europe.* International Journal of Climatology, 2016. **37**(1): p. 443-454.

362.   Kioumourtzis, G., C. Bouras, and A. Γκάμας, *Performance Evaluation of Ad Hoc Routing Protocols for Military Communications.* International Journal of Network Management, 2011. **22**(3): p. 216-234.

363.   Mahmood, B.A. and D. Manivannan, *GRB: Greedy Routing Protocol With Backtracking for Mobile Ad-Hoc Networks.* 2015.

364.   Lee, Y.H. and X.H. Mao, *Study of Outdoor-to-Indoor Channel in Urban Area.* 2014.

365.   Lee, S. and A. Lim, *An Empirical Study on Ad Hoc Performance of DSRC and Wi-Fi Vehicular Communications.* International Journal of Distributed Sensor Networks, 2013. **9**(11): p. 482695.

366.   Mylitary.com, *Decoding Urban Warfare Communication Systems: Tactics and Technology.* 2024, Mylitary.com.

367.   Wang, H., G. Newman, and Z. Wang, *Urban Planning as an Extension of War Planning.* Journal of Contemporary Urban Affairs, 2019. **3**(1): p. 1-12.

368.   Vardanyan, A., *The Impact of War Generations on Spatial Environment and Military Resistance Challenges in Urban Planning.* E3s Web of Conferences, 2023. **436**: p. 12010.

369.   Erenoğlu, R.C., O. Erenoğlu, and N. Arslan, *Accuracy Assessment of Low Cost UAV Based City Modelling for Urban Planning.* Tehnicki Vjesnik - Technical Gazette, 2018. **25**(6).

370.   Hu, D. and J. Minner, *UAVs and 3D Modeling to Aid Urban Planning and Preservation: A Systematic Review.* 2023.

371.   Doan, T., et al., *Volume Comparison of Automatically Reconstructed Multi-Lod Building Models for Urban Planning Applications.* Isprs Annals of the Photogrammetry Remote Sensing and Spatial Information Sciences, 2021. **V-4-2021**: p. 169-176.

372. Southall, R. and F. Biljecki, *The VI-Suite: A Set of Environmental Analysis Tools With Geospatial Data Applications*. Open Geospatial Data Software and Standards, 2017. **2**(1).
373. Smith, S.J., N. Tomic, and C. Gebhard, *The Father, the Son and the Holy Ghost: A Grounded Theory Approach to the Comparative Study of Decision-Making in the NAC and PSC*. European Security, 2017. **26**(3): p. 359-378.
374. Smith, S.J. and C. Gebhard, *EU–NATO Relations: Running on the Fumes of Informed Deconfliction*. European Security, 2017. **26**(3): p. 303-314.
375. Hakimzadeh, K., *The Issue of Decision Up-Creep in Network Centric Warfare*. 2003.
376. Lee, H. and B.P. Zeigler, *System Entity Structure Ontological Data Fusion Process Integrated With C2 Systems*. The Journal of Defense Modeling and Simulation Applications Methodology Technology, 2010. **7**(4): p. 206-225.
377. Vogelaar, A.L.W. and E.H. Kramer, *Mission Command in Dutch Peace Support Missions*. Armed Forces & Society, 2004. **30**(3): p. 409-431.
378. Gustavsson, P.M., et al., *Operations Intent and Effects Model*. The Journal of Defense Modeling and Simulation Applications Methodology Technology, 2010. **8**(1): p. 37-59.
379. Mahlock, L.M., *Coalition Air Command and Control Interoperability and Net Centric Warfare: Who Manages Air C2 Interoperability in the Netted Environment*. 2005.
380. Malici, A., *Discord and Collaboration Between Allies*. Journal of Conflict Resolution, 2005. **49**(1): p. 90-119.
381. Devetak, S., et al., *Performance Analysis of One Model of Communication and Information System in Military Operation*. Defence Science Journal, 2019. **69**(3): p. 290-297.
382. Etefia, B., M. Gerla, and L. Zhang, *Supporting Military Communications With Named Data Networking: An Emulation Analysis*. 2012.
383. Naqvi, M.A., et al., *Nature Inspired Flying Vehicles and Future Challenges in Aerospace*. International Journal of Automotive Engineering and Technologies, 2015. **4**(1): p. 40.
384. Seo, K.-M., W. Hong, and T.G. Kim, *Enhancing Model Composability and Reusability for Entity-Level Combat Simulation: A Conceptual Modeling Approach*. Simulation, 2017. **93**(10): p. 825-840.
385. Hofmann, M., *Challenges of Model Interoperation in Military Simulations*. Simulation, 2004. **80**(12): p. 659-667.
386. Queralta, J.P., et al., *Collaborative Multi-Robot Search and Rescue: Planning, Coordination, Perception, and Active Vision*. Ieee Access, 2020. **8**: p. 191617-191643.

387. Mallek, S., et al., *Contribution to the Definition of a New Type of More Technological Ambulance: Use Case of French Civil Security Ambulances.* Open Journal of Safety Science and Technology, 2016. **06**(04): p. 126-142.

388. Dawson, D., K. Vercruysse, and N. Wright, *A Spatial Framework to Explore Needs and Opportunities for Interoperable Urban Flood Management.* Philosophical Transactions of the Royal Society a Mathematical Physical and Engineering Sciences, 2020. **378**(2168): p. 20190205.

389. Killea, K.J., *The New A" in MAGTF".* 2001.

390. Muller, M.F., E.R. Loures, and O. Canciglieri, *Interoperability Assessment for Building Information Modelling.* 2015.

391. Vercruysse, K., D. Dawson, and N. Wright, *Interoperability: A Conceptual Framework to Bridge the Gap Between Multifunctional and Multisystem Urban Flood Management.* Journal of Flood Risk Management, 2019. **12**(S2).

392. Scrocca, M., I. Baroni, and I. Celino, *Urban IoT Ontologies for Sharing and Electric Mobility.* Semantic Web, 2023. **14**(4): p. 617-638.

393. Noy, B., *The Role of Zoning Policies on Sustainable Development and Climate Change.* 2022.

394. Martins, T.M., et al., *An Urban Ontology to Generate Collaborative Virtual Environments for Municipal Planning and Management.* 2012.

395. Spain, R.D., H.A. Priest, and J.S. Murphy, *Current Trends in Adaptive Training With Military Applications: An Introduction.* Military Psychology, 2012. **24**(2): p. 87-95.

396. Kilcullen, D. and G. Pendleton, *Future Urban Conflict, Technology, and the Protection of Civilians.* 2021.

397. Muhammedally, S., *Understanding Risks and Mitigating Civilian Harm in Urban Operations.* Canadian Army Journal, 2024: p. 46-63.

398. Grubb, A., et al., *From "Sad People on Bridges" to "Kidnap and Extortion": Understanding the Nature and Situational Characteristics of Hostage and Crisis Negotiator Deployments.* Negotiation and Conflict Management Research, 2018. **12**(1): p. 41-65.

399. Mertes, M., J. Mazei, and J. Hüffmeier, *"We Do Not Negotiate With Terrorists!" but What if We Did?* Peace and Conflict Journal of Peace Psychology, 2020. **26**(4): p. 437-448.

400. Avdija, A.S., *Special Weapons and Tactics Operations.* Policing an International Journal, 2018. **41**(5): p. 651-658.

401. Nnawulezi, U. and S.B. Magashi, *Evolving Roles of the International Institutions in the Implementation Mechanisms of the Rules of*

International Humanitarian Law. Kutafin Law Review, 2023. **9**(4): p. 684-712.

402.   Orzeszyna, K., *Convergence of International Humanitarian Law and International Human Rights Law in Armed Conflicts*. Studia Iuridica Lublinensia, 2023. **32**(3): p. 237-252.

403.   Sarno, M. and V.B.V. Hasselt, *Suicide by Cop: Implications for Crisis (Hostage) Negotiations*. Journal of Criminal Psychology, 2014. **4**(2): p. 143-154.

404.   Colonomos, A., *Hostageship: What Can We Learn From Mauss?* Journal of International Political Theory, 2018. **14**(2): p. 240-256.

405.   Clemmow, C., Z. Marchment, and P. Gill, *Analyzing Person-exposure Patterns in Lone-actor Terrorism*. Criminology & Public Policy, 2019. **19**(2): p. 451-482.

406.   Wang, W., et al., *Parallel Realization of Moving Target Imaging in 2D MIMO Through-Wall Radar Applications*. 2015.

407.   Perone, S.A., et al., *Psychological Support Post-Release of Humanitarian Workers Taken Hostage: The Experience of the International Committee of the Red Cross (ICRC)*. British Journal of Guidance and Counselling, 2018. **48**(3): p. 360-373.

408.   Rai, G., *States' Use of Psychological Warfare to Deter Threats*. The Journal of Intelligence Conflict and Warfare, 2018. **1**(1).

409.   Nevmerzhitsky, I.V., *Procedure and Principles of Information and Psychological Operations in the North-Atlantic Alliance (Based on Nato Documents)*. Collection of Scientific Works of the Military Institute of Kyiv National Taras Shevchenko University, 2019(65): p. 107-116.

410.   Afkar, A.L.S., *The Application of Psychological Operation (PSYOP): A Case Study on the Siege of Sauk*. Malaysian Journal of Social Sciences and Humanities (Mjssh), 2022. **7**(7): p. e001643.

411.   Alosman, M.I.M. and M. Sawalmeh, *Psychological Operations and Their Ethical Implications in Phil Klay's Redeployment*. Jordan Journal of Modern Languages & Literatures, 2023. **15**(4). p. 1413-1426.

412.   Cronin, B., *Reckless Endangerment Warfare*. Journal of Peace Research, 2013. **50**(2): p. 175-187.

413.   Muhammad Asif Safdar, N., et al., *Common Article 3 and Asymmetric Warfare in the Context of Cyber Operations*. 2024. **4**(1): p. 15-23.

414.   Safdar, M.A., *Impact of the Applicability of Common Article 3 on the Asymmetric Warfare*. 2023. **3**(1): p. 104-110.

415.   Clark, I., et al., *Crisis in the Laws of War? Beyond Compliance and Effectiveness*. European Journal of International Relations, 2017. **24**(2): p. 319-343.

416.   Huang, Q., et al., *Monitoring Urban Change in Conflict From the Perspective of Optical and SAR Satellites: The Case of Mariupol, a City in the Conflict Between RUS and UKR*. Remote Sensing, 2023. **15**(12): p. 3096.

417.   Bogdanowicz, Z.R., J.W.H. Price, and K. Patel, *Effect-Based Weapon-Target Assignment Optimization With Collateral Damage Under Control*. Applied Mathematical Sciences, 2016. **10**: p. 519-541.

418.   Solar, C., *Trust in the Military in Post-Authoritarian Societies*. Current Sociology, 2020. **70**(3): p. 317-337.

419.   Bakar, D.M., S. Syamsunasir, and A. Adriyanto, *Mediating Effect of Indonesia's Public Trust on National Security From the Sense of Safety, Knowledge, and Willingness to Participate in State Defense Efforts*. Security and Defence Quarterly, 2021. **36**(4): p. 69-88.

420.   Tiargan-Orr, R. and M. Eran-Jona, *The Israeli Public's Perception of the IDF*. Armed Forces & Society, 2015. **42**(2): p. 324-343.

421.   Garb, M., *Public Trust in the Military: The Slovenian Armed Forces in a Comparative Analysis*. Current Sociology, 2015. **63**(3): p. 450-469.

422.   Malešič, M. and M. Garb, *Public Trust in the Military From Global, Regional and National Perspectives*. 2018: p. 145-159.

423.   Bilgiç, A., G.H. Gjørv, and C. Wilcock, *Trust, Distrust, and Security: An Untrustworthy Immigrant in a Trusting Community*. Political Psychology, 2019. **40**(6): p. 1283-1296.

424.   Moldjord, C. and I.D. Hybertsen, *Training Reflective Processes in Military Aircrews Through Holistic Debriefing: The Importance of Facilitator Skills and Development of Trust*. International Journal of Training and Development, 2015. **19**(4): p. 287-300.

425.   Ott, D.L. and S. Michailova, *Cultural Intelligence: A Review and New Research Avenues*. International Journal of Management Reviews, 2016. **20**(1): p. 99-119.

426.   Minshew, L.M., et al., *Development of a Cultural Intelligence Framework in Pharmacy Education*. American Journal of Pharmaceutical Education, 2021. **85**(9): p. 8580.

427.   Azevedo, A., *Cultural Intelligence: Key Benefits to Individuals, Teams and Organizations*. American Journal of Economics and Business Administration, 2018. **10**(1): p. 52-56.

428.   Jenni, G.D.L., et al., *Military Perspectives on Public Relations Related to Environmental Issues*. Journal of Public Relations Research, 2015. **27**(4): p. 353-369.

429.   Vlajčić, D., et al., *The Role of Geographical Distance on the Relationship Between Cultural Intelligence and Knowledge Transfer*. Business Process Management Journal, 2018. **25**(1): p. 104-125.

430. Kistyanto, A., et al., *Cultural Intelligence Increase Student's Innovative Behavior in Higher Education: The Mediating Role of Interpersonal Trust.* International Journal of Educational Management, 2021. **36**(4): p. 419-440.

431. Moffett, L., et al., *Inside the Shadows: A Survey of UK Human Source Intelligence (HUMINT) Practitioners, Examining Their Considerations When Handling a Covert Human Intelligence Source (CHIS).* Psychiatry Psychology and Law, 2021. **29**(4): p. 487-505.

432. Hart, A., et al., *Developing Cultural Awareness Curricular Competencies for Humanitarian Non-Governmental Organization Staff.* Prehospital and Disaster Medicine, 2021. **36**(6): p. 669-675.

433. Barrón-Cedeño, A., et al., *Proppy: A System to Unmask Propaganda in Online News.* Proceedings of the Aaai Conference on Artificial Intelligence, 2019. **33**(01): p. 9847-9848.

434. Oates, S., D. Lee, and D. Knickerbocker, *Data Analysis of Russian Disinformation Supply Chains: Finding Propaganda in the U.S. Media Ecosystem in Real Time.* 2022.

435. Aggarwal, K. and A. Sadana, *NSIT@NLP4IF-2019: Propaganda Detection From News Articles Using Transfer Learning.* 2019.

436. Mikhalkova, E., et al., *UTMN at SemEval-2020 Task 11: A Kitchen Solution to Automatic Propaganda Detection.* 2020.

437. Gupta, P., et al., *Neural Architectures for Fine-Grained Propaganda Detection in News.* 2019.

438. Martino, G.D.S., et al., *Fine-Grained Analysis of Propaganda in News Article.* 2019.

439. Baly, R., et al., *Predicting Factuality of Reporting and Bias of News Media Sources.* 2018.

440. Chatfield, A.T., C.G. Reddick, and U. Brajawidagda, *Tweeting Propaganda, Radicalization and Recruitment.* 2015: p. 239-249.

441. Yaser, E., *Social Strategies of Terrorism.* International Journal of Social Science Research and Review, 2023. **6**(6): p. 76-85.

442. Tudorache, P. and L. Ispas, *The Relation Cultural Awareness – Tactical Multinational Military Structures' Effectiveness.* International Conference Knowledge-Based Organization, 2017. **23**(1): p. 325-330.

443. Clever, L., et al., *Behind Blue Skies: A Multimodal Automated Content Analysis of Islamic Extremist Propaganda on Instagram.* Social Media + Society, 2023. **9**(1).

444. Holmes-Eber, P., E. Tarzi, and B. Maki, *U.S. Marines' Attitudes Regarding Cross-Cultural Capabilities in Military Operations.* Armed Forces & Society, 2016. **42**(4): p. 741-751.

445.   Badawy, A. and E. Ferrara, *The Rise of Jihadist Propaganda on Social Networks*. SSRN Electronic Journal, 2017.

446.   Wood, J., C. Wong, and S. Paturi, *Addressing Food Security in Constrained Urban Environments*. 2021: p. 169-186.

447.   Slade, C., C. Baldwin, and T. Budge, *Urban Planning Roles in Responding to Food Security Needs*. Journal of Agriculture Food Systems and Community Development, 2016: p. 1-16.

448.   Meyer, K.D., et al., *Increasing Resilience of Material Supply by Decentral Urban Factories and Secondary Raw Materials*. Frontiers in Manufacturing Technology, 2023. **3**.

449.   Mahendra, A., et al., *Towards a More Equal City: Seven Transformations for More Equitable and Sustainable Cities*. 2021.

450.   Lammers, D., et al., *Airborne! <scp>UAV</scp> Delivery of Blood Products and Medical Logistics for Combat Zones*. Transfusion, 2023. **63**(S3).

451.   Hughes, K., et al., *Autonomous Aerial Supply Delivery: Parachute Release and Impact Study*. 2022.

452.   Meng, F.Y., et al., *Application and Research of Automatic Ammunition Resupply and Replenishment System for Artillery: A Review*. Journal of Physics Conference Series, 2023. **2478**(12): p. 122070.

453.   Johnson, R.J., *Post-Cold War United Nations Peacekeeping Operations: A Review of the Case for a Hybrid Level 2+ Medical Treatment Facility*. Disaster and Military Medicine, 2015. **1**(1).

454.   Chen, X., et al., *Application and Prospect of a Mobile Hospital in Disaster Response*. Disaster Medicine and Public Health Preparedness, 2020. **14**(3): p. 377-383.

455.   Newgard, C.D., et al., *Prospective Validation of the National Field Triage Guidelines for Identifying Seriously Injured Persons*. Journal of the American College of Surgeons, 2016. **222**(2): p. 146-158e2.

456.   Doutchi, M., et al., *Health Transformation Toward Universal Healthcare Coverage Amidst Conflict: Examining the Impact of International Cooperation in Niger*. Frontiers in Public Health, 2024. **12**.

457.   Maghfiroh, M.F.N. and S. Hanaoka, *Mobile Clinics: Medical Service Strategy for Disaster Healthcare Response Operation*. Journal of Industrial Engineering and Management, 2022. **15**(3): p. 470.

458.   Malembaka, E.B., et al., *The Use of Health Facility Data to Assess the Effects of Armed Conflicts on Maternal and Child Health: Experience From the Kivu, DR Congo*. BMC Health Services Research, 2021. **21**(S1).

459.   Poole, D.N., et al., *Damage to Medical Complexes in the Gaza Strip During the Israel–Hamas War: A Geospatial Analysis*. BMJ Global Health, 2024. **9**(4): p. e014768.

460. Gao, M.-Z., et al., *Designing Mobile Epidemic Prevention Medical Stations for the COVID-19 Pandemic and International Medical Aid.* International Journal of Environmental Research and Public Health, 2022. **19**(16): p. 9959.

461. Harrison, M., *Casualty Evacuation in Korea, 1950-53: The British Experience.* Korean Journal of Medical History, 2023. **32**(2): p. 503-552.

462. Leasiolagi, J., et al., *Proposed Specifications of a Mobile Operating Room for Far-Forward Surgery.* Canadian Journal of Surgery, 2018. **61**(6 Suppl 1): p. S180-S183.

463. Brown, J.B., et al., *Prehospital Lactate Improves Accuracy of Prehospital Criteria for Designating Trauma Activation Level.* Journal of Trauma and Acute Care Surgery, 2016. **81**(3): p. 445-452.

464. Mohajervatan, A., M. Ashabi, and F. Rezaei, *Investigating Emergency Medical System Challenges in Mass Casualty Incidents: A Case Report Study in Road Traffic Accidents in Northern Iran.* Health in Emergencies & Disasters Quarterly, 2024. **9**(3): p. 239-246.

465. Hunter, O., et al., *Science Fiction or Clinical Reality: A Review of the Applications of Artificial Intelligence Along the Continuum of Trauma Care.* World Journal of Emergency Surgery, 2023. **18**(1).

466. Roger Lane, L.L., & Himayu Shiotani, *Opportunities to Improve Military Policies and Practices to Reduce Civilian Harm From Explosive Weapons in Urban Conflict.* 2019, United Nations Institute for Disarmament Research.

467. Căruţaşu, V. and D. Căruţaşu, *Assessment of Maintenance Needs Required to Preserve the Operational Status of Combat Vehicles Participating in Military Operations.* International Conference Knowledge-Based Organization, 2015. **21**(3): p. 787-792.

468. Virca, I., V. Dascălu, and C. Grigoraş, *Research on Improving the Maintenance Activities for Military Vehicles.* International Conference Knowledge-Based Organization, 2015. **21**(3): p. 896-903.

469. Ochando, F.J., et al., *Data Acquisition for Condition Monitoring in Tactical Vehicles: On-Board Computer Development.* Sensors, 2023. **23**(12): p. 5645.

470. Barakat, S., S. Milton, and G. Elkahlout, *Reconstruction Under Siege: The Gaza Strip Since 2007.* Disasters, 2019. **44**(3): p. 477-498.

471. Smith, R.J., *Isolation Through Humanitarianism: Subaltern Geopolitics of the Siege on Gaza.* Antipode, 2016. **48**(3): p. 750-769.

472. Farhat, T., et al., *Responding to the Humanitarian Crisis in Gaza: Damned if You Do... Damned if You Don't!* Annals of Global Health, 2023. **89**(1).

473. Rose, W., et al., *Urban Logistics: Establishing Key Concepts and Building a Conceptual Framework for Future Research.* Transportation Journal, 2017. **56**(4): p. 357-394.

474. Watts, S., *Under Siege: International Humanitarian Law and Security Council Practice Concerning Urban Siege Operations.* SSRN Electronic Journal, 2014.

475. Morana, J., J. González-Feliu, and F. Semet, *Urban Consolidation and Logistics Pooling.* 2013: p. 187-210.

476. Rao, C., et al., *Location Selection of City Logistics Centers Under Sustainability.* Transportation Research Part D Transport and Environment, 2015. **36**: p. 29-44.

477. Li, J.Y. and Q. Xiao, *Grey Relational Analysis Method for Urban Logistics.* Applied Mechanics and Materials, 2013. **321-324**: p. 3065-3068.

478. Ville, S., J. González-Feliu, and L. Dablanc, *The Limits of Public Policy Intervention in Urban Logistics: Lessons From Vicenza (Italy).* European Planning Studies, 2013. **21**(10): p. 1528-1541.

479. Xian-wen, G., *Coupling Coordinated Development Model of Urban-Rural Logistics and Empirical Study.* Mathematical Problems in Engineering, 2019. **2019**(1).

480. Penney, G., et al., *Calculation of Critical Water Flow Rates for Wildfire Suppression.* Fire, 2019. **2**(1): p. 3.

481. Gusmão, A., *Urban Logistics Guidelines for the Efficiency of Goods Distribution in Brazilian Cities.* 2024.

482. Molloy, O., *How are Drones Changing Modern Warfare?* 2024, Australian Army - Department of Defence or the Australian Government.

483. Total Military Insight, *The Impact of AI in Urban Warfare Strategies and Tactics.* 2024, Total Military Insight.

484. SDI Santient Digital Inc., *The Most Useful Military Applications of AI in 2024 and Beyond.* 2024, SDI Santient Digital Inc.,.

485. Kimes, M., *Revolutionizing Military Operations with Decentralized AI: A Game Changer for Digital Interaction.* 2024, LinkedIn.

486. Department of the Army, *FM 2-91.4 Intelligence Support to Urban Operations.* 2008.

487. Agbossou, I., *Urban Augmented Reality for 3D Geosimulation and Prospective Analysis.* 2023.

488. Ssin, S., H. Cho, and W. Woo, *GeoACT: Augmented Control Tower Using Virtual and Real Geospatial Data.* Interaction Design and Architecture(s), 2021(48): p. 122-142.

489. Kallberg, J., et al., *The tactical considerations of augmented and mixed reality implementation.* Mil. Rev, 2022. **662**: p. 105.

490. Sheldon, J.R., *cyberwar,* in *Encyclopedia Britannica.* 2024.

491.   Bhaiyat, H. and S. Sithungu, *Cyberwarfare and Its Effects on Critical Infrastructure.* International Conference on Cyber Warfare and Security, 2022. **17**(1): p. 536-543.

492.   Guadagno, R.E. and K. Guttieri, *Fake News and Information Warfare.* 2019: p. 167-191.

493.   Payne, B. and E.L. Mienie, *The Impact of Cyber-Physical Warfare on Global Human Security.* International Journal of Cyber Warfare and Terrorism, 2019. **9**(3): p. 36-50.

494.   البدوي, ح., الحرب الرقمية والأمن السيبراني... خطر التهديدات يقابله تعزيز الدفاعات. 3 .2024(3): p. 153-180.

495.   Igakuboon, A.N., *An Appraisal of the Legal Framework for the Protection of Civilians in Cyber-Warfare Under International Humanitarian Law.* International Journal of Research and Scientific Innovation, 2022. **09**(07): p. 14-26.

496.   Valeriano, B., B. Jensen, and R.C. Maness, *Cyber Strategy.* 2018.

497.   Moore, D., *Charting Intangible Warfare.* 2022: p. 45-68.

498.   Schulze, M., *Cyber in War: Assessing the Strategic, Tactical, and Operational Utility of Military Cyber Operations.* 2020: p. 183-197.

499.   Harknett, R.J. and M. Smeets, *Cyber Campaigns and Strategic Outcomes.* Journal of Strategic Studies, 2020. **45**(4): p. 534-567.

500.   Healey, J., *The Implications of Persistent (And Permanent) Engagement in Cyberspace.* Journal of Cybersecurity, 2019. **5**(1).

501.   Kello, L., *The Meaning of the Cyber Revolution: Perils to Theory and Statecraft.* International Security, 2013. **38**(2): p. 7-40.

502.   Kim, S.-K., S.-p. Cheon, and J.-H. Eom, *A Leading Cyber Warfare Strategy According to the Evolution of Cyber Technology After the Fourth Industrial Revolution.* International Journal of Advanced Computer Research, 2019. **9**(40): p. 72-80.

503.   Imperva, *Cyber Warfare.* 2024, Imperva.

504.   Spezio, A.E., *Electronic Warfare Systems.* Ieee Transactions on Microwave Theory and Techniques, 2002. **50**(3): p. 633-644.

505.   Gupta, M., G. Hareesh, and A.K. Mahla, *Electronic Warfare:Issues and Challenges for Emitter Classification.* Defence Science Journal, 2011. **61**(3): p. 228-234.

506.   Military Equipment HQ, *The Strategic Role of Electronic Warfare in Urban Warfare.* 2024, Military Equipment HQ

507.   Hou, Y., et al., *Artificial Intelligence Technology Pushes Forward the Modernization of Firepower Weapon Equipment.* 2023: p. 5.

508.   Konaev, M., et al., *U.S. And Chinese Military AI Purchases.* 2023.

509. Zhang, J., et al., *A Survey on Joint-Operation Application for Unmanned Swarm Formations Under a Complex Confrontation Environment*. Journal of Systems Engineering and Electronics, 2023. **34**(6): p. 1432-1446.

510. Gaire, U.S., *Application of Artificial Intelligence in the Military: An Overview*. Unity Journal, 2023. **4**(01): p. 161-174.

511. Rashid, A.B., et al., *Artificial Intelligence in the Military: An Overview of the Capabilities, Applications, and Challenges*. International Journal of Intelligent Systems, 2023. **2023**: p. 1-31.

512. BĂLĂCeanu, I. and C.-M. BÎRsan, *The Characteristics of Unconventional Warfare*. Review of the Air Force Academy, 2018. **16**(1): p. 11-16.

513. Akpoto, Y.M., et al., *Extremity Injuries in Soldiers During the Conflict in Mali: Experience of Togo Level Two Hospital*. International Orthopaedics, 2015. **39**(10): p. 1895-1899.

514. Atkinson, C., *Hybrid Warfare and Societal Resilience: Implications for Democratic Governance*. Information & Security an International Journal, 2018. **39**(1): p. 63-76.

515. Mahmood Alvi, N.B., S.F. Habeeb, and M.U.D. Aftab, *Fifth-Generation Warfare and Challenges for Pakistan*. Journal of Management Practices Humanities and Social Sciences, 2022. **6**(6).

516. Kaur, M., R. Khanna, and A. Sharma, *Radar Cross Section Reduction Techniques using Metamaterials*. International Journal of Innovative Technology and Exploring Engineering (IJITEE), 2019. **8**(9S).

517. Gonçalves, C.P., *Cyberspace and Artificial Intelligence: The New Face of Cyber-Enhanced Hybrid Threats*. 2020.

518. Robinson, M., K. Jones, and H. Janicke, *Cyber Warfare: Issues and Challenges*. Computers & Security, 2015. **49**: p. 70-94.

519. Wessels, A., *Boer Guerrilla and British Counter-Guerrilla Operations in South Africa, 1899 to 1902*. Scientia Militaria South African Journal of Military Studies, 2011. **39**(2).

520. Besenyő, J., *Guerrilla Operations in Western Sahara: The Polisario Versus Morocco and Mauritania*. Connections the Quarterly Journal, 2017. **16**(3): p. 23-45.

521. Álvarez, M.M. and J.A. Gonzalez, *The Rhodesian War (1965-1980)*. 2023: p. 236-253.

522. Tsigo, E.B. and E. Ndawana, *Unsung Heroes? The Rhodesian Defence Regiment and Counterinsurgency, 1973–80*. International Journal of Military History and Historiography, 2019. **39**(1): p. 88-120.

523. Dietz, A.S. and E.A. Schroeder, *Integrating Critical Thinking in the U.S. Army: Decision Support Red Teams*. New Directions for Adult and Continuing Education, 2012. **2012**(136): p. 29-40.

524. Foran, H.M., et al., *Soldiers' Perceptions of Resilience Training and Postdeployment Adjustment: Validation of a Measure of Resilience Training Content and Training Process.* Psychological Services, 2012. **9**(4): p. 390-403.

525. Crameri, L., I. Hettiarachchi, and S. Hanoun, *A Review of Individual Operational Cognitive Readiness: Theory Development and Future Directions.* Human Factors the Journal of the Human Factors and Ergonomics Society, 2019. **63**(1): p. 66-87.

526. Hickey, J.P., B. Donne, and D. O'Brien, *Effects of an Eight Week Military Training Program on Aerobic Indices and Psychomotor Function.* Journal of the Royal Army Medical Corps, 2012. **158**(1): p. 41-46.

527. O'Toole, P. and S.R. Talbot, *Fighting for Knowledge: Developing Learning Systems in the Australian Army.* Armed Forces & Society, 2010. **37**(1): p. 42-67.

528. Madhavan, P., *Training Issues in Military and Defense.* Proceedings of the Human Factors and Ergonomics Society Annual Meeting, 2010. **54**(27): p. 2272-2273.

529. Lucian Valeriu Scipanov, D.R., Petrişor Pătraşcu, and Dănuţa Mădălina Scipanov, *Possible Solutions for Capitalizing E-Education Tools in Military Science.* 2023.

530. Woodruff, T., *Who Should the Military Recruit? The Effects of Institutional, Occupational, and Self-Enhancement Enlistment Motives on Soldier Identification and Behavior.* Armed Forces & Society, 2017. **43**(4): p. 579-607.

531. Alim, H., et al., *A Review on Issue and Consideration of Military Personnel Cognitive Readiness for Malaysian Army.* International Journal of Academic Research in Progressive Education and Development, 2023. **12**(4).

532. Carless, D., et al., *Psychosocial Outcomes of an Inclusive Adapted Sport and Adventurous Training Course for Military Personnel.* Disability and Rehabilitation, 2013. **35**(24): p. 2081-2088.

533. Levine, D.H., *Care and Counterinsurgency.* Journal of Military Ethics, 2010. **9**(2): p. 139-159.

534. Maekawa, W., *External Intelligence Assistance and the Recipient Government's Violence Against Civilians.* Conflict Management and Peace Science, 2022. **40**(5): p. 511-532.

535. Cayford, M. and W. Pieters, *The Effectiveness of Surveillance Technology: What Intelligence Officials Are Saying.* The Information Society, 2018. **34**(2): p. 88-103.

536. Трут, В.П., *The Tactics of Conducting Combat Operations of Soviet Troops During the Battle of Stalingrad.* Vestnik Volgogradskogo

Gosudarstvennogo Universiteta Serija 4 Istorija Regionovedenie Mezhdunarodnye Otnoshenija, 2018. **23**(1): p. 13-25.

537. Burden, D., *The Battles of Hue: Understanding Urban Conflicts Through Wargaming*. Journal of Strategic Security, 2023. **16**(3): p. 128-140.

538. Yarchi, M., *Does Using 'Imagefare' as a State's Strategy in Asymmetric Conflicts Improve Its Foreign Media Coverage? The Case of Israel*. Media War & Conflict, 2016. **9**(3): p. 290-305.

539. Takshe, A.A., I.v.d. Molen, and J.C. Lovett, *Examining the Lack of Legal Remedies for Environmental Damage in the 2006 Lebanon–Israel War*. Environmental Policy and Governance, 2012. **22**(1): p. 27-41.

540. Zein, H.E., *"The Utilised Frames by Militant Organisations in the Arab-Israeli Conflict: Case of Hezbollah's Military Arm"*. Scientific Bulletin of the Politehnica University of Timişoara Transactions on Modern Languages, 2020. **14**: p. 13-22.

541. Melki, J., *The Interplay of Politics, Economics and Culture in News Framing of Middle East Wars*. Media War & Conflict, 2014. **7**(2): p. 165-186.

542. Mansour-Ichrakieh, L., *The Impact of Israeli and Saudi Arabian Geopolitical Risks on the Lebanese Financial Market*. Journal of Risk and Financial Management, 2021. **14**(3): p. 94.

543. Pahlavi, P. and É. Ouellet, *Institutional Analysis and Irregular Warfare: Israel Defense Forces During the 33-Day War of 2006*. Small Wars and Insurgencies, 2012. **23**(1): p. 32-55.

544. Pfeifer, H., *Recognition Dynamics and Lebanese Hezbollah's Role in Regional Conflicts*. 2021.

545. Pratto, F., et al., *When Domestic Politics and International Relations Intermesh: Subordinated Publics' Factional Support Within Layered Power Structures*. Foreign Policy Analysis, 2013. **10**(2): p. 127-148.

546. Estriani, H.N., P.A. Dewanto, and H. Asyidiqi, *Asymmetric and Hybrid Warfare in Postmodern Times: Lesson From Hezbollah-Israeli War 2006*. Andalas Journal of International Studies (Ajis), 2023. **12**(1): p. 27.

547. DeVore, M.R., A. Stähli, and U. Franke, *Dynamics of Insurgent Innovation: How Hezbollah and Other Non-State Actors Develop New Capabilities*. Comparative Strategy, 2019. **38**(4): p. 371-400.

548. Wahab, H., *Syria's Sect-Coded Conflict: From Hezbollah's Top-Down Instrumentalization of Sectarian Identity to Its Candid Geopolitical Confrontation*. Contemporary Review of the Middle East, 2021. 8(2): p. 149-167.

549. Abadi, J., *Israel and Lebanon: Relations Under Stress*. Contemporary Review of the Middle East, 2020. **7**(1): p. 90-115.

550.   Gassama, S.K., M. Ebrahimi, and K. Yusoff, *Justum Ad Bellum &Amp; Israel's 2006 Attack on Lebanon: An Examination of Just War Principles.* Asian Social Science, 2018. **14**(8): p. 68.

551.   Sela, A., *From Revolution to Political Participation: Institutionalization of Militant Islamic Movements.* Contemporary Review of the Middle East, 2015. **2**(1-2): p. 31-54.

552.   Neiger, M. and E. Zandberg, *Communicating Critique: Toward a Conceptualization of Journalistic Criticism.* Communication Culture and Critique, 2010. **3**(3): p. 377-395.

553.   Eilam, E., *Arab Factor as a Part of the Iranian–Israeli Conflict: An Israeli Perspective.* Contemporary Review of the Middle East, 2016. **3**(4): p. 422-432.

554.   Lupovici, A., *The Limits of Securitization Theory: Observational Criticism and the Curious Absence of Israel.* International Studies Review, 2014. **16**(3): p. 390-410.

555.   Olesker, R., *National Identity and Securitization in Israel.* Ethnicities, 2013. **14**(3): p. 371-391.

556.   Delerue, F., *Civilian Direct Participation in Cyber Hostilities.* Idp Revista De Internet Derecho Y Política, 2014(19): p. 3.

557.   Houri, W.E. and D. Saber, *Filming Resistance.* Radical History Review, 2010. **2010**(106): p. 70-85.

558.   Schulzke, M., *Simulating Terrorism and Insurgency: Video Games in the War of Ideas.* Cambridge Review of International Affairs, 2014. **27**(4): p. 627-643.

559.   Cornum, R.L., M.D. Matthews, and M.E.P. Seligman, *Comprehensive Soldier Fitness: Building Resilience in a Challenging Institutional Context.* American Psychologist, 2011. **66**(1): p. 4-9.

560.   Tornero-Aguilera, J.F., et al. *Optimising Combat Readiness: Practical Strategies for Integrating Physiological and Psychological Resilience in Soldier Training.* in *Healthcare.* 2024. MDPI.

561.   Phocas, B., *Urban Combat Fitness: Preparing Today for Tomorrow's Fight.* 2022, From the Green Notebook.

INDEX